Textbook of Machine Learning and Data Mining

with Bioinformatics Applications

Hiroshi Mamitsuka

International Standard Book Number (ISBN): 978–4–9910445–0–2

Library of Congress Control Number: 2018956749

To Miko

and

Hiro

Preface

When the idea of writing a book came to my mind, originally the first idea was to focus on so-called semi-structured data, such as sequences, trees and graphs, which exactly I have been working on, more than other data types, and have more expertise. However once I started writing this book, I was getting more interested in covering the entire machine learning and data mining techniques, which are in fact connected to each other in many aspects. In particular, methods for semi-structured data sometimes can be similar to or the same as those for more basic data types. Also I became more interested in, simply speaking, writing a more introductory book, which can be read even by entry-level machine learners. This would be because writing this book made me remember my days I started working on machine learning and also machine learning itself just started its history. I thus tried writing a book comprehensible for even those who are working on other fields but have some interest in machine learning and/or data mining and also no stern background knowledge on them. As a result I might have focused more on what can be consistently used through different approaches and also different data types. In other words, many (differently regarded) approaches should have some common motivations and ideas behind them, even between those in different data types. These points are partially already known and written in papers, but not so much at the level of books, while they must be the things which make people understand machine learning approaches more clearly. Thus this book covers wide aspects of machine learning, shedding light on such aspects, which are shared by different methods. This would clarify how machine learning methods are similar to or different from each other.

This book is not only for researchers in the relevant fields but also a wide variety of people including students and engineers who are interested in artificial intelligence or more widely data science, regardless of their background. I hope that this book is helpful for a lot of readers to have some clearer picture of machine learning and data mining.

I started working on machine learning when I was working with a company, where at that time probably the most flexible place in Japan to start doing new research. The manager of the department I belonged was open-minded and had a great interest in artificial intelligence, thinking that the area is very promising in the future. I was in the "machine learning" group of the department, and people there, including me, were strongly encouraged by him to embark upon the new research area. I would like to thank him and people in the department for their support of my doing machine learning research.

After I moved to academia, I have enjoyed more freedom to do research. That is, I could do research, simply following my own preference. The university system is rather bureaucratic as well as other organizations in Japan, while in the Institute for Chemical Research, Kyoto University, thanks to my colleagues and particularly warm arrangement of senior professors, I have been comfortably working in the institute. Without the atmosphere of respecting unique, original ways of thinking in the institute, I did not come to the idea of writing a book.

I started writing this book when I was staying at and working closely with the Department of Computer Science, Aalto University, Finland. I would like to thank all colleagues and people in the department. Aalto University welcomed me warmly and kindly, offering all I need to do research and other work. The system of the department, school, university and even country are all well organized, also always exploring efficiency and its improvement. I'd like to thank people who established and are maintaining those systems. This book was not realized without all efficient systems of the country.

In this book, part of the research the author has conducted with colleagues are introduced. Performing the collaborative research has not been done without the tremendous effort by the collaborators. Particularly in those work, a lot of missing parts of the author's ability were complemented by the excellent insights and capability of the collaborators. I deeply thank those collaborators I raise below alphabetically: Naoki Abe, Kiyoko Aoki-Kinoshita, David duVerle, Timothy Hancock, Kosuke Hashimoto, Minoru Kanehisa, Masayuki Karasuyama, Atsuyoshi Nakamura, Canh Hao Nguyen, Yasuko Ono, Motoki Shiga, Hiroyuki Sorimachi, Ichigaku Takigawa, Koji Tsuda, Keiko Udaka, Nobuhisa Ueda, Raymond Wan, Takashi Yoneya, Sohiya Yotsukura and Shanfeng Zhu.

Hiroshi Mamitsuka
Uji, Japan / Espoo, Finland
August, 2018.

Contents

Chapter 1

Introduction

Currently machine learning (ML) and data mining (DM) are the largest part of artificial intelligence (AI) research, and AI itself would be one of the largest research area in computer science and also being emerged as an important part of the entire science and more generally our society. Behind the technologies of AI, there have been developed various approaches and methods of ML for a variety of applications, although the history of ML itself is not so long, comparing with other research fields, like basic sciences and engineering.

ML research started in 1970s to 80s, and the community was not becoming so large for a long time, even after the DM research started around 1980s to 90s. However after around 2005 to 2010, the research area related with ML and DM has become drastically huge, affecting a lot of impacts on a variety of science and engineering fields, and also our society and life. Originally the application of ML was very limited and closely related with scientific or academic research areas, such as natural language processing, speech recognition, computer vision, robotics and medical engineering. However due to the development of internet, the applications were becoming more diverse, such as web searching, recommendation, fraud detection, etc. Also ML has accelerated pure science research, including drug discovery process, biological experiment management and physical or chemical simulations. Furthermore by the recent explosive data increase, the applications of ML are not limited to science and engineering but also abundant fields of our society, such as autonomous driving, industrial applications, finance, gaming, cooperate management, construction and agriculture.

There would be a lot of ways to classify a variety of ML and DM methods into some sort of order. In this book, we first focus on input data types. We segment ML and DM methods into seven categories, from a viewpoint of input data types. More in detail we categorize data types into the following sevens: *vectors*, *sets*, *sequences (strings)*, *trees*, *graphs*, *nodes in a graph* and *integrated data*. The last integrated data means that data can be a combination of one or more datasets, such as sequences plus graphs. We will define and also explain each of these seven types in Chapter 2. Then each of the seven data types is described by one chapter: Chapters 3 to 9, in the above order. In each chapter, we further classify the methods for the corresponding data type, by either problem settings or their techniques. That is, problem settings are supervised, unsupervised and semi-supervised learning. This was done for *vectors* and *nodes in a graph* (Chapters 3 and 8). On the other hand, the techniques are frequent pattern mining, probabilistic models (statistical learning) and kernel learning. This way of classification by techniques was done for *sets*, *sequences*, *trees*, *graphs* (Chapters 4 to 7). Then we explain one or more standard methods for each classified setting, focusing on already established methods. Also in the last of each chapter, we raise one or more application examples of bioinformatics.

These data types, such as vectors and graphs, look totally different from each other, and also maybe because of this, ML methods for different data types would be thought to be different from each other. They however definitely have some common ideas, motivations and derivation, etc., especially techniques behind different methods. One example is the formulation of principal component analysis, kernel K-means clustering (both for vectors), spectral clustering (for graphs), and canonical correlation analysis (for integrated data), all take the form of so-called Rayleigh quotient and eventually their setting can be solved by a (generalized) eigenvalue problem. Also there is a reverse case, in which different methods have been developed for the same problem setting for the same data type, while they have been developed by different ideas. We will explain these cases on what point they are different and also how reasonably they have been developed. We think that describing such shared points of different methods or unique features of each method will clarify each corresponding method.

Theoretical work are usually classified mainly by their significance into different levels, such as *Theorems*, *Lemmas*, etc. In this book theoretical results are all presented as *Propositions*, since they are rather already established.

After the seven chapters corresponding to the seven data types, we extensively explain practical manners of evaluating the results obtained by applying ML methods to actual data. Even if ML methods can be analyzed theoretically, it would be reasonable to investigate how well predictions by ML methods can be succeeded, to understand the performance of the ML methods themselves and also the hardness of the problem or the data, to which the methods are applied.

The last chapter is Appendix, which describes the terms/methods used in this book. Although the corresponding sections of Appendix are cited in the main text as much as possible, readers can refer Appendix if coming across the terms not described well in the main text. The last part of Appendix is the detail of several derivations in the main text.

The idea of this book is to cover major and standard approaches of already established problem settings in ML and DM extensively, showing their relevance and differences. Then ML and DM areas covered by this book are major, while much more ML topics are not considered in this book, because of rather being new or too complicated for an introductory text book. These uncovered topics include reinforcement learning, transfer learning, query learning, learning to rank, statistical Bayesian learning for numerous settings, such as Bayesiann belief networks, etc. There exist good books and reviews already for each of these problem settings, and interested readers can refer to them.

Chapter 2

Concepts and Terminology

2.1 Machine Learning and Data Mining

We will describe a high-level abstract on machine learning and data mining (also additionally bioinformatics), introducing key terms to be used. These terms are explained later more. Readers, particularly entry-level machine learners, are recommended to go through this chapter without thinking to understand the terms fully within this chapter.

Two terms, Machine Learning and Data Mining, can be considered interchangeably, like that they are synonyms. We then use the term "machine learning" mainly, while data mining are more application-oriented, and we use either of these two terms, depending on situations.

Data, input of machine learning, are a set of records, each being called an *instance*, *example* or *sample*. We use the "instance" among them through this book. One instance has *features*, *variables* or *variates*, where these features are visible and unchanged (during learning) and so called *observable variables*.

Occasionally instances can be classified into more than one categories, typically two (binary). Here are two examples:

Example 1: Mobile phone customers of some company can be segmented into at least the following two: *current subscribers* and *former subscribers* (though current subscribers might be categorized into more details, due to the risk of leaving. Also promising subscribers among totally non-subscribers would be also a possible, important category for mobile phone companies).

Example 2: Regarding some disorder, *patients* and *healthy people* (also patients would be categorized into more, due to the seriousness of the disorder).

As such, when instances can be segmented into categories, each category is called a *class*, and each of the above *current subscribers*, *former subscribers*, *patients* and *healthy people*, etc. is called a *label*. The label is the name of a class, and if there are five labels, this means there exist five classes. Here are more examples:

Example 3: A customer is an instance, where the evaluation by one customer against some item is a label. Also even if a customer buys an item, this is already a label, to fill the corresponding part of the data.

Example 4: A gene is an instance, where its function can be a label.

In Example 3, the evaluation can be binary or some moderate-size number, like a five-stage, while in Example 4, the functions of genes cannot be even by such a small number but a larger number. These two examples imply that labels would not be necessarily assigned to all instances (or functions are not necessarily assigned to all genes), implying that assigning labels or functions to genes would be a possible problem setting. This means that a label is a value of one feature which should be predicted in machine learning.

Again all instances do not necessarily have labels, because obtaining labels is expensive. For example, in Example 3, we need to ask customers to give some evaluation against some item. Also we need some biological experiments to annotate functions to genes in life sciences, which needs cost and also time, which in some cases might not be good enough if it is really hard to find functions of genes.

Thus in machine learning we have a paradigm which do not assume classes and labels. This setting is called *unsupervised learning*, where the objective is to understand the data *distribution* or more generally to summarize data. A typical approach for summarizing data is *clustering*, where instances are grouped into several *clusters*. Clusters are different from observable variable like features and variables where their values can be trained from given data. Thus this variable is called a *latent variable*. Estimating latent variables means clustering.

On the other hand, we call learning using labels *supervised learning*, where inputs are features of instances and estimate a function to predict the label. This function is called a *hypothesis*, *model*, *classifier* or *predictor*. Note that model is not only the concept for supervised learning but also for estimating distribution in unsupervised learning. This function has *parameters*, where the values of parameters are trained (learned or estimated) by using input data, and the trained hypothesis is used to predict labels of unknown instances (or data). The first input data are called *training data* and the second is called *test data*.

Estimating a model from given training data means, for example, to keep the values given by the model closest to those of training data. Thus we can set up an *objective function*, for example, in supervised learning, being consistent with the error of the model from training data, i.e. an *error function* or *loss function*. We can then minimize the objective function, which turns into an optimization problem, like minimizing the error function. The optimization in terms of training data only causes overfitting of the model to the training data. Avoiding that, the model needs to be generalized, which is called *generalization* or *regularization*. In practice, we add one or more constraints to the objective function, where the constraints are, mathematically, terms, called *regularizers* or

Table 2.1: Data in life sciences.

	data types	examples
gene	vector	gene expression
	sequence (string)	nucleic acid sequence
	node in graph	gene regulatory network
protein	sequence (string)	amino acid sequence
	node in graph	protein-protein interaction network
chemical compound	(molecular) graph	chemical structure
(e.g. drug)	node in graph	metabolic pathway
glycan	tree	sequence structure

regularization terms. Thus to solve some problem, we can set up an objective function with regularizers, which we often call problem *formulation.* Usually in machine learning model formulation, model parameters to be estimated are in either the objective function or regularizers. Also model formulation sometimes has *hyperparameters*, which are given arbitrary by users or decided empirically but not trained from input data. For example, in model formulation, the coefficients of regularizers, called *regularization coefficients* are one typical hyperparameter.

We call values, which are clearly different from the data distribution, an *outlier.*

Now let's get back to Example 3, where the situation is an E-commerce site. It would not be the case that an customer has bought almost all items in the E-commerce site, while usually most customers bought only just a few goods in the site. Thus the data from E-commerce, i.e. a matrix of users (rows) and items (columns) has only a small number of elements, filled by some values, showing the goods evaluation by customers or just the user purchase history. More concretely the row vectors or users have only a few number of values, with others being *missing information.* This situation of data is called *sparse* data, and interestingly sometimes this data *sparsity* is useful to solve the problem efficiently even if the given matrix is huge.

2.2 Bioinformatics: Connections to Data Types

Bioinformatics is the study on data in life sciences, particularly molecular biology, so the focus being placed on molecules working in cells, such as *genes*, *proteins*, *chemical compounds*, etc. We can describe data formats of those molecules which are regarded as multiple ways even for one single molecule type, depending on situations.

Gene is the most attention-paid term in life sciences. Recent so-called *high-throughput* techniques allow to measure the expression of thousands of genes in cells simultaneously. The measurement can be performed under many different conditions, which results in numerical *vectors* of genes. At the same time, physically genes are *nucleic acids*, consisting of four *bases* (represented by four letters:

A, T, C, G), which are building blocks of *nucleic acid sequences*. Simply speaking they are *sequences* or *strings* of four letters. Proteins can be generated, following the expression of genes, by which the term "gene" is used in most cases, instead of protein. For example, gene function means the function of the corresponding protein. Physically proteins are also sequences, while building blocks of proteins are 20 different types of *amino acids*, by which sequences of proteins are *amino acid sequences* or strings of 20 letters. Another aspect of genes is gene regulation, which are eventually represented by a *graph*, called *gene regulatory network*, in which *nodes* are genes and an edge shows that a gene is regulated by another gene. Similarly one aspect of proteins is three-dimensional binding between two or more proteins, which are called *protein-protein interactions*, also usually being represented by a graph with nodes of proteins and edges of interactions.

Glycans are called the third molecule in cells, next to nucleic acids and proteins. Building blocks of glycans are *monosaccharides*, and glycan sequences are generated by the connection of monosaccharides. The difference of glycan sequences from gene sequences is that glycans allow to have branches (but no cycles) in their sequences, which results in glycan *trees*.

Nucleic acids of genes, amino acids of proteins and monosaccharides of glycans are all *chemical compounds*, and so molecules. The chemical structure of molecules are *graphs*, so-called *molecular graph*, meaning that chemical compounds can be presented by graphs. On the other hand, for example, *metabolic network* is a collection of chemical reactions in cells and shows the process of generating chemical compounds from other compounds. Thus chemical compounds can be nodes in the graph of metabolic network.

Table 2.1 shows the summary of molecules in cells and their possible data formats. As you can see from this table, bioinformatics data can be five types. When we think about general applications of machine learning and data mining, we can add one more data type, *sets*, to them, resulting in six types: vectors, sets, sequences (strings), trees, graphs and nodes in a graph. These data types are explained in more detail in the next section.

2.3 Six Types of Data

Data can be categorized into six types, partially according to the observation on the life science data. This book is organized according to the six data types, so each chapter being for each data type, as follows:

Chapter 3: Vectors

Chapter 4: Sets

Chapter 5: Sequences (strings)

Chapter 6: Trees

Chapter 7: Graphs

Chapter 8: Nodes in a graph

	Feature 1	Feature 2	Feature 3	Feature 4
Instance 1	A	B	C	D
Instance 2	A	C	C	C
...				

Figure 2.1: Instances, each being a vector, are a matrix.

Instance 1: $\{A, B, C, D\}$, Instance 2: $\{D, C, B, A\}$

Figure 2.2: Two instances, each being a set.

Chapter 9 is then on machine learning for data integration.

2.3.1 Vectors

When each instance has multiple features, the simplest and most frequent data in machine learning and data mining is vectors, where each instance has always a certain number of features. Fig. 2.1 shows an illustrative example of vectors (called *feature vectors*) for each instance, resulting in a matrix for multiple instances. Here are features, terminology and examples:

Building blocks: are feature values, which can be either discrete or continuous.

Fixed length and order: The length of a vector is fixed, and also a sort of feature values are ordered, meaning that for example, if Feature 1 in Fig. 2.1 is fixed to be a binary taking A or B, all first values of instances in this dataset must be A or B.

Example 1: demographic data: One instance is an individual, where features are age, gender, etc.

Example 2: gene expression: One instance is a gene, and features are experimental conditions, under which expression of the corresponding gene are measured. Each matrix element is the expression value of the corresponding gene and the corresponding experimental condition.

2.3.2 Sets

Each instance is a set, and input data is a set of sets. Sets are the most flexible machine learning data type. One set, i.e. an instance, can have an arbitrary number of elements, without any order. Fig. 2.2 shows an illustrative examples of sets. Here are features, terminology and examples:

Building blocks: One set has elements, which can be discrete or continuous values.

Instance 1: *DCBA*, Instance 2: *ADBCA*

Figure 2.3: Two instances, each being a sequence.

No order in elements: We focus on discrete values for elements. Elements have no order, and so for example, in Fig. 2.2, {A, B, C, D} and {D, C, B, A} are the same.

No fixed size of elements: Each instance has its own size. We can say that sets are more general than vectors, and vectors are a special case of sets, in the sense that the size and order of features in sets are both fixed in vectors.

Example: Market basket: One typical example is *market basket*, which is a set of items bought by one customer at a department store, a convenience store or an E-commerce site per time. By focusing on a certain number of items, one set can be a vector. In general the number of items is always very large, while each user buys only a few number of items, which makes the data very *sparse*. In fact elements of a matrix of users vs. items are mostly missing.

2.3.3 Sequences and Strings

When elements in one set are ordered, the set becomes a *sequence*. In particular elements are a finite number of letters, such as the alphabet, the sequence is a *string* (see below in some more detail). The size of elements in a set, i.e. the length of a sequence, is changeable over different instances.

Thus again the sequence is a special case of a set, while the length of each sequence is not fixed, meaning that a vector is a further special case of a sequence. Fig. 2.3 shows an illustrative example of sequences. Below we describe the definition and terminology of sequences.

Discrete element for sequences: Discrete elements of sequences are called *letters* or *characters*. The set of letters is called the *alphabet*. A sequence of letters is called a string.

Subsequence and substring: A consecutive part of a sequence is called a *subsequence*. Also a consecutive part of a string is called a *substring*. For example, in Instance 1 of Fig. 2.3, DC and DCB are substrings, while DCA, i.e. the first, second and fourth letters, is not a substring.

Example 1: natural language: Typical sequence examples are natural language, which are the main data of computational linguistics, natural language processing and speech recognition, etc., which are classical applications of machine learning.

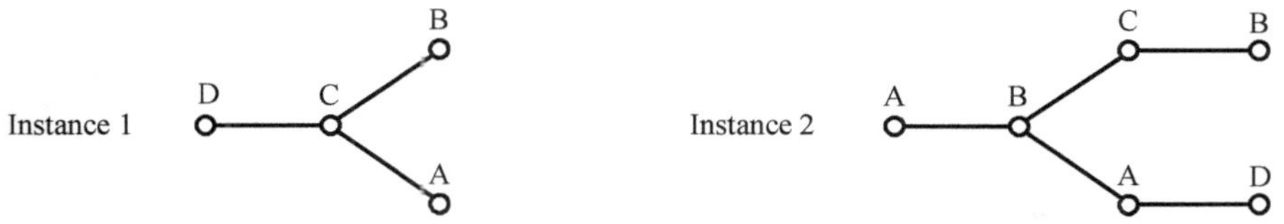

Figure 2.4: Two instances, each being a tree.

Example 2: gene sequence: A nucleic acid sequence can be a string of four letters (corresponding to four types of nucleic acids). Similarly an amino acid sequence can be a string of twenty letters (corresponding to twenty amino acids). The length of one sequence/string can be different.

2.3.4 Trees

Each instance is a tree, and so the input data is a set of trees. Description of trees can be easier if graphs are already defined, and so suppose that graphs are already defined, trees are graphs without any cycling edges (or cycles).

Fig. 2.4 shows illustrative examples of trees, where Instance 1 has a branch from C to B and also to A, and similarly Instance 2 has a branch from B to C and A. Here are more terminology and features of trees:

Building blocks: A tree consists of nodes and edges.

Cycles: No cycles of edges in trees. This discriminates trees from graphs.

Labels: A tree with labels is called a *labeled tree*. For example, in Fig. 2.4, A, B, C and D are labels. We consider labeled trees.

Root: In general, any node of a tree can be a *root*, while a tree is called a *rooted tree*, if one node of the tree is fixed as the root. For example, D of Instance 1 is fixed as the root, and A of Instance 2 is fixed as the root. We focus on rooted trees, and so we consider *labeled rooted trees*.

Leaves and internal nodes: In the rooted tree, except the root, we regard all nodes with only one edge as *leaves*. Nodes except leaves and the root are called *internal nodes*.

Ordered tree: For a rooted tree, nodes can be ordered from the root to leaves, and the tree is an *ordered tree*. Ordered trees are generated by allowing branches in sequences. In other words, sequences are a special case of ordered trees, by not allowing any branches.

Parent-children: In the rooted tree, for two nodes connected by an edge, the node closer to the root is called *parent* and the other node is called a *child*. Comparing with sequences, a feature of trees is a parent can have more than one children, while in sequences (and strings), a parent can have only one. Also the root cannot have a parent. In ordered trees, the parent to child direction is regarded as the order.

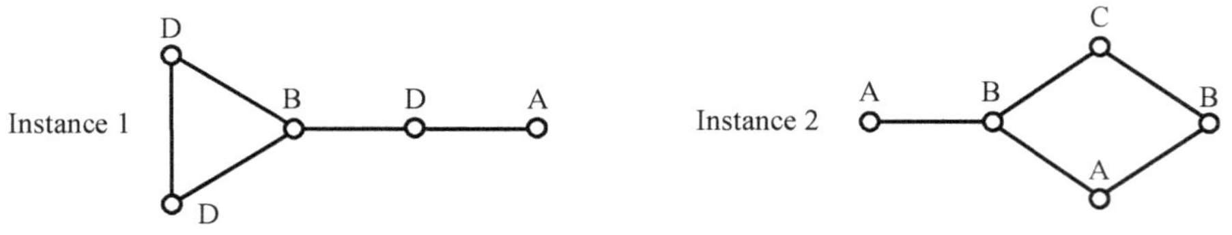

Figure 2.5: Two instances, each being a graph.

Siblings, ancestors and descendants: Nodes with the same parent are called *siblings*. For a child, its parent and any node of the parent side are all *ancestors*. Similarly for a parent, its children and any node of the children side are all *descendants*.

Depth: The number of edges from the root is called the *depth*. Usually the depth of the root is zero. The nodes with the same depth are called nodes at the same *layer*. On the other hand, the number of edges from a leaf is called the *height*. Usually the height of an leaf is zero.

Direction of edge: In ordered trees, usually the direction is from the root to leaves.

Subtree: We explain regular definition of *subtree* below, while in frequent subtree mining, subtree means any connected part of a tree. This definition is different from the regular definition (see Section 6.3 for more detail). In fact the regular definition is based on the definition of subsequence in sequences/strings, while the above definition (any part of a tree) is derived from the definition of subgraph in graphs.

A subtree, a part of a tree, has some node (of the original tree) as the root (of this subtree) and also all nodes and edges of the leaf side of the root. For example, in Instance 2 of Fig. 2.4, if the root is the most left-hand side A, CB and AD of the most right-hand side can be subtrees, while BAD cannot be a subtree, because if B is specified as the root of a subtree, the subtree must be B(AD)CB.

Example: glycan (carbohydrate sugar chain): Building blocks of glycans are around 10 to 15 types of monosaccharides, which can be letters (labels). Glycan is generated by the connection (binding) of these monosaccharides, and the connection allows branches, resulting in trees with labels of monosaccharides. Glycans will be explained more in Section 6.4.

2.3.5 Graphs

One instance is a graph and input data are a set of graphs. Fig. 2.5 shows illustrative examples of graphs. Comparing with trees, one important feature of graphs is that graphs have cycles. For example, in Instance 1 of Fig. 2.5, two D and B are connected each other in the left-hand side, which turns into a cycle. This is not allowed in a tree, and trees are a special case of graphs. Below are features, terminology and examples of graphs:

Building blocks: A graph consists of nodes and edges connecting nodes.

Degree: The number of edges connecting to one node is called the *degree* of the node.

Labels: A graph with labels is called a *labeled graph.* Instances in Fig. 2.5 are two labeled graphs.

Direction of edges: If edges have some direction, the graph is called a *directed graph,* while if no directions, the graph is called an *undirected graph.*

Cycles: In graphs, any connection is allowed, and so it is possible to start with some node and come back to the same node along with edges. This is called a *cycle.* Also the cycle allows to go from a node to another node by using more than one routes (edge connections). Again graphs with no cycles are trees, meaning that there is always only one route in a tree if going from one node to another.

Subgraph: A part of a graph is called a *subgraph.* If all nodes and edges are connected each other in a subgraph, the subgraphs is called a *connected subgraph.* We consider only connected subgraphs as subgraphs.

Spanning tree: A tree with all nodes of a undirected graph is called a *spanning tree.* This is an important concept to deal with a graph efficiently.

Isomorphism: If two graphs have the same structure in terms of nodes and edge connectivity, they are called *isomorphic.*

Example: chemical structure of chemical compounds: The types of (physical) elements of chemical compounds are limited and so can be labels. Then the chemical structure (called a *molecular graph*) of a chemical compound can be regarded as a labeled graph.

2.3.6 Nodes in a Graph

Input data is a graph, and one instance is a node in the graph. Fig. 2.6 shows a simple, illustrative example. Nodes are instances, meaning that nodes are all unique (Note that this does not mean that labels of nodes are all different). Also all nodes in one graph is a set of all instances.

Example 1: social network: A social network is a graph with nodes for individuals and edges for some relationship between the individuals connected by the corresponding edges. Note that nodes are all unique, and they can be assigned by a limited number of labels, such as a male or a female.

Example 2: gene regulatory network: A gene regulatory network is a collection of molecular-level biological knowledge on gene regulations, such as that gene A is regulated by gene C. Regarding gene regulations as binary relations, i.e. two nodes connected by one edge, a graph with all binary relations of gene regulations is a gene regulatory network, where nodes are all unique genes. Again in this case also, genes can be with a limited number of labels, such as gene functions.

2.3.7 Notes on Data Types

Structured and Semi-structured Data

The most typical data in machine learning, i.e. vectors, are called *structured data*, while others are called *semi-structured data*. We think that the term *structured* data are derived from the fact that vectors can be organized well into a matrix with lows for instances and columns for features. However, sometimes even in machine learning and data mining publications, these two terms, i.e. structured and semi-structured data, are used in different ways, such as structured data for graphs. Thus you can be careful when you come across the term "structured data".

Structured data or vectors were long-time the major data in machine learning, while semi-structured data are relatively new in machine learning, except for sequences and strings (which have been used in applications of machine learning, such as speech recognition, natural language processing and bioinformatics).

Inclusion Relations among Data Types

Let $\mathcal{S}_V$, $\mathcal{S}_S$, $\mathcal{S}_Q$, $\mathcal{S}_T$ and $\mathcal{S}_G$ be all possible data of vectors, sets, sequences, trees and graphs, respectively.

If elements of a set are ordered, the elements become a sequence (or a string), and so sequences are part of sets. Also the size (length) of elements in a set (sequence) is fixed over all instances, vectors are part of sets (sequences). This observation leads to the following inclusion relations:

$$\mathcal{S}_S \supseteq \mathcal{S}_Q \supseteq \mathcal{S}_V \tag{2.1}$$

Trees are graphs without cycles, indicating that trees are a special case of graphs. Ordered trees are a subset of trees. Sequences are ordered trees without any branches, meaning that sequences are a special case of ordered trees. Thus in summary the following relations come out:

$$\mathcal{S}_G \supseteq \mathcal{S}_T \supseteq \mathcal{S}_{OT} \supseteq \mathcal{S}_Q, \tag{2.2}$$

where $\mathcal{S}_{OT}$ is all data of ordered trees.

Data Transformation

In the above relations, when one data type, say A, can be defined by some constraint on another data type, say B, if the constraint is removed, A becomes B.

For example, if the order of elements in sequences is removed, the sequences can be sets. More practically, two sequences, ATCG and ACGT, can be the same set {A, C, G, T} if we remove the order of letters in sequences,

In general however this type of transformation are not performed, because clearly this transformation loses some important information. For example, in the above toy example, originally ATCG and ACGT are different, and the order of letters is definitely good information to discriminate them.

Thus simply speaking some appropriate machine learning method should be developed for each data type.

Machine Learning for Different Data Types

However machine learning algorithms for some data type with less constraints can be applied to another data type with more constraints rather easily sometimes. On the other hand, applying some algorithm with more constraints to those with less constraints is not easy or impossible. For example, an algorithm for graphs can be applied to ordered trees or sequences just by considering the order of nodes of simpler graphs, while an algorithm for trees or sequences cannot be applied to graphs so easily.

Also even if one algorithm can be applied to another data type, like that an algorithm for graphs can be applied to sequences, this would never be the best, because the algorithm for sequences can be designed as a more efficient one usually. Thus again, thinking about the method most proper for each data type would be an important aspect of machine learning and data mining.

2.4 Structure of This Book

Machine learning and data mining methods/ algorithms are summarized for each of the above six data types, and so there are six chapters on vectors, sets, sequences, trees, graphs and nodes on a graph. Then after the six chapters, we have one chapter which focuses on more than one data types, which are called by many ways in machine learning, such as *data fusion*, *multiview learning*, etc. On the other hand, we call all methods for data integration and machine learning *data-integrative machine learning* methods.

Chapter 3

Learning Vectors

The most typical data in machine learning is a vector for each instance, which is a matrix for all instances or input data. The matrix with rows for instances and features for columns have elements for the corresponding feature values. A key aspect of vectors or a matrix is all instances with a fixed number of features. Here are several examples:

Demographic data: Instances are individuals and features are personal information like thier age, gender, occupation, etc.

E-commerce or business data: Instances are users (customers), and features are if each item was bought by a user (or the number of times the user bought the items).

Life science data: We can raise the following two more concrete examples:

Gene expression
Instances are genes and features are different experimental conditions.

Drug sensitivity
Instances are patients and features are drugs that patients receive, where each element is the reaction result by the corresponding patient and drug.

A lot of machine learning methods have been proposed for this data type, and they are already matured. We explain three typical paradigms of machine learning: 1) unsupervised and 2) supervised learning and 3) feature leaning (such as feature selection), in this order, being followed by kernel learning, which is a framework covering over the three paradigms from a kernel viewpoint. Finally we will show several bioinformatics applications for this data type.

Notation in This Chapter

Input matrix with N instances and M features	$\boldsymbol{X} \in \mathcal{R}^{N \times M}$
Instances	$\boldsymbol{x}_i \ (i = 1, \ldots, N)$
Label of instance $\boldsymbol{x}_i$	y_i
Features	$\chi_m \ (m = 1, \ldots, M)$

(Note that $\boldsymbol{x}_i$ is the i-th instance and x_i is the i-th feature of instance $\boldsymbol{x}$. The m-th element of instance $\boldsymbol{x}_n$ and the n-th element of feature χ_m are both $\boldsymbol{X}_{nm}$)

3.1 Unsupervised Learning

The objective of unsupervised learning is to *summarize given data.* We will show several methods for this purpose.

3.1.1 Clustering – Objectives

Clustering is to group given instances into several groups, where each group has similar instances. The group is called a *cluster.* There are two types of clustering: 1) *disjoint clusters*: clusters have no overlap each other, and 2) *non-disjoint clusters*: clusters can be overlapped with each other. We focus on disjoint clusters. The objective of clustering is the following two:

Summarizing data: Again after grouping input data into several clusters, we can present the most (not only one but also more) representative/ typical instance from each cluster. This is one approach for summarizing the input data.

Label prediction: Given instances may have labels partially. We then do clustering given instances, and if you found a cluster with instances assigned by a certain label mostly, you can predict the corresponding label to (label) unknown instances in the same cluster. This is already label prediction.

There are two types of clustering:

1) **Partitional clustering** The input data are partitioned into groups (clusters) at once.

2) **Hierarchical clustering** Clusters have a hierarchical structure, like a cluster has two subclusters always until a subcluster is an instance. There are two possible ways for hierarchical clustering:

   2-1) **top-down:** The given data are divided into groups recursively until one group has only one instance.

   2-2) **bottom-up:** Starting with individual instances, two closest instances (or clusters) are repeatedly picked up and merged one by one iteratively into one cluster, until all instances are merged into one cluster covering all instances.

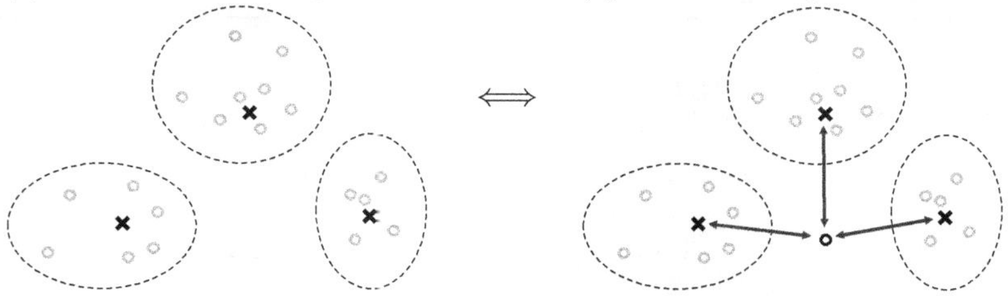

Figure 3.1: Schematic diagram of K-means.

We describe two methods of partitional clustering: K-means and finite mixture model, and then hierarchical clustering with the bottom-up type.

3.1.2 Clustering – K-means

K-means clustering is the most basic clustering method.

Background

For grouping instances, what we want to know first is the boundary of groups. In other words, if we know the locations of clusters, we can assign each instance to its closest cluster. Then, if we already know all instances in each cluster, we will be able to define clusters, by which the cluster boundary would be decided. Thus we can first assume that we already know clusters, and then assign instances to clusters. Then from the instances in each cluster, we can determine the cluster boundary. Thus again starting with some random initial cluster values, we can repeat the above two steps alternately. This *heuristics* is exactly the K-means clustering algorithm, which we will implement this more concretely below.

Algorithm

Now the problem is how to find the closest cluster of an instance or how to measure the distance between a cluster and an instance. If this is fixed, we can see the cluster boundary and assign any instance to the closest cluster. Thus for example, we assume that cluster k is the *means* (or *center*) $\boldsymbol{\mu}_k$ of all instances in the cluster:

$$\boldsymbol{\mu}_k \quad \leftarrow \quad \frac{\sum_{i|\boldsymbol{x}_i \in C_k} \boldsymbol{x}_i}{|C_k|}, \tag{3.1}$$

where C_k is the set of instances in the k-th cluster, $\boldsymbol{x} \in C_k$ is the instance $\boldsymbol{x}$ in cluster C_k and $|C_k|$ is the number of instances in cluster C_k.

Now an instance can be easily assigned to the closest cluster. Also we can have the boundary between two clusters which is defined as the line on which the distances from two neighboring clusters are equal. Fig. 3.1 shows a schematic picture of the algorithm implementing the above idea, which repeats two steps, i.e.

Algorithm 3.1: K-means.

```
1 Function K-means(X, K)
      Data: training data: X, #clusters: K
      Result: K cluster centers
2     Assign each instance randomly to one of K clusters.;
3     repeat
4         Compute the center of a cluster from examples assigned by using
          (3.1).;
5         Assign each of all examples to its closest cluster center.;
6     until convergence;
7     Output the latest cluster centers.;
```

a) computing the mean (or center) of each cluster, and b) assigning each instance to its closest cluster. **Algorithm 3.1** shows a pseudocode of this algorithm. In this algorithm, again the method repeats the two steps alternately: 1) computing the cluster center from given instances (line 4) and 2) assigning each instance to its closest cluster (line 5), which correspond to Fig. 3.1 (a) and (b), respectively.

Time Complexity

1. Cluster center computation

   Clusters are exclusive, and just only one cluster is for one instance. Thus the procedure should be first to scan over all instances to have all instances for each cluster, and then compute the means (center) for every cluster. Thus the number of iterations is $N + K$ times, and so the complexity is $O(NM)$, given $N >> K$.

2. Instance assignment to its closest cluster

   This step needs to consider all clusters for each instance, which is equivalent to the all combinations (pairs) of instances and clusters, resulting in $O(NMK)$.

Entirely the complexity of each iteration is $O(NMK)$.

Cluster Assignment Matrix

Let $\boldsymbol{Z} := (\boldsymbol{z}_1, \ldots, \boldsymbol{z}_K)$ be a *cluster assignment matrix*, with rows for instances and and columns for clusters. The vector $\boldsymbol{z}_k$ corresponds to the k-th cluster, with N elements, each corresponding to an instance. That is, $\boldsymbol{z}_k = (z_{1,k}, \ldots, z_{N,k})^\top$ and $z_{n,k} \in \{0, 1\}$. This means if instance $\boldsymbol{x}_n$ is in cluster k, $z_{n,k}$ is one; otherwise zero. In $\boldsymbol{Z}$, each row vector has only one element with 1 and others with zero.

Practically, for example, $\boldsymbol{Z}$ can be given as follows:

$$\boldsymbol{Z} = \begin{pmatrix} 0 & \cdots & 0 & 1 & 0 & \cdots & 0 \\ 0 & 1 & 0 & \cdots & 0 & \cdots & 0 \\ & & & \vdots & & & \\ 0 & \cdots & 0 & \cdots & 0 & 1 & 0 \end{pmatrix}. \tag{3.2}$$

When we think about two column vectors in cluster assignment matrix $\boldsymbol{Z}$, first since one instance cannot be in two clusters, two column vectors cannot have two 1s for any of the same row, meaning that the inner product of arbitrary two column vectors is always zero. On the other hand, the inner product of the same two column vectors shows the size of instances in the column or the size of the cluster (since again each column corresponds to a cluster). This can be summarized into the following:

$$\boldsymbol{z}_k^\top \boldsymbol{z}_{k'} = 0 \quad \text{if} \quad k \neq k', \tag{3.3}$$
$$\boldsymbol{z}_k^\top \boldsymbol{z}_k = |C_k| \quad \text{otherwise}. \tag{3.4}$$

Thus $\boldsymbol{Z}^\top \boldsymbol{Z}$ is a $(K \times K)$-diagonal matrix with the size of clusters for diagonal elements. Note that non-diagonal elements are just zero, as described above. Thus here is $\boldsymbol{Z}^\top \boldsymbol{Z}$:

$$\boldsymbol{Z}^\top \boldsymbol{Z} = \begin{pmatrix} |C_1| & 0 & \cdots & 0 \\ 0 & |C_2| & 0 & \vdots \\ \vdots & 0 & \ddots & 0 \\ 0 & \cdots & 0 & |C_K| \end{pmatrix}. \tag{3.5}$$

On the other hand, $\boldsymbol{Z}\boldsymbol{Z}^\top$ is a binary $(N \times N)$-symmetric matrix, in which if an element is 1, the two instances corresponding to the row and column, which specify this element, are in the same cluster; if zero, they are in different clusters. For example, if instance $\boldsymbol{x}_i$ and instance $\boldsymbol{x}_j$ are in the same cluster, the element of the i-th row and j-th column of $\boldsymbol{Z}\boldsymbol{Z}^\top$ is 1; 0 if they are not in the same cluster. Thus diagonal elements are all 1, and again other elements show the sharing of clusters. Below is an example:

$$\boldsymbol{Z}\boldsymbol{Z}^\top = \begin{pmatrix} 1 & 0 & 0 & 1 & 0 & 1 \\ 0 & 1 & 1 & 0 & 1 & 0 \\ 0 & 1 & 1 & 0 & 1 & 0 \\ 1 & 0 & 0 & 1 & 0 & 1 \\ 0 & 1 & 1 & 0 & 1 & 0 \\ 1 & 0 & 0 & 1 & 0 & 1 \end{pmatrix}. \tag{3.6}$$

You will be able to see that in this example, there are two, non-overlapping clusters, one being $C_1 = \{\boldsymbol{x}_1, \boldsymbol{x}_4, \boldsymbol{x}_6\}$ and the other being $C_2 = \{\boldsymbol{x}_2, \boldsymbol{x}_3, \boldsymbol{x}_5\}$.

There are no overlaps in this clustering, by which we can change the order of rows and columns so that the matrix can be transformed into a *block diagonal matrix* as follows:

$$\begin{pmatrix} 1 & 0 & 0 & 1 & 0 & 1 \\ 0 & 1 & 1 & 0 & 1 & 0 \\ 0 & 1 & 1 & 0 & 1 & 0 \\ 1 & 0 & 0 & 1 & 0 & 1 \\ 0 & 1 & 1 & 0 & 1 & 0 \\ 1 & 0 & 0 & 1 & 0 & 1 \end{pmatrix} \Rightarrow \begin{pmatrix} 1 & 1 & 1 & 0 & 0 & 0 \\ 1 & 1 & 1 & 0 & 0 & 0 \\ 1 & 1 & 1 & 0 & 0 & 0 \\ 0 & 0 & 0 & 1 & 1 & 1 \\ 0 & 0 & 0 & 1 & 1 & 1 \\ 0 & 0 & 0 & 1 & 1 & 1 \end{pmatrix}. \tag{3.7}$$

We note that (3.6) corresponds to a so-called *adjacency matrix*, which 1) shows similarity between instances and 2) is equivalent to a graph with unique nodes (see Chapter 8 for detail).

Objective Function of K-means

In fact we explained K-means without any *objective function*. However, when we build a machine learning algorithm for some problem, we need to understand what objective should be satisfied to solve the given problem. You may feel that the K-means algorithm was already presented, and so this problem of clustering was already solved. It however would be worth understanding what objective function is used in K-means and also optimized by the heuristics.

The objective function of K-means is the sum over all distances, each between one instance and its closest cluster. Thus suppose that clusters are represented by the means given by (3.1) and also the distance is the squared loss (squared Euclidean distance), the objective function is given as follows:

$$f(\boldsymbol{Z}) = \sum_{k=1}^{K}\sum_{i=1}^{N} \boldsymbol{Z}_{ik}||\boldsymbol{x}_i - \boldsymbol{\mu}_k||_2^2, \tag{3.8}$$

where $\boldsymbol{Z}_{ik}$ is the i-th row and k-th column of cluster assignment matrix $\boldsymbol{Z}$.

The optimization problem of K-means is to estimate the cluster assignment matrix $\boldsymbol{Z}$ so that the above objective function (3.8) is minimized. This problem is NP-hard, and so the K-means algorithm estimates $\boldsymbol{Z}$ by heuristics (i.e. the recursive alternating algorithm), which solves the problem approximately by finding the local optimum.

3.1.3 Clustering – Constrained K-means

Background: K-means

The objective function (3.8) can be transformed into the following:

$$J = \sum_{k=1}^{K}\sum_{j=1}^{N}\sum_{i=1}^{N} \boldsymbol{Z}_{ik}\boldsymbol{E}_{ij}\boldsymbol{Z}_{jk} \tag{3.9}$$

$$= \mathrm{trace}(\boldsymbol{Z}^{\mathsf{T}}\boldsymbol{E}\boldsymbol{Z}), \tag{3.10}$$

where the element of the i-th row and the j-th column of $\boldsymbol{E}$, i.e. $\boldsymbol{E}_{ij}$, can be given as follows:

$$\boldsymbol{E}_{ij} = \frac{1}{2N}(\boldsymbol{x}_i - \boldsymbol{x}_j)^2. \tag{3.11}$$

The derivation of (3.10) and (3.11) from (3.8) is shown in Section A.9.1. From (3.11), we can first see that $\boldsymbol{E}$ is a symmetric matrix with variance for each element (again see Section A.2.1). Also from (3.10), we can now see that K-means is an algorithm to optimize $\boldsymbol{Z}$ in terms that the cluster assignment of instances should be smooth over the variances.

The original K-means considers (3.8) only, meaning that other important issues in clustering, such as avoiding small clusters generated by *outliers* were not considered. However for example, obtaining clusters with an equal size would be generally important and also might be useful for many applications. Also sometimes *prior knowledge* on the size of clusters might be already set up before clustering instances.

Thus here let us consider how we can solve clustering with such cluster size constraints. The equation (3.5) shows that $\boldsymbol{Z}^\top \boldsymbol{Z}$ is a symmetric matrix with cluster sizes as diagonal elements. Thus we introduce a diagonal matrix $\boldsymbol{D}$ and give cluster size constraints in $\boldsymbol{D}$. Then to estimate $\boldsymbol{Z}$ under the constraints given in $\boldsymbol{D}$, we use $\boldsymbol{D}$ and (3.5), as follows:

$$\boldsymbol{Z}^\top \boldsymbol{Z} = \boldsymbol{D}. \tag{3.12}$$

Then we use (3.12) to optimize $\boldsymbol{Z}$ to satisfy the given constraints.

For example, if we do not have any prior knowledge, a possible constraint would make cluster sizes equal, and so we can set up diagonal elements of $\boldsymbol{D}$ equal as follows:

$$\boldsymbol{D} = \begin{pmatrix} |D| & 0 & \cdots & 0 \\ 0 & |D| & 0 & \vdots \\ \vdots & 0 & \ddots & 0 \\ 0 & \cdots & 0 & |D| \end{pmatrix} \tag{3.13}$$

$$= \begin{pmatrix} \frac{N}{K} & 0 & \cdots & 0 \\ 0 & \frac{N}{K} & 0 & \vdots \\ \vdots & 0 & \ddots & 0 \\ 0 & \cdots & 0 & \frac{N}{K} \end{pmatrix}. \tag{3.14}$$

where $|D|$ is the cluster size of a cluster in this clustering, and so $|D| = \frac{N}{K}$ where K is the number of clusters.

Similarly if we have some background prior knowledge already on the size of

Algorithm 3.2: Eigenvalue decomposition.

Function `Eigen_decomposition(`$\boldsymbol{X}, K$`)`
Data: training data (symmetrix matrix): $\boldsymbol{X}$, #eigenvectors: K
Result: matrix with first K eigenvectors: $\boldsymbol{M}$

Generate the first K eigenvectors by solving $\boldsymbol{X}\boldsymbol{z} = \lambda\boldsymbol{z}$.;
Output the eigenvectors as a matrix.;

Algorithm 3.3: (Cluster size) constrained K-means.

Function `Cluster_size_constrained_`K`-means(`$\boldsymbol{X}$, $\boldsymbol{D}$, K_1, K`)`
Data: training data: $\boldsymbol{X}$, diagonal matrix on cluster sizes: $\boldsymbol{D}$, #eigenvectors: K_1, #cluster centers: K
Result: K cluster centers

Compute $\boldsymbol{E}$ by (3.11).;
Run `Eigen_decomposition(`$\boldsymbol{E}, K_1$`)` to have first K_1 eigenvectors as $\boldsymbol{M}$.;
Run K`-means(`$\boldsymbol{M}, K$`)`.;
Output the resultant cluster centers.;

clusters, such as $|D_1^*|$, $|D_2^*|$, $|D_3^*|$, ..., $|D_K^*|$, we can use them as diagonals of $\boldsymbol{D}$:

$$\boldsymbol{D} = \begin{pmatrix} |D_1^*| & 0 & \cdots & 0 \\ 0 & |D_2^*| & 0 & \vdots \\ \vdots & 0 & \ddots & 0 \\ 0 & \cdots & 0 & |D_k^*| \end{pmatrix}. \tag{3.15}$$

Then we can add (3.12) to the optimization problem (3.10) of K-means. The optimization problem can be written as follows:

$$\min_{\boldsymbol{Z}} \quad \text{trace}(\boldsymbol{Z}^\top \boldsymbol{E}\boldsymbol{Z}) \text{ subject to } \boldsymbol{Z}^\top \boldsymbol{Z} = \boldsymbol{D}. \tag{3.16}$$

This problem has equality conditions only, by which we can use the method of Lagrange multipliers (See A.8.5 for detail). The Lagrangian can be written as follows:

$$f(\boldsymbol{Z}) = \text{trace}(\boldsymbol{Z}^\top \boldsymbol{E}\boldsymbol{Z}) - \lambda\ \text{trace}(\boldsymbol{Z}^\top \boldsymbol{Z} - \boldsymbol{D}). \tag{3.17}$$

We then set the gradient (derivative) of the Lagrangian with respect to $\boldsymbol{Z}$ equal to zero, which results in the following equation:

$$\boldsymbol{E}\boldsymbol{Z} = \lambda\boldsymbol{Z}. \tag{3.18}$$

This is a regular *eigenvalue problem*. The resultant values are obtained in *eigenspace*, and finally K-means can be ran over these values to estimate $\boldsymbol{Z}$.

Algorithm

Algorithm 3.3 shows a pseudocode of the above procedure of *constrained K-means clustering*. In this algorithm, we first compute $\boldsymbol{E}$ from $\boldsymbol{X}$ following (3.11) (line 2). We then solve the eigenvalue problem to have K_1 eigenvectors (line 3) for running K-means algorithm over these eigenvectors (line 4).

This algorithm uses the eigenvalue problem (or *eigenvalue decomposition*), which is important and often used in many places in machine learning. **Algorithm 3.2** shows a pseudocode of this algorithm.

Also the objective function (3.18) can be written as follows:

$$\text{trace}(\frac{\boldsymbol{Z}^\top \boldsymbol{E}\boldsymbol{Z}}{\boldsymbol{Z}^\top \boldsymbol{Z}}). \tag{3.19}$$

Rayleigh Quotient

This formulation is called *Rayleigh quotient* (see A.8.7). A lot of machine learning problems can be written by the Rayleigh quotient or more general *generalized Rayleigh quotient* (see also A.8.7).

3.1.4 Clustering – Finite Mixture Model

Background: Probabilistic Version of K-means

In K-means, as shown in (3.1), one cluster was represented by the means (center) over all instances assigned to this cluster. Instances in one cluster, however, would have some *distribution*, which can be different (broad, sharp or whatever), depending on clusters, even if the cluster center is the same. Then the distance between one instance and one cluster can be different depending on the distribution, implying that the cluster should be represented by its (instance) distribution rather than just one point. In this section, we think about clustering instances by clusters with distributions.

We remember K-means, particularly constrained K-means, which has two terms in its objective function (3.19), one for cluster centers and the other for cluster sizes. We can also introduce two distributions/probabilities, a probability of clusters: ρ_k for cluster k ($\sum_k \rho_k = 1$) and a probability that an instance is generated from a cluster: $\rho_{\boldsymbol{x}|k}$ for instance $\boldsymbol{x}$ given cluster k. The ρ_k, nothing to do with given data, is like cluster sizes in the sense that this probability connects to prior knowledge and can be uniform if there are no prior knowledge, like that cluster sizes should be balanced if no prior knowledge. On the other hand, the $\rho_{\boldsymbol{x}|k}$ is exactly the distribution of instances we thought in the previous paragraph, meaning that this is related with cluster centers in K-means. These two probabilities already show some structure, which we call a *model*, particularly a *probabilistic model*.

From these two probabilities, we can compute the joint probability of instance $\boldsymbol{x}$ and cluster k:

$$p(\boldsymbol{x}, k) = \rho_k \rho_{\boldsymbol{x}|k}. \tag{3.20}$$

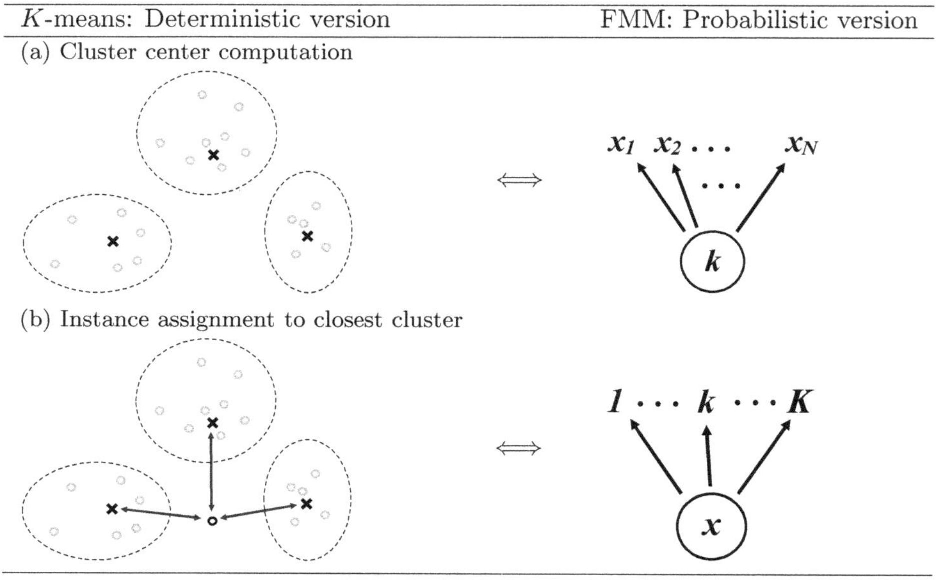

Figure 3.2: Correspondence between K-means and finite mixture model (FMM): (a) Cluster center computation in K-means is by $\rho_{\boldsymbol{x}|k}$ in FMM and (b) Instance assignment to closest cluster in K-means is by $\rho_{k|\boldsymbol{x}}$ in FMM.

This is called the *likelihood* of both instance and cluster k, i.e. $L(\boldsymbol{x}, k)$. Furthermore, the probability of instance $\boldsymbol{x}$, which is also called the likelihood $L(\boldsymbol{x})$ of $\boldsymbol{x}$ over all clusters, given the model, is given as follows:

$$L(\boldsymbol{x}) = p(\boldsymbol{x}) = \sum_k \rho_k \rho_{\boldsymbol{x}|k}. \tag{3.21}$$

This model is called a *finite mixture model* (FMM), and cluster k is called a *latent variable* or *hidden variable* (often written as $\boldsymbol{z}$), since this is not observable and can be estimated from data. We will describe how we can estimate the clusters, i.e. latent variable.

Algorithm: Estimating Parameters

If we try to set up K-means like algorithm for FMM, since we already have ρ_k, the prior knowledge, and $\rho_{\boldsymbol{x}|k}$, the instance distribution for a cluster, what we need further is the distance from each instance to cluster k. In other words, we would like to know which cluster each instance should go into. That is, we need to know the distribution of clusters given an instance. In fact, this distribution can be computed from the above two probabilities, as follows:

$$\hat{\rho}_{k|\boldsymbol{x}} \quad \leftarrow \quad \frac{L(\boldsymbol{x}, k)}{\sum_k L(\boldsymbol{x}, k)} = \frac{\rho_k \rho_{\boldsymbol{x}|k}}{\sum_k \rho_k \rho_{\boldsymbol{x}|k}}. \tag{3.22}$$

Algorithm 3.4: Parameter estimation algorithm of finite mixture model (EM algorithm).

```
1 Function Learning_finite_mixture_model(X, K)
      Data: training data: X, #clusters: K.
      Result: K probability distributions.
2     Assign each of all instances randomly to one of K clusters.;
3     Decide p_{k|x} (cluster assignment probability) randomly.;
4     repeat
5         M-step: compute ρ_{x|k}, k = 1..., K, parameters of K probability
          distributions, corresponding to clusters, by using the assigned
          instances.;
6         E-step: update ρ̂_{k|x} according to (3.22) and assign each instance to
          cluster, according to (3.23).;
7     until convergence;
8     Output the latest K probability distributions.
```

We then can assign each instance into cluster $k^*_{\boldsymbol{x}}$, which satisfies

$$k^*_{\boldsymbol{x}} \quad \leftarrow \quad \arg\max_k \hat{\rho}_{k|\boldsymbol{x}}. \tag{3.23}$$

Note that ρ_k is cluster distribution, which is *a priori* given, while $\hat{\rho}_{k|\boldsymbol{x}}$ is also cluster distribution but after given data. Thus ρ_k is called a *prior distribution*, and $\hat{\rho}_{k|\boldsymbol{x}}$ is called a *posterior distribution*. Also (3.22), which allows to compute a posterior probability from a priori probability, is called the *Bayes theorem*.

Thus to estimate clusters, we already have two important points: 1) instance distribution $\rho_{\boldsymbol{x}|k}$ for cluster k which corresponds to the cluster center in K-means, and 2) cluster distribution $\hat{\rho}_{k|\boldsymbol{x}}$ for instance $\boldsymbol{x}$ which corresponds to the distance from instance $\boldsymbol{x}$ to cluster k in K-means.

Thus now we can set up a K-means like algorithm, which repeats the following two steps alternately:

1. Compute distribution $\rho_{\boldsymbol{x}|k}$ from the instances given to cluster k.
2. Update $\hat{\rho}_{k|\boldsymbol{x}}$ according to (3.22) and then assign instance $\boldsymbol{x}$ to cluster $k^*_{\boldsymbol{x}}$ according to (3.23).

Algorithm 3.4 shows a pseudocode of this algorithm. As you can seen from the derivation of this algorithm, this algorithm is a probabilistic version of the K-means clustering algorithm. Also this algorithm is an implementation of the so-called *EM (Expectation and Maximization) algorithm*, a general framework for estimating parameters of probabilistic models by local optimization based on *maximum likelihood*, which in FMM, maximizes the likelihood of instance $\boldsymbol{x}$ given the model, i.e. (3.21). The EM algorithm repeats the two steps, called Maximization-step (M-step) and Expectation-step (E-step), alternately (see Section A.8.9), which exactly correspond to the two steps of lines 5 and 6 in **Algorithm 3.4**.

We note that in practice we can assume some (concrete) distribution, such as the normal distribution, for the instance distribution $\rho_{\boldsymbol{x}|k}$ for cluster k. Then the M-step of **Algorithm 3.4** would become the step to estimate parameters of such a practical distribution, for example, the mean and variance of the normal distribution. However it is possible to keep $\rho_{\boldsymbol{x}|k}$ as it is. In this case, we need to estimate $\rho_{\boldsymbol{x}|k}$ in the M-step, which can be as follows, according to the Bayes theorem again[1]:

$$\hat{\rho}_{\boldsymbol{x}|k} \quad \leftarrow \quad \frac{L(\boldsymbol{x},k)}{\sum_{\boldsymbol{x}} L(\boldsymbol{x},k)} = \frac{\rho_{\boldsymbol{x}|k}\rho_k}{\sum_{\boldsymbol{x}} \rho_{\boldsymbol{x}|k}\rho_k}. \tag{3.25}$$

Time Complexity

The algorithm for clustering (or parameter estimation) by FMM is a probabilistic version of the K-means clustering algorithm. This means that the computational complexity of this algorithm is the same as that of the K-means clustering algorithm.

Notes on Finite Mixture Model

Model description: We might not have mentioned the structure of FMM clearly, and so the structure can be written as in (3.21). The (3.21) shows the structure of FMM as follows: 1) each instance is generated at each cluster, according to $\rho_{\boldsymbol{x}|k}$, 2) this probability is weighed by the probability of cluster k and 3) summer up over all clusters, we have the likelihood of instance $\boldsymbol{x}$.

Two parameters: From the model, we can see that there are two parameters:

1. ρ_k: Cluster generation probability
2. $\rho_{\boldsymbol{x}|k}$: Instance generation probability, given cluster k. As mentioned above, practically, this can be a set of parameters of some distribution, like the normal distribution.

No constraints: In Section 3.1.3, we showed clustering of K-means with the constraint. This constraint can be used for a variety of purposes, such as controlling cluster sizes, like balanced clusters. Also this can be regularization, which allows the problem to be solved by eigenvalue decomposition. Similarly FMM has the prior probability, while there are no extension like the constrained K-means from K-means.

Probabilistic model: As mentioned earlier, FMM is a probabilistic model. Probabilistic models can be divided into *generative models* and *discriminative models*, which correspond to unsupervised and supervised learning,

[1]As well as $\rho_{\boldsymbol{x}|k}$, the prior distribution also can be updated using the current probability values, as follows:

$$\hat{\rho}_k \quad \leftarrow \quad \frac{\sum_{\boldsymbol{x}} \rho_k \rho_{\boldsymbol{x}|k}}{\sum_{\boldsymbol{x}} \sum_k \rho_k \rho_{\boldsymbol{x}|k}}. \tag{3.24}$$

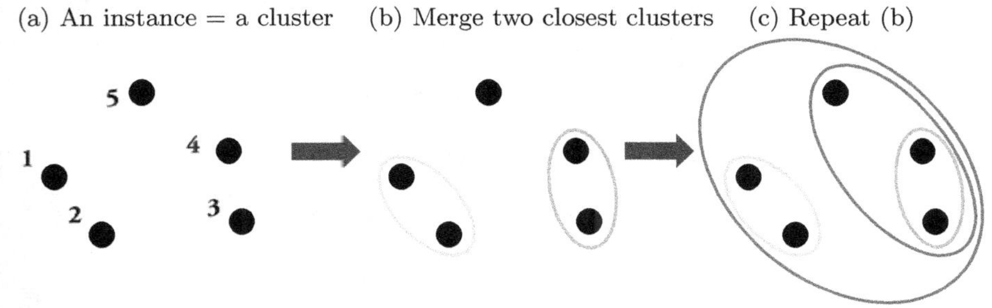

Figure 3.3: Schematic diagram of bottom-up hierarchical clustering procedure, and dendrogram.

respectively. In fact discriminative models use labels to estimate parameters of the models, while generative models do not use labels for parameter estimation. FMM is a generative model, because of unsupervised learning, parameters being estimated without labels.

3.1.5 Clustering – Hierarchical Clustering

Idea and Algorithm

K-means and FMM are typical partitional clustering, which divides instances into K clusters or groups at once, where clusters are all flat. However this magic number K is usually hard to decide. Also clusters are not necessarily so independent each other, and there would be at least partially inclusion relations among clusters which might be important to be discovered. *Hierarchical clustering* is advantageous in these two points.

The objective of hierarchical clustering is to understand the structure of the data space by a *binary tree* (which is called a *dendrogram*), where all instances are leaves of this tree. Leaves and internal nodes have the inclusion relation that the set of instances of a parent always has the set of instances of its child. That is, instances assigned to a child are all assigned to its parent. The root has all instances. This tree itself is a process of hierarchical clustering.

There are two major ways of building a dendrogram (binary tree) of hierarchical clustering: (a) *top-down* (*divisive*), which, starting with the root, partitions instances recursively, and (b) *bottom-up* (*agglomerative*), which, starting with leaves, merges instances recursively.

(a) Top-down: the number of possible partitioning of given instances is huge, which requests some heuristics for partitioning. A simple, practical way is to run partitional clustering, say K-means, recursively. That is, we first divide given data by K-means, and then each divided part is further divided by K-means. This can be repeated until no clusters. However repeatedly applying K-means definitely needs a large amount of computation time, and so the top-down approach is not major, while the bottom-up approach is the most standard for hierarchical

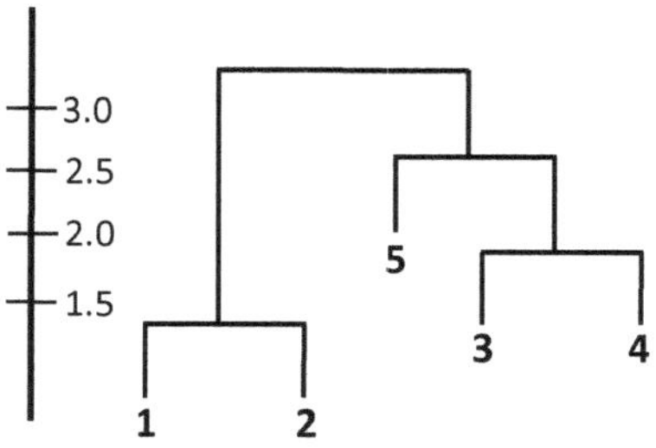

Figure 3.4: Dendrogram of hierarchical clustering.

Algorithm 3.5: Agglomerative (bottom-up) hierarchical clustering algorithm.

Function `Hierarchical_clustering(`$\boldsymbol{X}, K$`)`
  Data: training data: $\boldsymbol{X}$, #clusters: K
  Result: dendrogram
  Generate a cluster for each of all instances.;
  repeat
    Generate a distance matrix with distances among all possible clusters.;
    Merge two clusters with the shortest distance among those in the distance matrix.;
  until *All instances are in only one cluster*;
  Output the entire merging process as the dendrogram.;

---

clustering.

(b) Bottom-up: Fig. 3.3 (a-c) shows the procedure. In this figure, instances can be distributed on a two-dimensional space (meaning only two features). The algorithm first (a) set all instances as clusters (initialization), then (b) the closest clusters (instances) are merged, and (c) merging the closest clusters of (b) is repeated until all instances into one.

Fig. 3.4 shows the tree (dendrogram), showing the process of merging instances. This dendrogram shows that starting with all instances in different clusters, instances 1 and 2 are merged at the first step, and then instances 3 and 4 are merged into a cluster, which is merged with instance 5, and finally this cluster of 3, 4 and 5 is merged with that of 1 and 2. This is the same procedure as Fig. 3.3 (a-c).

Thus a dendrogram shows not only the process of merging but also the order of the merging. A dendrogram is the hierarchical clustering result itself. Also by using the dendrogram, we can generate partitional clusters, by stopping merging clusters when the number of clusters reaches K. For example, in Fig. 3.4, two clusters {1,2}, {3,4,5} can be obtained for the scale of 3.0, and three clusters {1,2}, {3,4}, {5} can be obtained for the scale of 2.0. We can change K by changing the value of the scale.

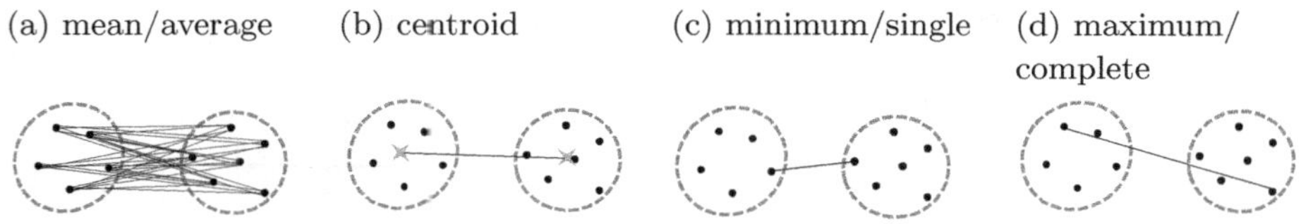

Figure 3.5: Four representative methods for measuring the distance between two clusters.

Algorithm 3.5 shows a pseudocode of the procedure of bottom-up hierarchical clustering. Mainly this algorithm repeats generating a distance matrix between instances and current clusters (line 4) and then merging two closest clusters (or instances) (line 5), until all instances are in one big cluster. This algorithm selects two closest clusters at each step, which means locally optimizing every time. This type of algorithm is called a *greedy algorithm*.

Distances in Hierarchical Clustering

Hierarchical clustering needs computing the distance between clusters (or instances). This distance is called the *linkage*. A cluster has more than one instances, and even if we define the distance between two instances, we can have more than one way of measuring the linkage of two clusters. Here suppose that the distance between two instances is already defined, we introduce four typical linkages below (also Fig. 3.5 shows visible pictures of these linkages):

(a) mean/ average linkage: The mean over the distances of all instance pairs.

(b) centroid linkage: The distance between the center (means) of each cluster.

(c) minimum/ single-linkage: The shortest distance among those of all possible instance pairs

(d) maximum/ complete linkage: The longest distance among those of all possible instance pairs.

(c) Minimum and (d) maximum linkages both use only one instance pair, which is vulnerable against noise and/or outliers. Thus (a) mean linkage and (b) centroid would be more preferable, particularly (a) mean linkage would be reasonable since the objective function is clearly set, as shown in (3.29).

Time Complexity

We focus on the mean linkage ((a) in Fig. 3.5). We then assume that the distance between two instances is computed already for all instance pairs (In fact this needs already $O(N^2)$).

The algorithm starts with assigning the distance of two instances to the distance of two clusters for all clusters. We then merge the closest two clusters into

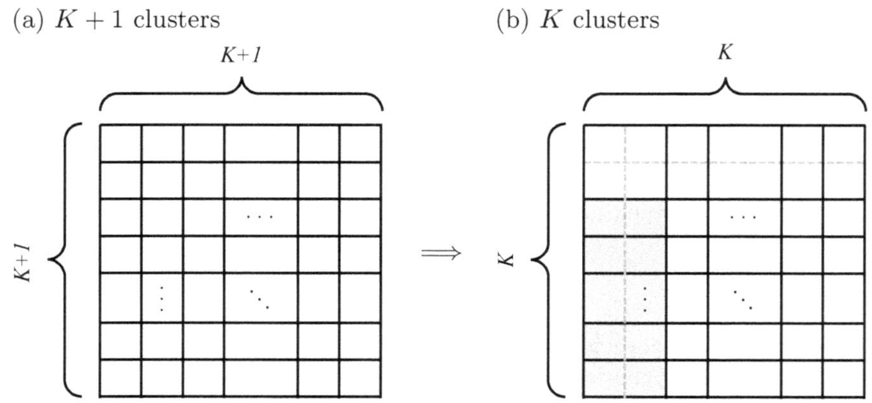

Figure 3.6: Distance matrices among (a) $K+1$ clusters and (b) K clusters: obtained by merging (the first) two out of $K+1$ clusters into one cluster.

one repeatedly. Suppose that we now have $K+1$ clusters, and are trying to generate K clusters out of $K+1$ clusters. All distances among $K+1$ clusters can be written as a $((K+1)\times(K+1))$-matrix, shown in (a) of Fig. 3.6, where each element of (a) has the distance between the corresponding two clusters. We then pick up the two clusters with the shortest distance (suppose that the first and second instances, without loss of generality) and merge them into K clusters. Fig. 3.6 (b) shows the resultant K clusters have the $(K \times K)$-distance matrix, where the dotted lines show that the first and second clusters in (a) are merged into the first instance of K clusters.

Then there are three points:

1. Both matrices are symmetric, and so we can just think about the half (upper triangle or lower triangle) of each matrix.

2. We do not have to think about the distance between instances any more. Instead we can just use distance between clusters. For example, suppose that two clusters, $k_{old,1}$ and $k_{old,2}$ in $K+1$, are merged into a cluster, k_{new}, the distance between some cluster k and cluster k_{new}, i.e. $d_{k,k_{new}}$ can be computed just by using only cluster distance as follows:

$$d_{k,k_{new}} = d_{k,k_{old,1}} + d_{k,k_{old,2}}, \tag{3.26}$$

   where $d_{k,k_{old,1}}$ ($d_{k,k_{old,2}}$) are the distance between k and $k_{old,1}$ ($k_{old,2}$).

3. From $K+1$ clusters to K clusters, the change is only the first two rows (and columns), and other elements are all kept the same, meaning that these elements are just kept as they are. Thus the distance we have to update is between each cluster and the new cluster only. Note that they are only $K-1$ elements, which are shaded in (b) of Fig. 3.6.

This means we need update only $K-1$ elements on merging $K+1$ clusters to K clusters. This is repeated from $K = N$ to $K = 1$:

$$\sum_{K=N}^{K=1} (K-1) \tag{3.27}$$

Overall the complexity would be in the order of N^2, i.e. $O(N^2)$.

The time complexity of partitional clustering (K-means and FNN) was $O(NMK)$ for one iteration Comparing with that, we have to say that this complexity of hierarchical clustering is relatively large, though hierarchical clustering is thought to be computatinally light.

Notes on Hierarchical Clustering

Visualization: In the original matrix of input data, two instances merged in hierarchical clustering are not necessarily neighbors. We however can change the order of instances, so that all instances are neighboring leaves of the two-dimensional dendrogram obtained by hierarchical clustering. This makes more similar instances closer, leading more visual understanding of input data by the dendrogram. This is the biggest reason why hierarchical clustering is popular and used, particularly in biology-related fields.

Objective function: When we use the mean linkage, the objective function to be minimized in case of K clusters is given as the following, which is the same as (3.10) of K-means:

$$\begin{aligned} J &= \sum_{k=1}^{K}\sum_{j=1}^{N}\sum_{i=1}^{N} \boldsymbol{Z}_{ik}\boldsymbol{E}_{ij}\boldsymbol{Z}_{jk} & (3.28)\\ &= \mathrm{trace}(\boldsymbol{Z}^\top \boldsymbol{E}\boldsymbol{Z}). & (3.29)\end{aligned}$$

However, the difference of hierarchical clustering from K-means is as follows: In K-means, the cluster assignment matrix $\boldsymbol{Z}$ can be freely estimated to optimize (3.29), unless any constraints exist. On the other hand, in hierarchical clustering, the dendrogram is given. In other words, $K+1$ clusters are already given. Then K clusters are optimized just by merging two clusters out of $K+1$ clusters, so that (3.29) is minimized. This means that optimization is done within a limited number of candidates. So simply if $K+1$ clusters are not an optimum set of $K+1$ clusters, K clusters can never be optimum. Reversely, if $K+1$ clusters are an optimum, the optimum K clusters might be rather easier to estimate than all possible choices of K clusters.

This is simply general, serious characteristics of a greedy algorithm, which just selects a local optimum always.

3.1.6 Biclustering and Visualization

Biclustering is clustering of not only instances (row vectors) but also features (column vectors) at the same time, finding groups, which are specified by not only the row side but also the column side. These groups are called *biclusters*. Particularly biclustering uses hierarchical clustering for the both sides. As mentioned in the previous section, we can change the order of instances, according to (the leaves of) the dendrogram of hierarchical clustering. When we apply hierarchical clustering to features at the same time, a dendrogram can be generated over features by hierarchical clustering. Then the key point of biclustering is both the order of instances and the order of features can be changed at the same time. In other words, the order of instances (row vectors) and that of features (column vectors) are independent, leading to the order of instances and that of features, which both are consistent with the dendrograms of both sides. This is useful to visually capture the similarity of instances and that of features at the same time, particularly in terms of knowledge discovery. We will show real data examples of biclustering in Section 3.5.1.

Again biclustering or hierarchical clustering is well-used in many applications, because of clustering results and process can be illustratively given by these clustering methods. That is, *visualization* might be sometimes more weighed or prioritized than time complexity, prediction accuracy, etc.

3.1.7 Probabilistic Model (Generative Model)

Bayesian Belief Network

Biclustering focused on not only relations among instances but also those among features. Similarly we now learn the relations among features in a more detailed way than just grouping.

We build a probabilistic model of instance $\boldsymbol{x}$, generated from its features, starting with the assumption that features contribute to the instance *independently*:

$$L(\boldsymbol{x}) = p(x_1, \ldots, x_M) \tag{3.30}$$
$$= \prod_m p(x_m). \tag{3.31}$$

Practically, however, features are not independent of each other, and it would be more reasonable to assume some *dependency* among features. We thus define the structure of the generative model by finding the internal, dependency among features. For example, the second feature would depend upon the third and fifth features, and the fourth feature depends upon the first and fifth features. If these dependencies are deterministic (complete), it would be rather easy, while practically these dependencies would be stochastic/probabilistic. *Bayesian belief network* (BBN) is a technique to represent this type of probabilistic dependencies [54]. Fig. 3.7 shows the structure of a BBN which is a directed graph with nodes for variables (now features) and directed edges for dependencies. Fig. 3.7 shows that for example, there are three nodes, x_2, x_3 and x_5 for three features,

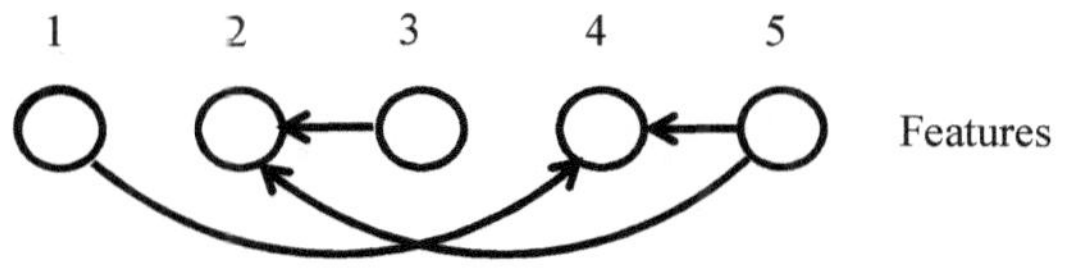

Figure 3.7: An example of Bayesian belief network over features.

where directed edges from x_3 and x_5 to x_2 shows that probability $p(x_2|x_3, x_5)$ is significant. Now we can give the likelihood of instance $\boldsymbol{x}$ by using BBN as generative model, as follows:

$$L(\boldsymbol{x}) = p(x_1, \ldots, x_5) \tag{3.32}$$
$$= p(x_1)p(x_2|x_3, x_5)p(x_3)p(x_4|x_1, x_5)p(x_5). \tag{3.33}$$

More generally, likelihood of instance $\boldsymbol{x}$ can be written by BBN as follows:

$$L(\boldsymbol{x}) = \prod_m p(x_m|\Gamma(x_m)), \tag{3.34}$$

where $\Gamma(x_m)$is a set of features on which x_m depends.

BBN cannot include all dependencies, because of probabilistic model for which probabilities cannot be computed for a directed cycle. That is, BBN has a constraint of no cycles, and the network satisfying this constraint is called a *directed acyclic graph* (DAG).

Learning Bayesian Belief Network

We just briefly overview the learning of Bayesian belief network [54] which has two steps:

1. Learning the structure of DAG

   Generate a probabilistic model, by extracting all possible largest dependencies out of data, keeping the constraint of DAG.

2. Learning parameter values

   Estimate probability parameters for dependencies, once after the structure of DAG is determined.

BBN has nodes (variables) for features and acquire the relation (dependencies) among features from given data. Dependencies are more direct than clustering, while dependencies in BBN are limited, due to the constraint of DAG. Also similar to biclustering, BBN can be constructed over instances as well as features, resulting in that we can see the dependencies over columns and also rows for the given data matrix.

More generally probabilistic models, including BBN, are called *graphical models*. A graphical model shows the dependencies of variables in instances. For

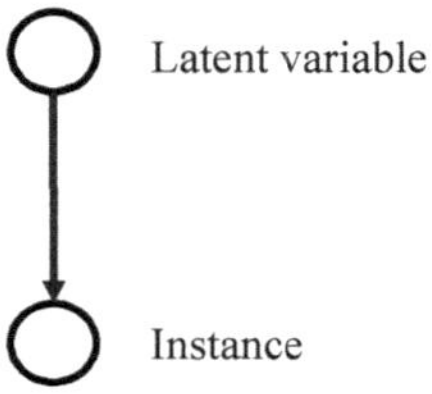

Figure 3.8: Graphical model for finite mixture model.

example, as shown by (3.21), in FMM, instance $\boldsymbol{x}$ is generated from latent variable $\boldsymbol{z}$ (k in (3.21)), corresponding to clusters. Fig. 3.8 shows the graphical model of FMM.

3.1.8 Matrix Factorization

Background: Matrix Factorization?

So far we could see that K-means provides cluster centers, hierarchical clustering gives the process of clustering as dendrogram, and Bayesian belief network shows dependencies among features. All three summarize the data and show the summary visually. In this section we summarize the entire given data, i.e. the given matrix or a set of vectors, somewhat more directly.

For this objective, what we can do first is to remove the same, redundant rows (or columns), although this would be usually done in data preprocessing. The next step is to make the matrix *full-rank* by reducing the rows (or columns) which can be represented by liner combinations of other rows (or columns).

The process so far does not lose any information. Here we try to approximate the original input matrix ($\boldsymbol{X}$ of rank M) by lower-rank matrices with a smaller rank, say K, than N and M. As a simple example, we can start with eigenvalue decomposition (see Section A.4.3), where we assume that the given input is a square matrix ($\boldsymbol{X}$ keeps $N = M$). After decomposing the input matrix $\boldsymbol{X}$ and taking the K largest eigenvalues, by using eigenvectors, $\boldsymbol{u}_i$ ($i = 1, \dots, K$), and eigenvalues, α_i ($i = 1, \dots, K$), $\boldsymbol{X}$ can be represented as follows:

$$\boldsymbol{X}\begin{pmatrix} \boldsymbol{u}_1 \cdots \boldsymbol{u}_K \end{pmatrix} = \begin{pmatrix} \boldsymbol{u}_1 \cdots \boldsymbol{u}_K \end{pmatrix}\begin{pmatrix} \alpha_1 & 0 & 0 \\ 0 & \ddots & 0 \\ 0 & 0 & \alpha_K \end{pmatrix}, \tag{3.35}$$

or just by matrices only,

$$\boldsymbol{X}\boldsymbol{U} = \boldsymbol{U}\boldsymbol{\Lambda}, \tag{3.36}$$

where $\boldsymbol{U}$ has eigenvectors as column vectors[2] and $\boldsymbol{\Lambda}$ is a diagonal matrix with

[2]Eigenvectors have as orthogonal system, means

$$\boldsymbol{U}^\top \boldsymbol{U} = \boldsymbol{I} \tag{3.37}$$

Thus the transpose of $\boldsymbol{U}$ is the inverse matrix of $\boldsymbol{U}$.

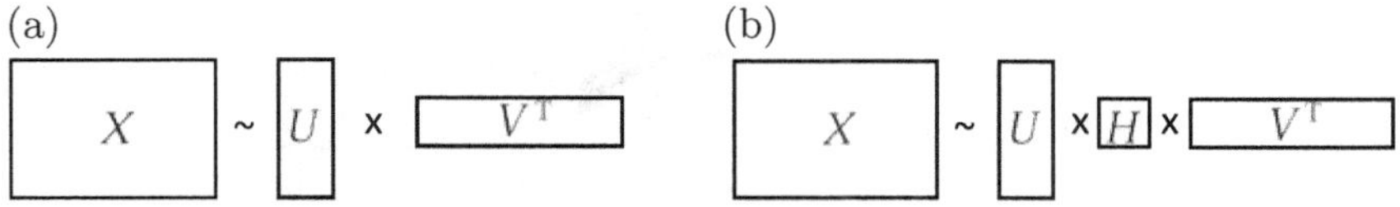

Figure 3.9: Matrix factorization into (a) two and (b) three low-rank matrices.

eigenvalues for diagonals.

Eigenvector matrix $\boldsymbol{U}$ is an orthogonal matrix, and then $\boldsymbol{X}$ is approximated by *low-rank matrix* $\boldsymbol{U}$ as follows:

$$\boldsymbol{X} \sim \boldsymbol{U}\boldsymbol{\Lambda}\boldsymbol{U}^{\top}. \tag{3.38}$$

Mathematically this looks the same as eigenvalue decomposition (see Section A.4.3), while we take only the K largest eigenvalues and so $\boldsymbol{X}$ is approximated by $\boldsymbol{U}$ and $\boldsymbol{\Lambda}$.

Letting $\boldsymbol{V} = \boldsymbol{\Lambda}\boldsymbol{U}^{\top}$, the following transformation would be further possible:

$$\boldsymbol{X} \sim \boldsymbol{U}\boldsymbol{V}^{\top} \tag{3.39}$$

$$= \begin{pmatrix} \boldsymbol{u}_1 \cdots \boldsymbol{u}_K \end{pmatrix} \begin{pmatrix} \boldsymbol{v}_1^{\top} \\ \vdots \\ \boldsymbol{v}_K^{\top} \end{pmatrix}. \tag{3.40}$$

This approximation is possible, even if $\boldsymbol{X}$ is not a square matrix. In the literature of *factor analysis*, column vectors of $\boldsymbol{V}^{\top}$ are called *factors*, and $\boldsymbol{U}$ is called the *factor loading matrix*. This is rather interchangeable, because the transpose of $\boldsymbol{U}\boldsymbol{V}^{\top}$ is $\boldsymbol{V}\boldsymbol{U}^{\top}$, for which the column vectors of $\boldsymbol{U}^{\top}$ are factors and $\boldsymbol{V}$ is the column loading matrix. Thus simply we just call vectors of low-rank matrices *factors*. Decomposing $\boldsymbol{X}$ into two low-rank matrices is called *low-rank approximation* and *matrix factorization*. Fig. 3.9 shows schematic pictures of matrix factorization, where the input matrix $\boldsymbol{X}$ is decomposed into two or three low-rank matrices, which we call, *bi-factorization* and *tri-factorization*, respectively[3].

(a) matrix bi-factorization: two low-rank matrices:

$$\boldsymbol{X} \sim \boldsymbol{U}\boldsymbol{V}^{\top}. \tag{3.41}$$

(b) matrix tri-factorization: three low-rank matrices:

$$\boldsymbol{X} \sim \boldsymbol{U}\boldsymbol{H}\boldsymbol{V}^{\top}. \tag{3.42}$$

$\boldsymbol{H}$ is usually a square matrix. The above $\boldsymbol{\Lambda}$ is also a square matrix.

In bi-factorization, $\boldsymbol{U}$ and $\boldsymbol{V}^{\top}$ are $(N \times K)$ and $(K \times M)$ matrices, respectively. Also in tri-factorization, $\boldsymbol{U}$, $\boldsymbol{H}$ and $\boldsymbol{V}$ can be $(N \times K_1)$, $(K_1 \times K_2)$ and $(K_2 \times M)$ matrices (again regularly $K_1 = K_2 = K$).

[3]Decomposition into more than three matrices is possible, while the results become more a blackbox as the number of decomposed matrices is larger, regardless of higher computational complexity. Thus usually we use bi- or tri-factorization.

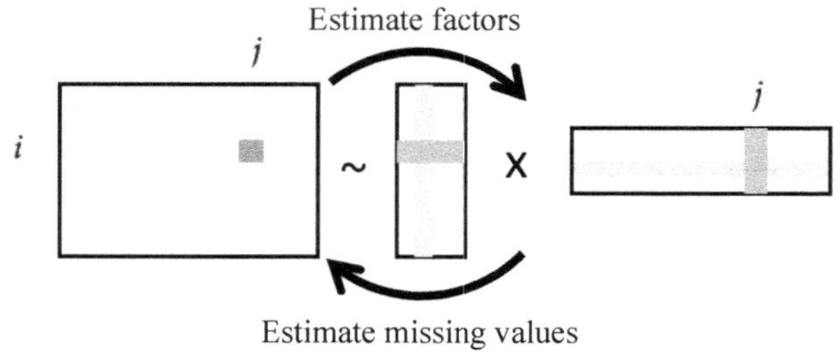

Figure 3.10: Matrix completion of $\boldsymbol{X}$ by matrix factorization.

We will next see how we can estimate obtain these low-rank matrices, $\boldsymbol{U}$ and $\boldsymbol{V}$, focusing on bi-factorization.

Algorithm: Estimating Low-rank Matrices

If given $\boldsymbol{X}$ is a square matrix, one way of estimating low-rank matrices would be eigenvalue decomposition, equivalent to singular value decomposition (SVD). However SVD assumes the complete data, which would not be used for data with missing values, and so the traditional approach of SVD would not be a practical choice.

As shown in Fig. 3.9 (a), approximating $\boldsymbol{X}$ by two low-rank matrices, $\boldsymbol{U}$ and $\boldsymbol{V}$ means minimizing the error between $\boldsymbol{X}$ and $\boldsymbol{U}\boldsymbol{V}^\top$. If the error is measured by the sum of all elements of the squared loss, which is called the *matrix L^2 norm* (see Section A.6), the minimization problem can be given as follows:

$$\min_{\boldsymbol{U},\boldsymbol{V}} ||\boldsymbol{X} - \boldsymbol{U}\boldsymbol{V}^\top||_2^2. \tag{3.43}$$

This formulation has two parameters, $\boldsymbol{U}$ and $\boldsymbol{V}$, which cannot be solved analytically at the same time. Thus a possible solution is, for example, the *alternating least square* (ALS) algorithm (see Section A.8.3), which repeats a step of that we fix one parameter, making the problem convex, and estimate the other parameter, and do the same manner by changing the parameters. We set the partial derivative of the objective function with respect to $\boldsymbol{U}$ or $\boldsymbol{V}$ equal to zero, and obtain the following update rules (For derivation, see the formulation with the regularization term):

$$\boldsymbol{U} \leftarrow \boldsymbol{X}\boldsymbol{V}(\boldsymbol{V}^\top\boldsymbol{V})^{-1}, \tag{3.44}$$
$$\boldsymbol{V}^\top \leftarrow (\boldsymbol{U}^\top\boldsymbol{U})^{-1}\boldsymbol{U}^\top\boldsymbol{X}, \tag{3.45}$$

where we assume that $\boldsymbol{V}^\top\boldsymbol{V}$ and $\boldsymbol{U}^\top\boldsymbol{U}$ are *regular matrices*[4] and can have the inverse matrix[5].

[4] A regular matrix has an inverse matrix, and called also as *invertible matrix* or *non-singular matrix*. See Section A.4.2 for more detail.

[5] For a non-zero vector $\boldsymbol{a}$,

$$\boldsymbol{a}^\top\boldsymbol{U}^\top\boldsymbol{U}\boldsymbol{a} = (\boldsymbol{U}\boldsymbol{a})^\top\boldsymbol{U}\boldsymbol{a} = ||\boldsymbol{U}\boldsymbol{a}||^2 \geq 0, \tag{3.46}$$

Just satisfying the L^2 norm over the error between $\boldsymbol{X}$ and $\boldsymbol{U}\boldsymbol{V}^\top$ is usually still too flexible, leaving a lot of possibilities in solutions of $\boldsymbol{U}$ and $\boldsymbol{V}$, which might make hard to interpret the results. Thus usually we give constraints on the range of $\boldsymbol{U}$ and $\boldsymbol{V}$. A normal and well-accepted constraint is the L^2 norm over $\boldsymbol{U}$ ($\boldsymbol{V}$) which means that again the squared sum over all elements of $\boldsymbol{U}$ ($\boldsymbol{V}$) must be a constant. Note that the L^2 norm is the most standard constraint, and this type of constraint over parameters is called *regularization*[6]. The objective function is reformulated as follows:

$$\min_{\boldsymbol{U},\boldsymbol{V}} ||\boldsymbol{X} - \boldsymbol{U}\boldsymbol{V}^\top||_2^2 + \lambda(||\boldsymbol{U}||_2^2 + ||\boldsymbol{V}||_2^2), \tag{3.47}$$

where λ is called the *regularization coefficient*, i.e. a hyperparameter. Due to the equality constraint, we can use the method of Lagrange multipliers. The Lagrange function this time is given as follows:

$$L(\boldsymbol{U},\boldsymbol{V}) = ||\boldsymbol{X} - \boldsymbol{U}\boldsymbol{V}^\top||_2^2 + \lambda(||\boldsymbol{U}||_2^2 + ||\boldsymbol{V}||_2^2). \tag{3.48}$$

Then the partial derivative of the Lagrange function is given as follows.

$$\frac{\partial f(\boldsymbol{U},\boldsymbol{V})}{\partial \boldsymbol{U}} = -2\boldsymbol{X}\boldsymbol{V} + 2\boldsymbol{U}\boldsymbol{V}^\top\boldsymbol{V} + 2\lambda\boldsymbol{U}, \tag{3.49}$$

$$\frac{\partial f(\boldsymbol{U},\boldsymbol{V})}{\partial \boldsymbol{V}} = -2\boldsymbol{X}^\top\boldsymbol{U} + 2\boldsymbol{V}\boldsymbol{U}^\top\boldsymbol{U} + 2\lambda\boldsymbol{V}. \tag{3.50}$$

We set them equal to zero, and can have the following update rules (see Section A.9.2 for derivation):

$$\boldsymbol{U} \leftarrow \boldsymbol{X}\boldsymbol{V}(\boldsymbol{V}^\top\boldsymbol{V} + \lambda\boldsymbol{I}_K)^{-1}, \tag{3.51}$$

$$\boldsymbol{V}^\top \leftarrow (\boldsymbol{U}^\top\boldsymbol{U} + \lambda\boldsymbol{I}_K)^{-1}\boldsymbol{U}^\top\boldsymbol{X}. \tag{3.52}$$

Algorithm 3.6 shows a pseudocode of the above algorithm.

The L^2 norm is the most typical norm, while we here raise two constraints which are often used in machine learning or matrix factorization.

1. **Sparse constraint**
   Sparse learning keeps only a small number of elements non-zero in low-rank matrices (most of elements are zero). So when we interpret the results, we can just focus on the non-zero elements, which makes the interpretation easier. This is implemented by using a lower order norm, such as the L^1 norm, than the L^2 norm, for regularization terms (see Section A.6). We will explain sparse learning in Section 3.3.4 also.

---

where if (3.46) is non-zero, $\boldsymbol{U}^\top\boldsymbol{U}$ is a *positive definite matrix* and nonsingular matrix, which has an inverse matrix. This means that what we assume here is (3.46) is non-zero, and $\boldsymbol{U}^\top\boldsymbol{U}$ is a positive definite matrix. See **Proposition 3.1** also regarding the case of the regularization term.

[6]The L^2 norm term in (3.47) simply says $(||\boldsymbol{U}||_2^2 + ||\boldsymbol{V}||_2^2)$, but note that this means for example $||\boldsymbol{U}||_2^2$ is a constant. Thus precisely the L^2 norm term should be $(||\boldsymbol{U}||_2^2 - \text{Const.}) + (||\boldsymbol{V}||_2^2) - \text{Const.})$. However in optimization this is the same as (3.47). Thus a simple way is taken, and this manner is used throughout this book for the L^2 norm regularization.

Algorithm 3.6: Alternating least square (ALS: additive update) for solving (3.47).

```
Function Alternating_least_square_for_matrix_factorization(X, K)
    Data: Training data: X, rank: K (low-rank matrices)
    Result: U, V
    Initialize U and V.;
    repeat
        Update U by (3.51).;
        Update V by (3.52).;
    until convergence;
    Output U and V.;
```

2. Non-negative constraint

This constraint can be applied to the non-negative input, and keeps the element values of lower-rank matrices non-zero. Thus non-negative values can be regarded as weights, so simply larger values being more important, which makes understand the learning results easier.

Update rules, such as (3.51) and (3.52), need the inverse matrix. This step does not guarantee that updated matrices are non-negative. Also *additive or subtractive update rules* cannot guarantee the non-negative results as well, even if the input is non-negative.

Thus update rules which guarantee non-negative outputs use *multiplicative updates*. That is, the current parameter value is always multiplied by some positive number, and the parameter value is bigger if the number is larger than one; otherwise it becomes smaller.

Matrix factorization with non-negative constraints has the same problem setting as (3.47), and the objective function is also the same as (3.48), while the difference is that low-rank matrices must be always non-negative.

$$\min_{\boldsymbol{U}\geq 0, \boldsymbol{V}\geq 0} ||\boldsymbol{X} - \boldsymbol{U}\boldsymbol{V}^\top||_2^2 + \lambda(||\boldsymbol{U}||_2^2 + ||\boldsymbol{V}||_2^2). \tag{3.53}$$

The direction of minimizing the objective function (3.48) is consistent with the negative value of the partial derivative of this function (so the negative value of the partial derivative was added in additive update of (3.51) and (3.52)). Then, in the partial derivative of (3.49) and (3.50), we use the positive terms (e.g. $\boldsymbol{U}\boldsymbol{V}^\top\boldsymbol{V}$) to make the parameter value small and negative terms (e.g. $-\boldsymbol{X}\boldsymbol{V}$ and $-\lambda\boldsymbol{U}$) to make the parameter value larger. That is, positive terms are for the denominator, and negative terms are for the numerator.

In summary, the partial derivatives of $\boldsymbol{U}$ and $\boldsymbol{V}$ are (3.49) and (3.50), re-

Algorithm 3.7: Multiplicative update rules for non-negative matrix factorization of (3.53).

Function `Non-negative_matrix_factorization`($\boldsymbol{X}, K$)
  Data: training data: $\boldsymbol{X}$, rank: K (low-rank matrices)
  Result: $\boldsymbol{U} \geq 0, \boldsymbol{V} \geq 0$
  Initialize $\boldsymbol{U}$ and $\boldsymbol{V}$.;
  repeat
    Update $\boldsymbol{U}$ by (3.54).;
    Update $\boldsymbol{V}$ by (3.55).;
  until *convergence*;
  Output $\boldsymbol{U}$ and $\boldsymbol{V}$.;

spectively, and then update rules are given as follows:

$$\boldsymbol{U}_{i,k} \leftarrow \boldsymbol{U}_{i,k} \frac{(\boldsymbol{X}\boldsymbol{V})_{i,k}}{(\boldsymbol{U}\boldsymbol{V}^\top\boldsymbol{V})_{i,k}}, \quad (3.54)$$

$$\boldsymbol{V}_{j,k} \leftarrow \boldsymbol{V}_{j,k} \frac{(\boldsymbol{X}^\top\boldsymbol{U})_{j,k}}{(\boldsymbol{V}\boldsymbol{U}^\top\boldsymbol{U})_{j,k}}. \quad (3.55)$$

Multiplicative updates are well-used for estimating low-rank matrices in matrix factorization. **Algorithm 3.7** shows the pseudocode of the algorithm with multiplicative updates.

Time Complexity

Estimation of $\boldsymbol{U}$ and $\boldsymbol{V}$ are symmetric, and we can just consider either of $\boldsymbol{U}$ or $\boldsymbol{V}$, and here we focus on $\boldsymbol{U}$ only. We consider the following four matrix computation, which can be processed sequentially: 1. $\boldsymbol{X}\boldsymbol{V}$, 2. $\boldsymbol{V}^\top\boldsymbol{V}$, 3. $(\boldsymbol{V}^\top\boldsymbol{V} + \lambda\boldsymbol{I}_K)^{-1}$, 4. $\boldsymbol{X}\boldsymbol{V} \times (\boldsymbol{V}^\top\boldsymbol{V} + \lambda\boldsymbol{I}_K)^{-1}$.

1. $\boldsymbol{X}\boldsymbol{V}$

   $\boldsymbol{X}$ is $(N \times M)$-matrix, and $\boldsymbol{V}$ is $(M \times K)$-matrix. So the computation process is repeated M times at $N \times K$ elements, meaning that $O(NKM)$. However suppose that $\boldsymbol{X}$ is a sparse matrix, we store non-zero elements and consider them only. Let M_{nz} be the maximum number of non-zero elements in $\boldsymbol{X}$, and then $O(NKM_{nz})$.

2. $\boldsymbol{V}^\top\boldsymbol{V}$

   $\boldsymbol{V}$ is $(M \times K)$-matrix, and then $O(K^2M)$.

3. $(\boldsymbol{V}^\top\boldsymbol{V} + \lambda\boldsymbol{I}_K)^{-1}$

   This is $(K \times K)$-matrix, and then the complexity of computing its inverse matrix is $O(K^3)$.

4. $\boldsymbol{X}\boldsymbol{V} \times (\boldsymbol{V}^\top\boldsymbol{V} + \lambda\boldsymbol{I}_K)^{-1}$

   Two matrices are $(N \times K)$-matrix and $(K \times K)$-matrix, and so $O(K^2N)$.

We can assume $K << N$ and $K << M$. K is rather a constant. So entirely $O(NKM_{nz})$.

Connection to Clustering

Constraint over low-rank matrices and regularization allow to reduce the flexibility of parameters, i.e. low-rank matrices, to make resultant matrices more comprehensible. Also we note that this constraint can control the resultant low-rank matrices. To show this point, we now put the following constraints on low-rank matrices $\boldsymbol{U} \in \mathcal{R}^{N\times K}$.

1. **Element values must be binary (zero or one):** $\boldsymbol{U} \in \{0,1\}^{N\times K}$

2. $\boldsymbol{U}^\top\boldsymbol{U} = \boldsymbol{D}$,

where $\boldsymbol{D}$ is a diagonal matrix, with a uniform number, N/K, for each of the diagonals.

We can then formulate the problem setting as follows:

$$\min_{\boldsymbol{U}\in\{0,1\}^{N\times K},\boldsymbol{V}} ||\boldsymbol{X} - \boldsymbol{U}\boldsymbol{V}^\top||_2^2 + \lambda_U||\boldsymbol{U}^\top\boldsymbol{U} - \boldsymbol{D}||_2^2 + \lambda_V||\boldsymbol{V}||_2^2. \tag{3.56}$$

In fact the above two constraints make $\boldsymbol{U}$ equivalent to the constraints of cluster assignment matrix $\boldsymbol{Z}$ in Section 3.1.2 of constrained K-means clustering. As shown by (3.2), cluster assignment matrix $\boldsymbol{Z}$ has only one value of 1 for each row vector (instance) and others of zero, meaning that values in $\boldsymbol{Z}$ are all binary. Also $\boldsymbol{Z}^\top\boldsymbol{Z}$ must have the same value for all elements of diagonals, the constraint making the cluster size the same over all clusters, exactly like constrained K-means clustering.

Thus by solving (3.56), the resultant low-rank matrices by matrix factorization provides the cluster assignment matrix $\boldsymbol{U}$, implying that clustering can be implemented by matrix factorization.

Rayleigh Quotient

For example, the motivation of eigenvalue decomposition is to approximate $\boldsymbol{X}$ by vector $\boldsymbol{u}$, as follows:

$$\hat{\boldsymbol{u}} = \arg\min_{\boldsymbol{u}} ||\boldsymbol{X} - \boldsymbol{u}\boldsymbol{u}^\top||_2^2. \tag{3.57}$$

The objective function to be minimized is as follows:

$$J = \boldsymbol{u}^\top\boldsymbol{X}\boldsymbol{u}. \tag{3.58}$$

Multiple $\boldsymbol{u}$ can be generated as follows:

$$\boldsymbol{U} = (\boldsymbol{u}_1, \ldots, \boldsymbol{u}_K). \tag{3.59}$$

We can give $\boldsymbol{U}$ the constraint of orthonormal system, and eigenvalue problem can be formulated as follows:

$$\min_{\boldsymbol{U}} \operatorname{trace}(\boldsymbol{U}^\top \boldsymbol{X}\boldsymbol{U}) \text{ subject to } \boldsymbol{U}^\top\boldsymbol{U} = \boldsymbol{I}_K. \tag{3.60}$$

Also this can be represented by Rayleigh quotient:

$$\min_{\boldsymbol{U}} R(\alpha, \boldsymbol{U}) \quad \Rightarrow \quad \min_{\boldsymbol{U}} \frac{\boldsymbol{U}^\top \boldsymbol{X}\boldsymbol{U}}{\boldsymbol{U}^\top\boldsymbol{U}}. \tag{3.61}$$

Thus we can see that matrix factorization and Rayleigh quotient can show an equivalent objective function.

Notes on Matrix Factorization

Approximation of $\boldsymbol{X}$ by $\boldsymbol{U}$ and $\boldsymbol{V}^\top$: The $\boldsymbol{U}$ and $\boldsymbol{V}^\top$ are trained so that $\boldsymbol{X}$ can be approximated by two low-rank matrices, i.e. $\boldsymbol{U}$ and $\boldsymbol{V}^\top$.

The generated low-rank matrices capture the factor (essence) of given data. Thus, for example, if $\boldsymbol{X}$ is noisy, low-rank approximation can remove the noise.

Only approximation to $\boldsymbol{X}$ is rather too flexible for $\boldsymbol{U}$ and $\boldsymbol{V}^\top$. We can add more constraints on the problem formulation, which controls $\boldsymbol{U}$ and $\boldsymbol{V}^T$ more. As shown already, these constraints allow matrix factorization to implement even clustering.

Missing values in $\boldsymbol{X}$: If $\boldsymbol{X}$ has missing values, simply we do not use the corresponding cells of $\boldsymbol{X}$ in matrix factorization, and we use the filled cells only to estimate $\boldsymbol{U}$ and $\boldsymbol{V}^\top$. Thus missing values are not problems in estimating the matrices

Once after low-rank matrices are estimated, we can estimate the missing values by using the obtained low-rank matrices, i.e. $\boldsymbol{U}$ and $\boldsymbol{V}^\top$.

In fact, matrix factorization has been used for item *recommendation* out of user-item matrices which have a huge number of items. However users do neither buy many items nor evaluate many items, resulting in a lot of missing values. Thus a user-item matrix has many missing values and is very sparse. Again once low-rank matrices are obtained, such missing values can be estimated by using the low-rank matrices. In this sense, matrix factorization matches item recommendation very well.

Fig. 3.10 shows this concept of learning without missing values and estimating missing values after training. This way of using matrix factorization is called *matrix completion.*

Space complexity reduction: When the original matrix is approximated by two low-rank matrices, The number of elements of the two low-rank matrices is now $N \times K + K \times M = K(N + M)$. In fact K is a constant (and smaller then N and M). So the space complexity is the linear order. On the other hand, the original space complexity is $O(NM)$, and thus matrix factorization reduces the space complexity significantly.

3.2 Supervised Learning

Supervised learning is classification if labels are discrete; otherwise, it is regression. Historically regression is from statistics, particularly multivariate analysis, while techniques for classification have been developed in machine learning. The objective of supervised learning is to build a model h, which can predict the unknown label of an instance from its features. Simply speaking, supervised learning is to estimate function h which satisfy the following, between instance $\boldsymbol{x}$ and its label y,

$$y = h(\boldsymbol{x}) \tag{3.62}$$

More in detail, the structure of model h is already decided by each approach, and so the objective is to estimate parameters of h, for example, linear coefficients if h is a linear function. An important point is given data is limited in size, and so model h should be trained carefully not to *overfitting* given data and to be general. This concept is called *generalization* or *regularization*.

The most general setting of classification is binary labels, such as positive and negative examples. In fact all supervised learning methods can be used for the binary classification problem. For example, a regression method can be used by predicting y and set a cut-off value to assign the prediction result to either of the two classes, such as positives and negatives.

Classification and regression are different in terms of label y, but we can generate the model with the same representation power for each of the two problem settings. For example, decision tree is for classification, and regression tree is for regression. The algorithms for these two models are both recursively partitioning the given space of training data, meaning that they are the same algorithm, except the partitioning criteria. This is true of other models in supervised learning. Also binary classification is the main problem setting in a lot of applications, including bioinformatics.

Overall for supervised learning of vectors, we just consider binary classification problem.

3.2.1 K-nearest Neighbors

Background and Algorithm

We start with a method called *K-nearest neighbors* (KNN) which is not necessarily a machine learning approach, since KNN has no models and so no parameters to estimate from the training data. The algorithm is simple, and given a test instance, takes the K nearest neighbors and the majority label of these K instances is used for prediction. This approach is simply called *majority voting.* **Algorithm 3.8** shows a pseudocode of this approach.

Time Complexity

The idea of the algorithm is very simple, while the complexity is very high, because all N training instances needs to be checked for each test instance, if we have no

Algorithm 3.8: K-nearest neighbor (learning and prediction), where $L[k]$ is the k-th element in list L and the label takes a binary value in $\{+1, -1\}$.

```
Function K_nearest_neighbor_prediction(X, y, x)
    Data: training data: X, training data labels: y, instance for
          prediction: x
    Result: estimated labels: ŷ
    Initialize L, a list of closest instances.;
    for i ← 1 to N do
        Compute distance d_i between x_i and x.;
        if d_i < d_L[K] then
            Replace the K-th element of the list L with i.;
            Record d_i.;
            Sort the elements of L in the ascending order of the distances.;
Output ŷ ← sign(Σ_{i|i∈L} y_i)
```

well-considered data structure. Also the algorithm needs compute the distance between each training instance and the given test instance. So the time complexity is at least $O(NM)$.

If we use some model in machine learning, training needs some computational cost, while prediction is easy, which just depends on the size of parameters. In fact this is an advantage of using some model in machine learning, while KNN does not have this merit. Thus KNN does not work particularly for large data.

Notes on KNN

K is the hyperparameter. If this is too small, the results are easily affected by noise, while the performance will be rather blunt if K is too large.

- Advantage

  The idea is intuitive and well-accepted.

- Disadvantage

  1. No learning results because of no models (though some kind of cause for prediction can be shown by neighbor instances).
  2. Particularly weak for sparse or scarce data
  3. High computational complexity, and then low scalability.

KNN is one approach of so-called *case-based reasoning* (CBR), which has the same properties of KNN.

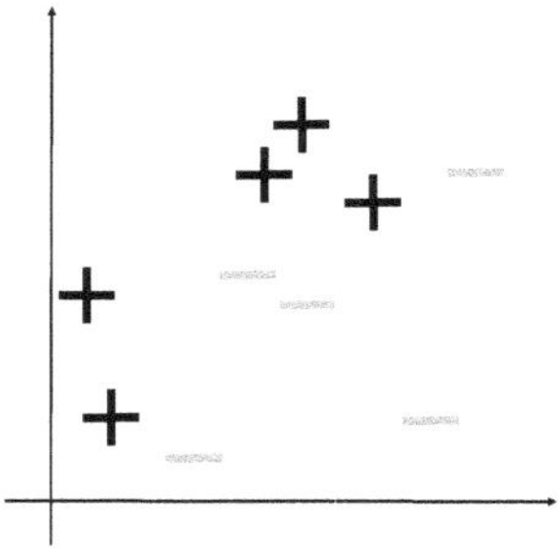

Figure 3.11: Toy data with five positives, five negatives and only two features.

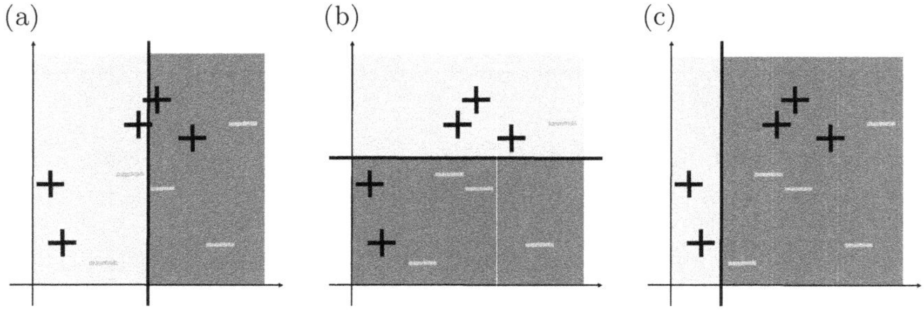

Figure 3.12: Three possible partitions of the toy data in Fig.3.11.

3.2.2 Decision Stump

Background

Decision stump, the simplest classification method, selects a feature value, which can classify the given training instances most (among all features) consistent with the label of training instances. This feature value, called *decision stump*, is used as the prediction rule of decision stump.

Fig. 3.11 is ten toy training instances, five positives (+) and five negatives (−), with two features, meaning ten points on two-dimensional data space. Decision stump selects one value of either of the two features to divide the data space so that ten examples should be clearly separated. In fact there are a lot of possibilities to select the feature value, and to understand the idea behind decision stump clearly, we consider the three possibilities shown in Fig. 3.12.

In each possibility, the data space is divided into two subspaces by a feature value, one being for positives and the other for negatives. This is the predictive rule of decision stump. We then count the number of instances correctly predicted. For example, in (a), by the decision stump, the left-hand side subspace is predicted as positives and the right-hand side as negatives, turning into six examples (three positives in the left-hand side and three negatives in the right-hand side) correctly predicted. Similarly the number of correctly predicted instances is seven for both (b) and (c). Thus, (b) and (c) are better than (a), while we are not sure which of (b) and (c) is better. To examine these (b) and (c), we can use the idea of *information gain.*

Information of information gain means *entropy*, which is given for data $\boldsymbol{X}$ with N instances (N_p positives and N_n negatives), as follows:

$$\text{Ent}(N_p, N_n) = -\frac{N_p}{N}\log_2(\frac{N_p}{N}) - \frac{N_n}{N}\log_2(\frac{N_n}{N}). \quad (3.63)$$

Then suppose that $\boldsymbol{X}$ is divided by feature χ_m into N_1 instances (N_{1p} positives and N_{1n} negatives) and N_2 instances (N_{2p} positives and N_{2n} negatives). Information gain can be defined as follows:

$$IG(\boldsymbol{X}, \chi_m) := \text{Ent}(N_p, N_n) - (\text{Ent}(N_{1p}, N_{1n}) + \text{Ent}(N_{2p}, N_{2n})). \quad (3.64)$$

Note that $N_1 + N_2 = N$, $N_{1p} + N_{1n} = N_1$ and $N_{2p} + N_{2n} = N_2$.

We can explain this by using an illustrative example. Fig. 3.11 shows ten examples with two classes, five examples for each class. Thus entropy of Fig. 3.11 is given as follows.

$$\text{Ent}(5, 5) = -\frac{1}{2}\log_2(\frac{1}{2}) - \frac{1}{2}\log_2(\frac{1}{2}) \quad (3.65)$$

$$= \frac{1}{2} + \frac{1}{2} = 1. \quad (3.66)$$

Thus, the entropy with no division is 1 bit. Next we can consider (b) of Fig. 3.12:

$$\text{Ent}(2,4) + \text{Ent}(3,1) = -\frac{6}{10}(\frac{2}{6}\log_2\frac{2}{6} + \frac{4}{6}\log_2\frac{4}{6}) - \frac{4}{10}(\frac{3}{4}\log_2\frac{3}{4} + \frac{1}{4}\log_2\frac{1}{4}) \quad (3.67)$$

$$= 0.875. \quad (3.68)$$

This division has entropy of 0.875 bits, and then information gain of (b) is 0.125 (=1.0 - 0.875).

Similarly we can consider the entropy of (c):

$$\text{Ent}(2,0) + \text{Ent}(3,5) = -\frac{2}{10}(\frac{2}{2}\log_2\frac{2}{2} + \frac{0}{2}\log_2\frac{0}{2}) - \frac{8}{10}(\frac{3}{8}\log_2\frac{3}{8} + \frac{5}{8}\log_2\frac{5}{8}) \quad (3.69)$$

$$= 0.764. \quad (3.70)$$

(c) is 0.764 bits, and information gain of (c) is 0.236bits (= 1.0 -0.764).

This means that (c) has more information gain than (b), and the feature value of (c) should be chosen. Thus once again decision stump picks up one feature value, which partitions the data space into two subspaces, which are most consistent with labels of training instances, in terms of some criterion, for example information gain.

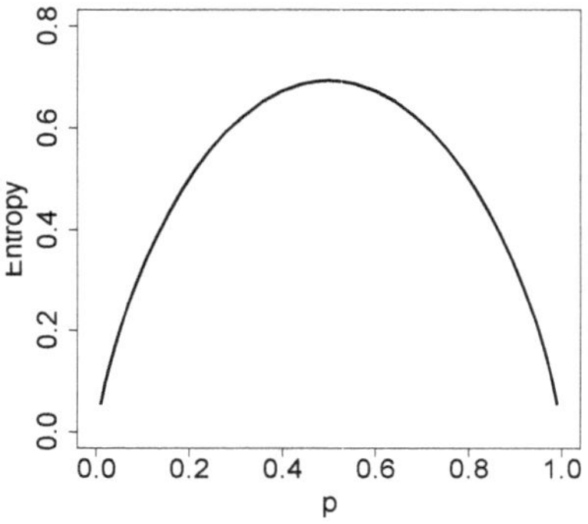

Figure 3.13: Entropy (y-axis): $-p \log p - (1-p)\log(1-p)$ where $0 \leq p \leq 1$ (p is x-axis).

Algorithm 3.9: Decision stump.

```
Function Decision_stump(X, y)
    Data: training data: X, training data labels: y
    Result: partitioned data: X_1 and X_2, and partition criterion: D
    for m ← 1 to M do
        Partition X into two by IG in (3.64).;
        if this partition is better than before then
            Save this partition as X_1 and X_2 (and also D).;
    Output (X_1, y) and (X_2, y), and partition D.;
```

Here we will take a more detailed look at the property of information gain. Fig. 3.13 shows entropy, $f(p)$, where $0 \leq p \leq 1$. In this figure, p (x-axis) has the range of 0 to 1, and y-axis shows entropy ($-p\log(p) - (1-p)\log(1-p)$). Note that p shows the balances of the partition. That is, $p = 0$ or 1 means the partition with only positives or only negatives. For $p = 0.5$, the partition is balanced, meaning the number of positives is equal to that of negatives. From this figure, entropy is larger with more unbiased distribution and smaller with more biased distribution. Information gain shows the decrease of entropy, meaning that information gain is larger with the division is more biased.

Thus information gain is not the only criterion for evaluating the partition of decision stump. As shown in Fig. 3.13, like entropy, any *convex* (upward) function can be a partition criterion for decision stump. Thus possible criteria include Area under the ROC curve (AUC) (see Section 10.1.2) [38], Gini index (or Gini coefficient), which is equivalent to AUC and minimum description length (MDL) [72], etc., which are all upward convex functions.

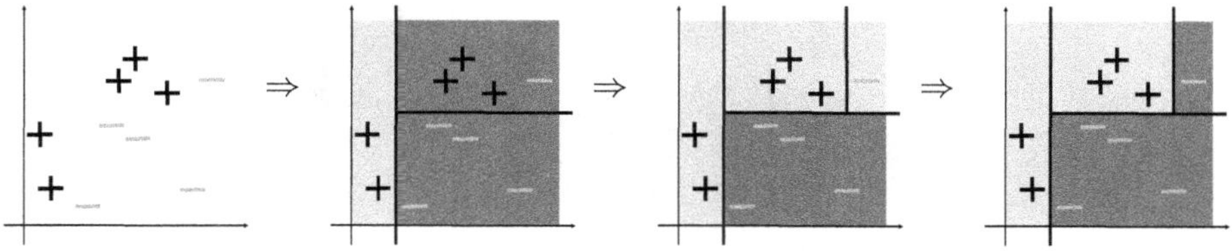

Figure 3.14: Decision tree: recursive partitioning over data space by decision stump.

Algorithm

Algorithm 3.9 shows a pseudocode of the learning algorithm of decision stump. In this pseudocode, $IG(\boldsymbol{X}, \chi_m)$ is a function of returning the value, which achieves the largest information gain, of m-th feature when $\boldsymbol{X}$ was partitioned. This feature value is a rule to apply to predict the label of any unknown instance.

Time Complexity

Regarding time complexity, for learning, we have to check all features to find the best feature value. For each feature, even if values are continuous, the number of times we examine the value is at most the number of instances. Thus the time complexity of finding the decision stump is totally $O(MN)$. For prediction, we can just apply the decision stump to the test instance, and the time complexity is $O(1)$.

3.2.3 Decision Tree

Decision Stump to Decision Tree: Recursive Partitioning

Decision stump was a feature vector which partitions the data space most consistent with labels, while only one feature will not be good enough for partitioning the data space. We then apply decision stump to the partitioned subspace repeatedly to have more detailed subspaces until each subspace has only one label. This procedure is called *recursive partitioning*.

Fig. 3.14 shows the best recursive partitioning of the toy data given in Fig. 3.11, where decision stump was applied three times to have subspaces with only one label. This recursive partitioning generates *decision tree*, which has the root with all instances and each leave with the corresponding instances, the tree showing the process of recursive partitioning from the root to leaves. Fig. 3.15 shows the tree corresponding to the recursive partitioning of Fig. 3.14. Decision tree visually shows the corresponding recursive partitioning of data space, and the path from the root to a leaf is one classification rule. For example, this decision

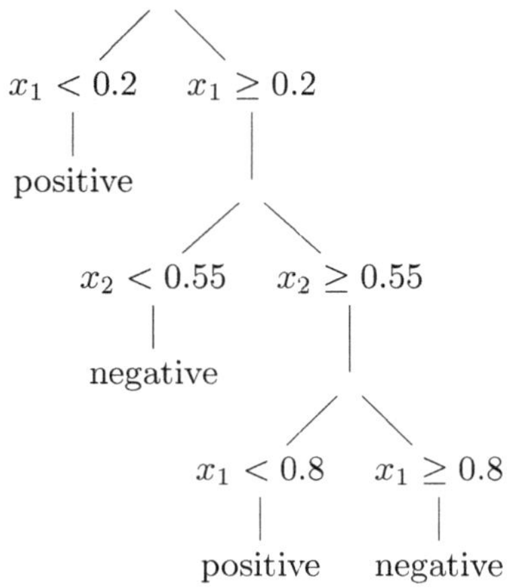

Figure 3.15: Decision tree, corresponding to the recursive partitioning of Fig. 3.14.

tree has the following four paths:

$$x_1 < 0.2 \quad \rightarrow \quad \text{positive.} \tag{3.71}$$

$$x_1 \geq 0.2 \quad \& \quad x_2 < 0.55 \quad \rightarrow \quad \text{positive.} \tag{3.72}$$

$$x_1 \geq 0.2 \quad \& \quad x_2 \geq 0.55 \quad \& \quad x_1 < 0.8 \quad \rightarrow \quad \text{positive.} \tag{3.73}$$

$$x_1 \geq 0.2 \quad \& \quad x_2 \geq 0.55 \quad \& \quad x_1 \geq 0.8 \quad \rightarrow \quad \text{negative.} \tag{3.74}$$

These four rules are the four subspaces shown in the most right-hand side of Fig. 3.14. That is, four rules are the left, right-bottom, middle and right-up of the last partitioning of Fig. 3.14. Again decision tree is a set of rules obtained by recursive partitioning of given data[7].

As shown in 3.14, recursive partitioning runs until each partitioned subspace has only one label. However, obviously, subspaces with only one label mean overfitting training data. To avoid overfitting, in reality, after all subspaces with only one label are generated, decision tree is usually made smaller by removing the edges from the leaves, to some degree. This process is called (tree) *pruning*. This is one way of *generalization*, and for pruning we can use information gain or entropy again.

[7] Note that these rules are sometimes redundant. For example, in the fourth rule, the first condition ($x_1 \geq 0.2$) is satisfied by the last condition ($x_1 \geq 0.8$). So this rule is as follows:

$$x_2 \geq 0.55 \quad \& \quad x_1 \geq 0.8 \quad \rightarrow \quad \text{Negatives} \tag{3.75}$$

Algorithm 3.10: Recursive data partitioning.

Function `Recursive_partitioning(`$\boldsymbol{X}$, $\boldsymbol{y}$`)`
  Data: training data: $\boldsymbol{X}$, training data labels: $\boldsymbol{y}$
  Result: partitioned data: $\boldsymbol{X}_1$, $\boldsymbol{X}_2$
  if *only one instance in* $\boldsymbol{X}$ *or only one label for all instances in* $\boldsymbol{X}$ **then**
    stop ;
  Run `Decision_stump(`$\boldsymbol{X}$, $\boldsymbol{y}$`)` to obtain the partition criterion D and the partitioned data ($\boldsymbol{X}_1$, $\boldsymbol{y}_1$) and ($\boldsymbol{X}_2$, $\boldsymbol{y}_2$).;
  Attach D to the blanching node for $\boldsymbol{X}$ in the decision tree.;
  Generate nodes of $\boldsymbol{X}_1$ and $\boldsymbol{X}_2$ being linked to the node of $\boldsymbol{X}$.;
  Run `Recursive_partitioning(`$\boldsymbol{X}_1$, $\boldsymbol{y}_1$`)`.;
  Run `Recursive_partitioning(`$\boldsymbol{X}_2$, $\boldsymbol{y}_2$`)`;

Algorithm 3.11: Decision tree: recursive partitioning + pruning.

Function `Decision_tree(`$\boldsymbol{X}$, $\boldsymbol{y}$`)`
  Data: training data: $\boldsymbol{X}$, training data labels: $\boldsymbol{y}$
  Result: optimal data partitioning
  Complete partitioning (one instance for one unique cluster) by running `Recursive_partitioning(`$\boldsymbol{X}$, $\boldsymbol{y}$`)`.;
  Prune the completed partitioning by generalization.;
  Output the entire partition results (which is the decision tree).;

Algorithm

Again the procedure of generating decision tree is recursively partitioning given training data $\boldsymbol{X}$, into subspaces with only one label, and then the generated tree is pruned. **Algorithm 3.10** shows a pseudocode of recursive purloining for generating decision tree. **Algorithm 3.11** shows a pseudocode of the entire decision tree algorithm, which starts running the recursive partitioning of **Algorithm 3.10**.

Time Complexity

The algorithm runs decision stump recursively over partitioned data. That is, $\boldsymbol{X}$ is divided into $\boldsymbol{X}_1$ and $\boldsymbol{X}_2$ by decision stump, and then decision stump is run over $\boldsymbol{X}_1$ and $\boldsymbol{X}_2$. Then letting N_1 and N_2 be the number of instances of $\boldsymbol{X}_1$ and $\boldsymbol{X}_2$, respectively, $N_1 + N_2 = N$. Thus suppose that any instance appears at most only once at each layer, the time complexity at each layer is always $O(NM)$, even if decision stump runs many times at one layer. Thus let T_D be the maximum tree depth, and the complexity is $O(NMT_D)$.

Prediction of Unknown Instance

An unknown test instance can be applied to the root of decision tree, and goes down to a leaf, along with the path, depending on the feature value the test instance has. Then the label of the leaf the test instance arrived is given as the predicted value. For example, we have the trained decision tree of Fig. 3.15 and a test instance with (x_1, x_2)=(0.5,0.75). At the root, the rule is if x_1 is less than 0.2, the instance has to go to the left; otherwise right. Then the value of the test instance is 0.5, and the instance can go to the right automatically. The next rule is if x_2 is less than 0.55, so the instance ($x_2 = 0.75$) can go to the left; otherwise right. Thus again the test instance can go to the right. Then the last rule is if x_1 is less than 0.8, the test instance can go to the left; otherwise right. So finally the test instance goes to the left and the label of this node, i.e. "positive", is assigned. The complexity is just the length of the path, i.e. the depth of the tree, so $O(T_D)$。

Overall a unique feature of decision tree is that recursive partitioning of given data automatically leads to a decision tree, from which we have a sort of rules, and one rule can be assigned to a test instance for prediction, using the label of the leaf.

---

decision stump $\rightarrow$ decision tree

3.2.4 Naive Bayes Classifier and Bayesian Belief Network Classifier

Decision Stump to Probabilistic Decision Stump

Decision stump used one feature value to generate a decisive classification rule. For example, if m-th feature, x_m, satisfies some condition ($c(x_m) = 1$), the instance is a positive ($y = 1$); otherwise a negative ($y = 0$):

$$\text{if } c(x_m) = 1 \text{ then } y = 1, \tag{3.76}$$
$$\text{otherwise } \quad y = 0. \tag{3.77}$$

For the entire training examples, we can generate a 2×2 contingency table by checking if each instance satisfies $c(x_m) = 1$ and its label, as below:

	$c(x_m) = 1$	$c(x_m) \neq 1$
$y = 1$	N_{TP}	N_{FP}
$y = 0$	N_{FN}	N_{TN}

For example, N_{TP} is the number of instances which satisfy $c(x_m) = 1$ and $y = 1$.

Thus from this table, we can compute the probability that the instance is a positive (or a negative) given that the instance satisfy $c(x_m) = 1$ (or not satisfy that), by using Bayes theorem, as follows:

$$p(y = 1|x_m) = \frac{p(x_m|y = 1)p(y = 1)}{p(x_m)}. \tag{3.78}$$

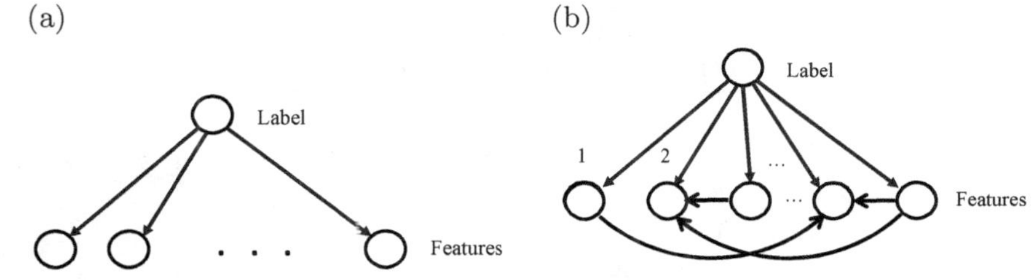

Figure 3.16: (a) Naive Bayes classifier, (b) An example of Bayesian belief network classifier.

For $c(x_m) = 1$,

$$p(y=1|x_m) = \frac{\frac{N_{TP}}{N_{TP}+N_{PP}} \frac{N_{TP}+N_{PP}}{N}}{\frac{N_{TP}+N_{PN}}{N}} = \frac{N_{TP}}{N_{PN}+N_{TP}}. \tag{3.79}$$

For $c(x_m) \neq 1$,

$$p(y=1|x_m) = \frac{\frac{N_{PP}}{N_{TP}+N_{PP}} \frac{N_{TP}+N_{PP}}{N}}{\frac{N_{PP}+N_{TN}}{N}} = \frac{N_{PP}}{N_{PP}+N_{TN}}. \tag{3.80}$$

This indicates that we can evaluate the rule given by decision stump in more detail by using the probability above, generating a probabilistic model for decision stump. Note that we can have the probabilities above and also $p(x_m|y=1)$.

Probabilistic Decision Stump to Naive Bayes Classifier

Decision stump selected only one feature, which can separate training instances the most consistent with labels. As mentioned in the above, then decision stump can be developed to probabilistic decision stump.

Then, instead of one feature, we can use all features for classification with probabilities, i.e. a probabilistic model with all features:

$$p(y=1|\boldsymbol{x}) = p(y=1|x_1,\ldots,x_M) \tag{3.81}$$
$$= \frac{p(y=1,x_1,\ldots,x_M)}{p(x_1,\ldots,x_M)} \tag{3.82}$$
$$\propto p(y=1,x_1,\ldots,x_M). \tag{3.83}$$

For simplicity, we assume that all features are independent.

$$p(y=1,x_1,\ldots,x_M) \propto p(y=1)p(x_1,\ldots,x_M|y=1) \tag{3.84}$$
$$= p(y=1)\prod_m p(x_m|y=1). \tag{3.85}$$

Note that each probability of the right-hand side was already derived for probabilistic decision stump. This probabilistic model is called a *naive Bayes classifier*.

Algorithm 3.12: Naive Bayes classifier, keeping the definition of $IG(\boldsymbol{X}, \chi_m)$ the same as that in **Algorithm 3.9**.

```
Function Naive_Bayes(X, y)
    Data: training data: X, training data labels: y
    Result: hypothesis: h

    for m ← 1 to M do
        Divide data into two partitions by IG(X, χ_m) and generate
        (2×2)-contingency table.;
        Compute p(x_m|y = 1) and p(x_m|y = 0) by using the generated
        (2×2)-contingency table.;
    Output hypothesis h, which can be computed according to (3.88).;
```

Algorithm 3.12 shows a pseudocode of a procedure of generating a naive Bayes classifier.

Regarding time complexity, a naive Bayes classifier needs to examine all features, which is however the same as decision stump, which has to scan all feature (and their values) as well to find the best feature value. Thus the time complexity of naive Bayes classifier is $O(NM)$. Naive Bayes classifier assumes that each feature is independent of each other and the contribution from each feature to predicting the label is equal. Fig. 3.16 (a) is a graphical model of naive Bayes classifier[8]

The odds that given $\boldsymbol{x}$, the probability of $y = 1$ to the probability of $y = 0$ are given as follows:

$$\frac{p(y=1|\boldsymbol{x})}{p(y=0|\boldsymbol{x})} = \frac{p(y=1,\boldsymbol{x})}{p(y=0,\boldsymbol{x})} \tag{3.86}$$

$$= \frac{p(y=1)\prod_m p(x_m|y=1)}{p(y=0)\prod_m p(x_m|y=0)} \tag{3.87}$$

$$= \frac{p(y=1)}{p(y=0)}\prod_m \frac{p(x_m|y=1)}{p(x_m|y=0)}. \tag{3.88}$$

Using the odds, we can predict the label of a test instance.

Naive Bayes Classifier to Bayesian Belief Network Classifier

Naive Bayes classifier of (3.85) assumes the independence among features, while in reality these features are not independent and there must be dependencies among features:

$$p(y=1, x_1, \ldots, x_M) \propto p(y=1)\prod_m p(x_m|\Gamma(x_m), y=1), \tag{3.89}$$

[8]A graphical model shows dependencies among variables, i.e. features. In a graphical model, for two variables connected by a directed edge, the variable of the destination of the edge depends upon the variable at the origin of the directed edge. Thus if variables have no edges to other variables, they are independent.

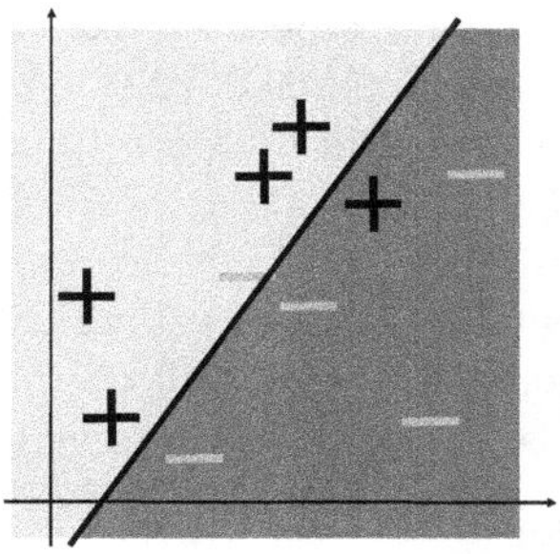

Figure 3.17: Linear regression.

where $\Gamma(x_m)$ means the set of features, on which feature x_m depends.

Also as well as naive Bayes classifier, we can compute the odds that given $\boldsymbol{x}$, the probability of $y = 1$ to the probability of $y = 0$, as follows:

$$\frac{p(y=1|\boldsymbol{x})}{p(y=0|\boldsymbol{x})} = \frac{p(y=1)\prod_m p(x_m|\Gamma(x_m), y=1)}{p(y=0)\prod_m p(x_m|\Gamma(x_m), y=0)} \tag{3.90}$$

$$= \frac{p(y=1)}{p(y=0)}\prod_m \frac{p(x_m|\Gamma(x_m), y=1)}{p(x_m|\Gamma(x_m), y=0)}. \tag{3.91}$$

This model with dependencies is called a *Bayesian belief network classifier.* Fig. 3.16(b) shows a schematic example of Bayesian belief network classifier with dependencies among features.

We explained Bayesian belief network already in unsupervised learning (Section 3.1.7). Comparing 3.16 (b) with Fig. 3.7, Bayesian belief network classifier is equivalent to Bayesian belief network, except labels.

As we have explained in the preceding pages, the model has been expanded in the following order:

---

decision stump → naive Bayes classifier → Bayesian belief network classifier

---

About the model representation power, there are several models proposed with the representation power between naive Bayes classifier and Bayesian belief network classifier. For example, tree-augmented naive Bayes classifier [20] places the constraint that the entire dependencies must not be more complex than a tree.

3.2.5 Linear (Ridge) Regression

Decision Tree to Ridge Regression

Recursive partitioning in decision tree uses feature values only, meaning that partitioning is always perpendicular to the dimension of a feature, in the input data space. Thus as shown in the most right-hand side of Fig. 3.14, data partitioning generates always a stepwise boundary between positives and negatives, even if all instances can be clearly separated. On the other hand, in reality, the boundaries between positives and negatives would be more smooth.

A simple solution to make these boundaries smooth is, as shown in Fig. 3.17, linear regression. That is, simply speaking, we can multiply some weight over each feature value and output the sum over them as the predicted label. Letting $\boldsymbol{w} = (w_1, \ldots, w_M)$ be the weight or *regression coefficient* of linear regression, the linear regression from input $\boldsymbol{x}$ to y can be given as follows:

$$\begin{aligned} y &= h(\boldsymbol{x}) && (3.92) \\ &= w_1 \cdot x_1 + w_2 \cdot x_2 + \cdots + w_M \cdot x_M + \epsilon && (3.93) \\ &= \boldsymbol{x}^\top \boldsymbol{w} + \epsilon. && (3.94) \end{aligned}$$

Also the linear regression over entire given data, i.e. from vectors $\boldsymbol{X}$ to $\boldsymbol{y}$, is given as follows:

$$\begin{aligned} \boldsymbol{y} &= h(\boldsymbol{X}) && (3.95) \\ &= \boldsymbol{X}\boldsymbol{w} + \epsilon. && (3.96) \end{aligned}$$

Estimating Regression Coefficients of Ridge Regression

We can minimize the residual $\boldsymbol{\epsilon}$ between the true $\boldsymbol{y}$ and predicted $\boldsymbol{y}$, to estimate $\boldsymbol{w}$.

$$\min \boldsymbol{\epsilon}^2. \quad (3.97)$$

For example, we can use the least square or the L^2 norm for the residual:

$$\min_{\boldsymbol{w}} ||\boldsymbol{y} - \boldsymbol{X}\boldsymbol{w}||_2^2. \quad (3.98)$$

We can solve this analytically to estimate $\boldsymbol{w}$, and suppose that $\boldsymbol{X}^\top \boldsymbol{X}$ is a non-singular matrix:

$$\hat{\boldsymbol{w}} = (\boldsymbol{X}^\top \boldsymbol{X})^{-1} \boldsymbol{X}^\top \boldsymbol{y}. \quad (3.99)$$

However totally similar to matrix factorization, $\boldsymbol{X}^\top \boldsymbol{X}$ is not necessarily a regular matrix, and to avoid overfitting training data, we place some regularization on $\boldsymbol{w}$ (a standard manner is the L^2 norm). We then formulate the problem as follows:

$$\min_{\boldsymbol{w}} (\frac{1}{2} ||\boldsymbol{y} - \boldsymbol{X}\boldsymbol{w}||_2^2 + \frac{\lambda}{2} ||\boldsymbol{w}||_2^2). \quad (3.100)$$

This regression is called *ridge regression*, and we can solve this problem analytically as follows:

$$\hat{\boldsymbol{w}} = (\boldsymbol{X}^\top \boldsymbol{X} + \lambda \boldsymbol{I}_M)^{-1} \boldsymbol{X}^\top \boldsymbol{y}, \quad (3.101)$$

where $\boldsymbol{I}_M$ is $(M \times M)$-matrix with one for diagonals and zero for other elements.

Algorithm 3.14 shows a pseudocode of learning and prediction algorithm of ridge regression.

Algorithm 3.13: Linear (ridge) regression (learning and prediction).

Function `Linear_regression(`$\boldsymbol{X}$, $\boldsymbol{y}$, $\boldsymbol{x}$`)`
  Data: training data $\boldsymbol{X}$, label: $\boldsymbol{y}$, unknown example: $\boldsymbol{x}_{new}$
  Result: trained weights:$\hat{\boldsymbol{w}}$
  Learning: estimate $\hat{\boldsymbol{w}}$ by using (3.101).;
  Prediction: predict unknown label by $\hat{\boldsymbol{w}}^\top \boldsymbol{x}_{new}$.;
  Output trained weights $\hat{\boldsymbol{w}}$.;

Time Complexity of Ridge Regression

Parameter estimation algorithm can be solved analytically, and then we consider the complexity of computing (3.101). We can focus on the following three parts, not related with $\boldsymbol{y}$, since $\boldsymbol{y}$ is a vector and computationally light.

1. $\boldsymbol{X}^\top \boldsymbol{X}$

   This is $(N \times M)$-matrix by $(M \times N)$-matrix, resulting in $O(M^2N)$.

2. $(\boldsymbol{X}^\top \boldsymbol{X} + \lambda \boldsymbol{I}_M)^{-1}$

   This is the inverse matrix of $(M \times M)$-matrix, and the complexity is $O(M^3)$.

3. $(\boldsymbol{X}^\top \boldsymbol{X} + \lambda \boldsymbol{I}_M)^{-1} \boldsymbol{X}^\top$

   Suppose that the inverse matrix is already computed (complexity is $O(M^3)$), this is $(M \times M)$-matrix by $(M \times N)$-matrix, resulting in $O(M^2N)$.

So totally $\max\{O(M^2N), O(M^3)\}$. The complexity is linear to the number of instances, which is rather light in computation, though it is heavier than that of decision tree.

Reason for Regularization

We have used regularization terms in our formulation in matrix factorization and also linear regression. We here raise two reasons why regularization terms are needed in machine learning once again:

Avoid overfitting: If we think about only the objective function, parameters might be overfit training data. To avoid this, we add regularization terms to make learning generalize.

Make learning/computation possible: Analytical solution of linear regression and alternating least square solution of matrix factorization have to have the inverse of $\boldsymbol{X}^\top \boldsymbol{X}$ without regularization. Thus we have to place an assumption that this is a regular matrix.

However, if we place a regularization term, particularly the L^2 norm over the parameters, this part is $(\boldsymbol{X}^\top \boldsymbol{X} + \lambda \boldsymbol{I})$ for ridge regression. We then do not need any assumption on the matrices of this formulation.

Table 3.1: Update rules of ALS for matrix factorization vs. update rules of linear regression.

Regularization terms	Update rule of $\boldsymbol{V}^\top$ in ALS of matrix factorization	Estimation of $\boldsymbol{\beta}$ in linear regression
No	$\boldsymbol{V}^\top \leftarrow (\boldsymbol{U}^\top \boldsymbol{U})^{-1} \boldsymbol{U}^\top \boldsymbol{X}$	$\hat{\boldsymbol{w}} = (\boldsymbol{X}^\top \boldsymbol{X})^{-1} \boldsymbol{X}^\top \boldsymbol{y}$
L^2 norm	$\boldsymbol{V}^\top \leftarrow (\boldsymbol{U}^\top \boldsymbol{U} + \lambda \boldsymbol{I}_K)^{-1} \boldsymbol{U}^\top \boldsymbol{X}$	$\hat{\boldsymbol{w}} = (\boldsymbol{X}^\top \boldsymbol{X} + \lambda \boldsymbol{I}_M)^{-1} \boldsymbol{X}^\top \boldsymbol{y}$

Proposition 3.1. $(\boldsymbol{X}^\top \boldsymbol{X} + \lambda \boldsymbol{I})$ *is a positive semidefinite matrix, meaning a nonsingular matrix.*

Proof. For non-zero vector $\boldsymbol{a}$,

$$\begin{aligned} \boldsymbol{a}^\top (\boldsymbol{X}^\top \boldsymbol{X} + \lambda \boldsymbol{I}) \boldsymbol{a} &= \boldsymbol{a}^\top \boldsymbol{X}^\top \boldsymbol{X} \boldsymbol{a} + \lambda \boldsymbol{a}^\top \boldsymbol{a} && (3.102) \\ &= (\boldsymbol{X}\boldsymbol{a})^\top \boldsymbol{X}\boldsymbol{a} + \lambda \boldsymbol{a}^\top \boldsymbol{a} && (3.103) \\ &= ||\boldsymbol{X}\boldsymbol{a}||^2 + \lambda ||\boldsymbol{a}||^2 \geq \lambda ||\boldsymbol{a}||^2 > 0 && (3.104) \end{aligned}$$

□

Thus $(\boldsymbol{X}^\top \boldsymbol{X} + \lambda \boldsymbol{I})$ is guaranteed to be a nonsingular matrix, and we do not have to place any assumption on this matrix. Practically this is a big reason for the regularization term [42].

Generalized Linear Model

We used the least square to minimize, meaning that we assumed that the residual follows a normal distribution. We can assume another distribution for the residual, and the model for any type of distribution is called *generalized linear model.* We do not go into this direction in this book.

Comparison of Linear Regression and ALS of Matrix Factorization

As mentioned briefly, the parameter estimation of linear (ridge) regression is very similar to the update rule of alternating least square (ALS) of matrix factorization. Table 3.1 shows the similarity between them in comparison, and this is very reasonable. The objective functions of both, i.e. (3.43) in matrix factorization and (3.98) in linear regression, is both the L^2 norm or the squared loss. Also ALS is an algorithm, which fixes one parameter and estimates the other. When one, for example, $\boldsymbol{U}$, is fixed, $\boldsymbol{V}$ is estimated from $\boldsymbol{U}$ and given data $\boldsymbol{X}$, which is almost the same as analytical solution of linear regression: parameter $\boldsymbol{w}$ is estimated from $\boldsymbol{X}$ and given label $\boldsymbol{y}$. Thus the difference is simply the parameter of matrix factorization is matrices, while that of linear regression is a vector. The key point is that matrix factorization (note that matrix factorization can perform clustering) is unsupervised learning and linear regression is supervised learning, while we can set up a similar objective function, resulting in similar (or the same) parameter estimation updates finally.

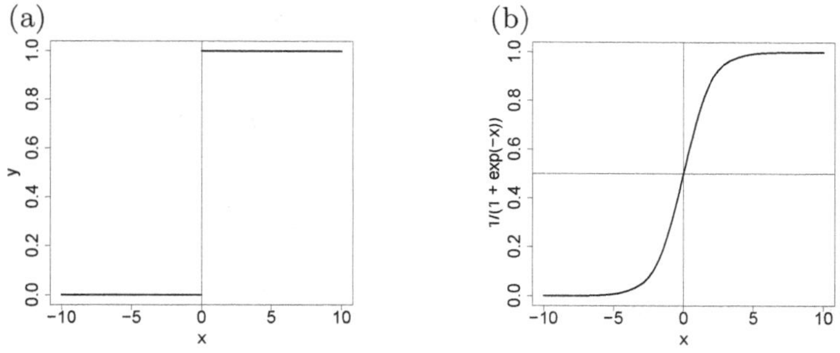

Figure 3.18: (a) Step function and (b) Sigmoid function.

3.2.6 Logistic Regression

Linear Regression to Logistic Regression

In linear regression, as shown by (3.94), the output of $h(\boldsymbol{x}) = \boldsymbol{x}^\top \boldsymbol{w} + \epsilon$ can be any continuous value, while our problem setting is binary classification. Then, to estimate the binary label, we now consider to assign the output of linear regression to either of $y = 1$ (positive) or $y = 0$ (negative). To implement this, we can use a function, which has the output of linear regression ($h(\boldsymbol{x})$) as input and outputs 1 if the input is larger than some cut-off value c (for example 0); otherwise 0. A simple function to implement this is a step function:

$$\text{if} \quad h(\boldsymbol{x}) > c \quad \text{then} \quad y = 1, \tag{3.105}$$

$$\text{otherwise} \quad y = 0. \tag{3.106}$$

For example, $c = 0$ is an unfair cut-off value for both positives and negatives. Fig. 3.18 (a) is a schematic example of a step function ($h(\boldsymbol{x})$) of $c = 0$. The classifier (or predictor) with $h(x)$ of a linear regression and also with a step function of $c = 0$ is called as *perceptron*.

However, a step function is too extreme, because originally if the output of $h(\boldsymbol{x})$ is larger, the output of the additional function should be closer to 1, and if smaller, closer to 0. This smooth function can be implemented by a sigmoid (logistic) function (see Section A.2.3)[9]:

$$\sigma(x) \quad = \quad \frac{1}{1 + \exp^{-x}}. \tag{3.107}$$

Fig. 3.18 (b) shows an example of sigmoid function ($\sigma(x)$), which is a monotone increasing function in the range of zero to one, where the input x takes an arbitrary numerical value. We can interpret the sigmoid function is a probability function given x, since the output takes a value between zero and one.

[9]A general sigmoid function is $\sigma(x) = \frac{1}{1+\exp^{-a \cdot x}}$, but here we use $a = 1$

We can use some regression function $h(\boldsymbol{x})$ for x:

$$p(y=1|\boldsymbol{x}) = \sigma(h(\boldsymbol{x})) = \frac{1}{1+\exp^{-h(\boldsymbol{x})}}. \tag{3.108}$$

$$p(y=0|\boldsymbol{x}) = 1-\sigma(h(\boldsymbol{x})) = \frac{\exp^{-h(\boldsymbol{x})}}{1+\exp^{-h(\boldsymbol{x})}}. \tag{3.109}$$

Particularly we can use a linear regression $h(\boldsymbol{x}) = \boldsymbol{x}^\top \boldsymbol{w}$:

$$p(y=1|\boldsymbol{x}) = \sigma(h(\boldsymbol{x})) = \frac{1}{1+\exp(-(\boldsymbol{x}^\top \boldsymbol{w}+\epsilon))}. \tag{3.110}$$

This regression is called *logistic regression*. Logistic regression is the most well-used regression in many application fields with multivariate analysis.

Parameter Estimation (Gradient Descent for Maximum Likelihood)

The difference of logistic regression from linear regression is only the output, which is limited within $[0,1]$, and so these two models are equivalent in representation power. However, the parameter of logistic regression, $\boldsymbol{w}$, cannot be obtained analytically and so in general is estimated by local optimization, particularly gradient descent (see Section A.8.4). For simplicity we write the output for $y=1$ and input $\boldsymbol{x}$ as $o(\boldsymbol{x})$. Again regression equations for positives and negatives are as follows:

$$p(y=1|\boldsymbol{x}) = o(\boldsymbol{x}) = \frac{1}{1+\exp(-(\boldsymbol{x}^\top \boldsymbol{w}+\epsilon))}. \tag{3.111}$$

$$p(y=0|\boldsymbol{x}) = 1-o(\boldsymbol{x}) = \frac{\exp(-(\boldsymbol{x}^\top \boldsymbol{w}+\epsilon))}{1+\exp(-(\boldsymbol{x}^\top \boldsymbol{w}+\epsilon))}. \tag{3.112}$$

As we mentioned earlier, the output of sigmoid function can be regarded as a probability, and so we can use the likelihood (log-likelihood) as the objective function to be maximized (maximum likelihood):

$$\begin{aligned} L(\boldsymbol{x}_i, \boldsymbol{w}) &= \sum_{i=1}^{N} \log p(y_i|\boldsymbol{x}_i) && (3.113)\\ &= \sum_{i=1}^{N} \log p(y=1|\boldsymbol{x}_i)^{y_i} p(y=0|\boldsymbol{x}_i)^{1-y_i} && (3.114)\\ &= \sum_{i=1}^{N} \log o(\boldsymbol{x}_i)^{y_i}(1-o(\boldsymbol{x}_i))^{1-y_i}. && (3.115) \end{aligned}$$

That is, for each instance $\boldsymbol{x}$:

$$\begin{aligned} L_i(\boldsymbol{w}) &= \log o(\boldsymbol{x}_i)^{y_i}(1-o(\boldsymbol{x}_i))^{1-y_i} && (3.116)\\ &= y_i \log o(\boldsymbol{x}_i) + (1-y_i)\log(1-o(\boldsymbol{x}_i)). && (3.117) \end{aligned}$$

Algorithm 3.14: Logistic regression (learning and prediction).

Function Logistic_regression($\boldsymbol{X}$, $\boldsymbol{y}$, $\boldsymbol{x}_{new}$)
    Data: training data: $\boldsymbol{X}$, training data labels: $\boldsymbol{y}$, unknown example: $\boldsymbol{x}_{new}$
    Result: predicted value for the unknown examples, trained $\hat{\boldsymbol{w}}$
    Assign random vales to $\hat{\boldsymbol{w}}$.;
    Learning: **repeat**
        update $\hat{\boldsymbol{w}}$ due to (3.120) ($\hat{\boldsymbol{w}}^{(new)} \leftarrow \hat{\boldsymbol{w}}^{(old)} + \eta \sum_{i=1}^{N} (y_i - o(\boldsymbol{x}_i))\boldsymbol{x}_i$).;
    until *convergence*;
    Prediction: predict the label of input $\boldsymbol{x}_{new}$, due to (3.110) ($p(y = 1|\boldsymbol{x}_{new}) = \frac{1}{1+\exp(-(\boldsymbol{x}_{new}^{\top}\hat{\boldsymbol{w}}+\epsilon))}$). ;
    Output the trained $\hat{\boldsymbol{w}}$.;

Then the gradient of $L_i(\boldsymbol{w})$ can be given as follows (see Section A.9.4 for derivation):

$$\frac{dL_i(\boldsymbol{w})}{d\boldsymbol{w}} = (y_i - o(\boldsymbol{x}_i))\boldsymbol{x}_i. \tag{3.118}$$

We can use the gradient descent, and the update rule for $\boldsymbol{w}$ can be obtained as follows:

$$\boldsymbol{w}^{(new)} \leftarrow \boldsymbol{w}^{(old)} + \eta \sum_{i=1}^{N} \frac{dL_i(\boldsymbol{w})}{d\boldsymbol{w}} \tag{3.119}$$

$$= \boldsymbol{w}^{(old)} + \eta \sum_{i=1}^{N} (y_i - o(\boldsymbol{x}_i))\boldsymbol{x}_i. \tag{3.120}$$

Algorithm 3.14 shows a pseudocode of the gradient descent algorithm for estimating the linear coefficients of logistic regression.

Time Complexity

For each iteration of the gradient descent, $(y_i - o(\boldsymbol{x}_i))\boldsymbol{x}_i$ needs only $O(M)$, and so the complexity of $\sum_{i=1}^{N}(y_i - o(\boldsymbol{x}_i))\boldsymbol{x}_i$ is $O(NM)$.

Gradient Descent for Least Square

A standard parameter estimation method for logistic regression is based on gradient descent (local optimization) for maximum likelihood, while as shown in linear regression and matrix factorization, we can use the squared loss (or L^2 norm) between true label y and output $o(\boldsymbol{x})$ of logistic regression:

$$e = \frac{1}{2}\sum_{i} ||o(\boldsymbol{x}_i) - y_i||_2^2. \tag{3.121}$$

The gradient descent of instance $\boldsymbol{x}$ is given as

$$\frac{de}{d\boldsymbol{w}} = \frac{de}{do(\boldsymbol{x}_i)} \frac{do(\boldsymbol{x}_i)}{dh(\boldsymbol{x}_i)} \frac{dh(\boldsymbol{x}_i)}{d\boldsymbol{w}}, \quad (3.122)$$

where

$$\frac{de}{do(\boldsymbol{x}_i)} = (o(\boldsymbol{x}_i) - y_i), \quad (3.123)$$

$$\frac{do(\boldsymbol{x}_i)}{dh(\boldsymbol{x}_i)} = o(\boldsymbol{x}_i)(1 - o(\boldsymbol{x}_i)), \quad (3.124)$$

$$\frac{dh(\boldsymbol{x}_i)}{d\boldsymbol{w}} = \boldsymbol{x}_i. \quad (3.125)$$

Thus the update rule of $\boldsymbol{w}$ is given as follows:

$$\boldsymbol{w}^{(new)} \leftarrow \boldsymbol{w}^{(old)} + \eta \sum_{i=1}^{N} \frac{de}{d\boldsymbol{w}} \quad (3.126)$$

$$= \boldsymbol{w}^{(old)} + \eta \sum_{i=1}^{N} (o(\boldsymbol{x}_i) - y_i) o(\boldsymbol{x}_i)(1 - o(\boldsymbol{x}_i))\boldsymbol{x}_i. \quad (3.127)$$

Comparing with the update rule of maximum likelihood, i.e. (3.120), the above update rule needs more computation at each step, and is expected to be disadvantage in computation time practically.

Comparison of Logistic Regression with Naive Bayes Classifier

Given $\boldsymbol{x}$, the odds of the probability of $y = 1$ to the probability of $y = 0$ in logistic regression are written as follows:

$$\frac{p(y = 1|\boldsymbol{x})}{p(y = 0|\boldsymbol{x})} = \prod_m \exp(x_m \cdot w_m + \epsilon). \quad (3.128)$$

Under the same condition, as shown in (3.88), the odds by naive Bayes classifier is given as follows:

$$\frac{p(y = 1|\boldsymbol{x})}{p(y = 0|\boldsymbol{x})} = \frac{p(y = 1)}{p(y = 0)} \prod_m \frac{p(x_m|y = 1)}{p(x_m|y = 0)}. \quad (3.129)$$

We assume that $p(x_m|y = 1)$ and $p(x_m|y = 0)$ in the right-hand side are continuous, and these distributions of the m-th features (for positives and negatives) are both normal distributions:

$$p(x_m|y = 1) = \exp(\frac{(x_m - \mu_{m1})^2}{\sigma_{m1}}), \quad p(x_m|y = 0) = \exp(\frac{(x_m - \mu_{m0})^2}{\sigma_{m0}}), \quad (3.130)$$

where μ_{m1} and σ_{m1} are the mean and variance of the m-th feature of positives, and similarly, μ_{m0} and σ_{m0} are the mean and variance of the m-th feature of negatives.

We assume that the variances of the m-th feature for positives and negatives are equal to each other ($\sigma_{m0} = \sigma_{m1} = \sigma_m$):

$$\frac{p(x_m|y=1)}{p(x_m|y=0)} = \exp(\frac{(x_m - \mu_{m1})^2 - (x_m - \mu_{m1})^2}{\sigma_m}) \quad (3.131)$$

$$= \exp(\frac{(\mu_{m1} - \mu_{m0})x_m + \mu_{m1}^2 + \mu_{m0}^2}{\sigma_m}). \quad (3.132)$$

Thus now the odds by naive Bayes classifier are:

$$\frac{p(y=1|\boldsymbol{x})}{p(y=0|\boldsymbol{x})} = \frac{p(y=1)}{p(y=0)} \prod_m \exp(\frac{(\mu_{m1} - \mu_{m0})x_m + \mu_{m1}^2 + \mu_{m0}^2}{\sigma_m}). \quad (3.133)$$

We further assume that the priors for the two classes are the same ($p(y=1) = p(y=0)$):

$$\frac{p(y=1|\boldsymbol{x})}{p(y=0|\boldsymbol{x})} = \prod_m \exp(\frac{(\mu_{m1} - \mu_{m0})x_m + \mu_{m1}^2 + \mu_{m0}^2}{\sigma_m}). \quad (3.134)$$

Let $w_m = \frac{\mu_{m1} - \mu_{m0}}{\sigma_m}$ and $\epsilon = \frac{\mu_{m1}^2 + \mu_{m0}^2}{\sigma_m}$:

$$\frac{p(y=1|\boldsymbol{x})}{p(y=0|\boldsymbol{x})} = \prod_m \exp(w_m \cdot x_m + \epsilon). \quad (3.135)$$

We now see that the odds by naive Bayes classifier, shown by (3.135) is totally the same as (3.128), which is the odds by logistic regression. Our assumption here on naive Bayes classifier is that the distributions of the m-th distribution for both class ($p(x_m|y=1)$ and $p(x_m|y=0)$) are two normal distributions with the same variance and $p(y=1) = p(y=0)$. Thus we can say that logistic regression is a special case of naive Bayes classifier, but roughly speaking, these two models have an equivalent representation power [10].

Another point is for logistic regression, if we assume the normal distributions with the same variance for positives and negatives at each feature, the odds by logistic regression can be written as follows:

$$\frac{p(y=1|\boldsymbol{x})}{p(y=0|\boldsymbol{x})} = \prod_m \frac{p(x_m|y=1)}{p(x_m|y=0)}. \quad (3.136)$$

Originally odds by naive Bayes classifier was given as follows:

$$\frac{p(y=1,\boldsymbol{x})}{p(y=0,\boldsymbol{x})} = \frac{p(y=1)}{p(y=0)} \prod_m \frac{p(x_m|y=1)}{p(x_m|y=0)}. \quad (3.137)$$

Thus, they have equivalent representation power, while naive Bayes classifier show joint probabilities, while logistic regression presents conditional probabilities. Once again in general probabilistic models can be classified into two types:

[10] A part of this description can be found in [64].

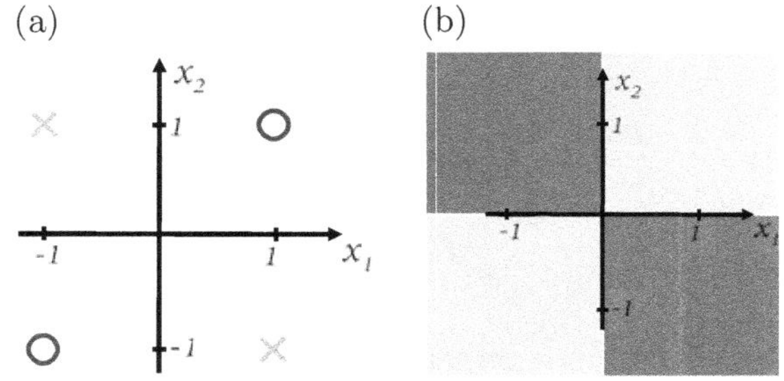

Figure 3.19: (a) Exclusive OR (XOR) (b) a XOR space.

Table 3.2: XOR.

y	x_1	x_2
1	-1	-1
0	-1	1
0	1	-1
1	1	1

generative model, which is the above first one, and *discriminative model*, which is the above second one. Thus now we can see that naive Bayes classifier and logistic regression are typical examples of the generative model and discriminative model, respectively. In other words or generally, we can just say that the generative model and discriminative model are for unsupervised learning and supervised learning.

In fact in supervised learning, this difference is not substantial. As shown the above, we can just change that if we consider $\frac{p(y=1)}{p(y=0)}$ or not.

Linear and Logistic Regression can Represent Exclusive OR (XOR)?

Both linear and logistic regression are based on the combinations of all features, keeping the same representation power. A simple example that these regression are unable to represent is Table 3.2, which has four instances with two labels. Fig. 3.19 (a) shows these four instances by ◯ for $y = 1$ and $\times$ for $y = -1$. This data is called *Exclusive OR (XOR)*.

Obviously these four instances cannot be classified by linear representation. Taking a more general look, by replacing each instance of Fig. 3.19 (a) with an instance with arbitrary values over the two dimensional space, we can see that as shown in Fig. 3.19 (b), a general XOR means the spaces taken by +1 (those by -1) are located at diagonals, being separated by the other spaces. Clearly it would be hard to discriminate these spaces by not only linear function but also other functions, like polynomials and more complex polnomial-based functions.

Thus what kind of models can deal with this XOR data? This question would

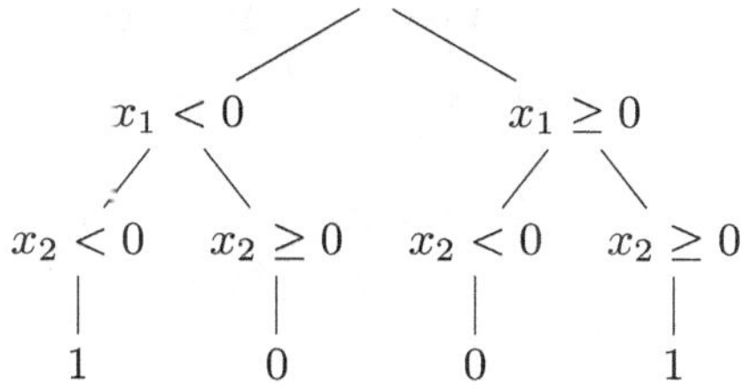

Figure 3.20: Decision tree, representing XOR.

be worth considering to understand the current classifiers (supervised learning models). First, on the data side, the key feature of this data (Fig. 3.19 (b)) is that the spaces with one label are clearly segmented (or divided) into subspaces. For example, the spaces with the label of 1 are $(x_1 > 0, x_2 > 0)$ and $(x_1 < 0, x_2 < 0)$. Similarly $(x_1 > 0, x_2 < 0)$ and $(x_1 < 0, x_2 > 0)$ for the label of -1.

Then on the classifier side, for example, decision tree was recursive partitioning data, implying that decision tree can capture the segmented places in data space. That is, for Fig. 3.19 (b), decision tree partitions the two-dimensional data space by $x_1 = 0$ first, and then the two partitioned spaces further by $x_2 = 0$, resulting in four subspaces, each being assigned by the corresponding label. This would make decision tree represent the four regions with correct labels in Table 3.2. Fig. 3.20 shows this decision tree.

On the other hand, a linear function or any more complex polynomial function (for simplicity we just write a linear function) cannot deal with this type of segmented subspaces just by one function. There would be only two solutions for the problem of dealing with this segmented subspaces:

Combinations of multiple hypotheses (functions): As we have seen the data space, we have to have multiple linear functions, each dealing with each segmented subspace. That is, we have to represent each of the segmented subspace by a hypothesis and combine them as the entire function. In other words, similar to clustering, each subspace (corresponding to a cluster in clustering) should be modeled by a hypothesis or function, and combine them, learning with each of them. Thus in the subsequent sections, we will explain the combination of logistic regression and then explain more on the combination of general hypotheses, which is called an *ensemble learning.*

Projection to different data spaces: Thus if the given data space is hard to classify instances, we can generate another data space in which instances are easier to classify and project the given instances into that data space. We will explain this approach after ensemble learning.

3.2.7 Layered Neural Network and Deep Learning

Again an approach of combining multiple hypotheses is called *ensemble learning*. In general *layered neural network* is not regarded as an ensemble learning method. However, from a viewpoint of logistic regression, which is a component of layered neural network, neural network is one type of ensemble learning, because of using multiple logistic regression parallelly at each layer.

Representing XOR by Combining Logistic Regression

Logistic regression has multiple numerical values as input and outputs a value (label) within [0,1]. We combine logistic regression, according to the following two steps (Fig. 3.21 (a)).

Step 1: multiple logistic function: We train multiple logistic regression for the same input. Although the inputs are the same, we use different initial values for the learning algorithm (of gradient descent) for different logistic regression, by which the learned results of multiple logistic regression are different.

Step 2: one logistic function: We then train one logistic regression, which use the outputs of the above Step 1 as input to summarize the outputs of Step 1.

We explain points behind each of these two steps below:

Step 1: multiple logistic regression for multiple data subspaces: The idea behind generating multiple logistic regression is to learn a unique subspaces (if there are) by each logistic regression (hopefully). We have no prior knowledge about subspaces and so just use random initial values to assign each model to a different data space. As we saw already, linear regression cannot deal with multiple (segmented) subspaces, and logistic regression has to just focus on the subspace, to which logistic regression is assigned by random initial parameter values. Then if initial values for parameter estimation are diverse enough, multiple logistic regression might be distributed over different subspaces, like each subspace by one logistic regression. Thus ensemble learning might be able to take care of the segmented data spaces like XOR or some more complex space.

Step 2: One logistic regression good enough for binary labels: The idea of Step 2 is to wrap up the outputs of multiple logistic regression of Step 1. Of course we can use more than one logistic regression for Step 2, while our setting is binary classification, for which one output (which can deal with both positives and negatives) is good enough.

We use logistic regression in two layers. Entirely this method (Fig. 3.21 (a) shows the network structure) is called three-*layered neural network*, three-layered *feed-forward neural network* or *artificial neural network*.

The most left-hand side of (a) is the layer with M input nodes, corresponding to M input features. This layer is called as *input layer*. Each of the K logistic

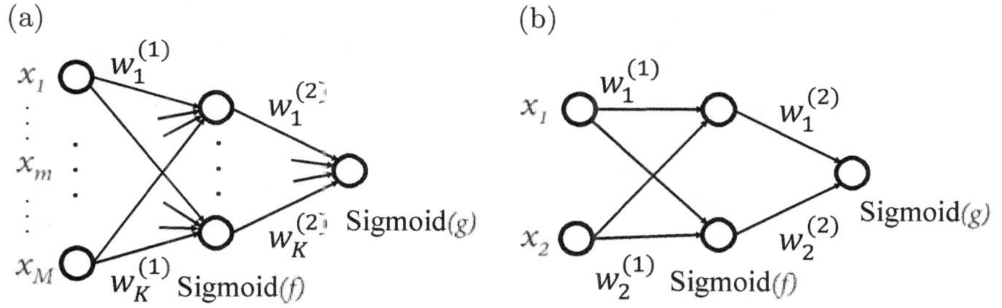

Figure 3.21: (a) Ensemble of logistic regression, (b) Two features and two nodes at the intermediate layer.

functions in Step 1 has M input features in the input layer as input and outputs at a node in the center layer. The center layer is called *intermediate layer.* Thus the intermediate layer has K nodes for K logistic regression. Logistic regression at Step 2 has K nodes of the intermediate layer as the input and the output at the node in the most right-hand side layer. This layer is called the *output layer.*

Once again we can check the number of nodes at each layer:

Input layer: M nodes for M features.

Intermediate layer: K nodes for K logistic regression. This number can be changed, regardless of input data.

Output layer: For binary classification problem, 1 node for the range of [0,1], corresponding to $y = 0$ to $y = 1$.

No process is done in the input layer, and each procedure is done between layers: weights are estimated between layers and the sigmoid function is assigned at a layer. We describe each function of this network, according to the two steps.

Step 1: Output of the input layer – Output of the intermediate layer:

The output of the input layer is linearly weighted and outputted to the intermediate layer as the input of a sigmoid function. The output of the sigmoid function, i.e. each output of a node in the intermediate layer, is the output of Step 1,

Step 2: Output of the intermediate layer–Output of the output layer:

The output of the intermediate layer is linearly weighted and outputted to the output layer as the input of a sigmoid function. The output of the sigmoid function is the output of Step 2. That is the total output of the layered network.

We describe the input and output of each layer by using equations below.

Table 3.3: Ensemble logistic regression (Fig. 3.21 (b)) for XOR.

y	x_1	x_2	$i_1^{(1)}(\boldsymbol{x})$	$o_1^{(1)}(\boldsymbol{x})$	$i_2^{(1)}(\boldsymbol{x})$	$o_2^{(1)}(\boldsymbol{x})$	$i^{(2)}(\boldsymbol{x})$	$o^{(2)}(\boldsymbol{x})$
1	-1	-1	$-\epsilon$	0	$+3\epsilon$	1	0	1/2
0	-1	1	$+\epsilon$	1	$+\epsilon$	1	$-\epsilon$	0
0	1	-1	$+\epsilon$	1	$+\epsilon$	1	$-\epsilon$	0
1	1	1	$+3\epsilon$	1	$-\epsilon$	0	0	1/2

- The input of the k-th logistic regression (the weighted sum over the output of the input layer), i.e. also the input of the k-th node in the intermediate layer:

$$i_k^{(1)}(\boldsymbol{x}) = \boldsymbol{x}^\top \boldsymbol{w}_k^{(1)} + \epsilon. \tag{3.138}$$

- The output of the k-th logistic regression, i.e. also the output of the k-th node in the intermediate layer. Then in the intermediate layer, the input goes through a sigmoid function:

$$o_k^{(1)}(\boldsymbol{x}) = \sigma(i_k^{(1)}(\boldsymbol{x})). \tag{3.139}$$

$$= \frac{1}{1+\exp(-i_k^{(1)}(\boldsymbol{x}))}. \tag{3.140}$$

- The input of the output layer (weighted sum over the outputs of the above K logistic regression:

$$i^{(2)}(\boldsymbol{x}) = \sum_k^K w_k^{(2)} o_k^{(1)}(\boldsymbol{x}) + \epsilon. \tag{3.141}$$

- The output of the output layer. Finally this goes into a sigmoid function:

$$o^{(2)}(\boldsymbol{x}) = \sigma(i^{(2)}(\boldsymbol{x})) \tag{3.142}$$

$$= \frac{1}{1+\exp(-i^{(2)}(\boldsymbol{x}))}. \tag{3.143}$$

We now check if this neural network can represent the XOR in Fig. 3.19. First we use only two input features, according to Fig. 3.19. Thus as shown in Fig. 3.21 (b), the number of nodes in the input layer is only 2.

We can choose an arbitrary number of nodes in the intermediate layer, but for simplicity we use two nodes in the intermediate layer, meaning two logistic regression at Step 1.

Suppose that the coefficients of the first and second linear regression are $\boldsymbol{w}_1^{(1)} = (+\epsilon, +\epsilon)$ and $\boldsymbol{w}_2^{(1)} = (-\epsilon, -\epsilon)$, respectively. The values of Step 1, $f_i^{(1)}(\boldsymbol{x})$ are shown in Table 3.3. If we take a large value for ϵ, $p_i^{(1)}(h = 1|\boldsymbol{x})$ $(i = 1, 2)$ after sigmoid functions are approximated by only 0 or 1, as shown in Table 3.3. Then from $p_1^{(1)}(h=1|\boldsymbol{x})$ and $p_2^{(1)}(h=1|\boldsymbol{x})$, we can see that we first take the intersection of these two and then their reverse. This for example can be implemented by $\boldsymbol{w}^{(2)} = (-\epsilon, -\epsilon)$. The result of this is shown in the last column of Table 3.3, which is consistent with y. This means that the combination of logistic regression, i.e. layered neural network, can implement XOR, if the above $\boldsymbol{w}^{(1)}$ and $\boldsymbol{w}^{(2)}$ are obtained from data.

We need to learn these coefficients from data. Although we just use only two logistic regression (only two nodes in the intermediate layer) in Fig. 3.21 (b), we can increase this number, to allow to deal with a larger number of subspaces.

Estimating Parameters for Layered Neural Network via Gradient Descent

We needed to use the gradient descent for estimating parameters of logistic regression. Layered neural network uses logistic regression, and so we again use the gradient descent for estimating parameters of layered neural network. The gradient descent for layered neural network is named as *back-propagation* (BP).

BP repeats the following two steps:

Feed-forward (FF) step: We compute the output from the input, using the current parameter values. We can just follow the above computation, from (3.138) to (3.143).

Back-propagation (BP) step: Using the error between the true $\boldsymbol{y}$ and the final output computed by FF, we update $\boldsymbol{w}^{(k)}$ by the gradient descent in the descendant order of k (i.e. from the output layer to the input layer).

The FF step would be straight-forward and easily understood, while the BP step would need more explanation. First we would make sure the following four preliminary points for implementing BP.

1. Three layered neural network has the following two parameters, which are estimated from the output layer to the intermediate layer:
   (a) Weights in the output layer: $\boldsymbol{w}^{(2)}$
   (b) Weights in the intermediate layer: $\boldsymbol{w}_k^{(1)} (k = 1, \ldots, K)$
2. The final output (of the output layer): $\hat{y} = o^{(2)}(\boldsymbol{x}) = p(y=1|\boldsymbol{x})$.
3. The derivative of $f(x) = \frac{1}{1+\exp(-x+\epsilon)}$ with respect to x is always
$$\frac{df(x)}{dx} = f(x)(1 - f(x)). \tag{3.144}$$
4. The error between the estimated $\hat{y}$ and true y is the squared loss (L^2 norm):
$$\begin{aligned} e &= \frac{1}{2}||\hat{y} - y||_2^2 &\quad (3.145)\\ &= \frac{1}{2}||o^{(2)}(\boldsymbol{x}) - y||_2^2. &\quad (3.146)\end{aligned}$$

We then estimate the two parameters by the gradient descent:

$w_k^{(2)}$: We take the partial derivative of the error with respect to $w_k^{(2)}$ (see Section A.9.4 for the derivation of (3.148)).

$$\frac{\partial e}{\partial w_k^{(2)}} = \frac{\partial e}{\partial o^{(2)}(\boldsymbol{x})} \frac{\partial o^{(2)}(\boldsymbol{x})}{\partial i^{(2)}(\boldsymbol{x})} \frac{\partial i^{(2)}(\boldsymbol{x})}{\partial w_k^{(2)}} \tag{3.147}$$

$$= (o^{(2)}(\boldsymbol{x}) - y) o^{(2)}(\boldsymbol{x})(1 - o^{(2)}(\boldsymbol{x})) o_k^{(1)}(\boldsymbol{x}). \tag{3.148}$$

The above equation is just for one instance $\boldsymbol{x}$. Considering all entire instances, $\boldsymbol{X}$, the update rule of $\boldsymbol{w}^{(2)}$ for the gradient descent is as follows:

$$w_k^{(2):new} \leftarrow w_k^{(2):old} + \eta \frac{\partial e}{\partial w_k^{(2)}} \tag{3.149}$$

$$= w_k^{(2):old} + \eta \sum_{\boldsymbol{x} \in \boldsymbol{X}} (o^{(2)}(\boldsymbol{x}) - y) o^{(2)}(\boldsymbol{x})(1 - o^{(2)}(\boldsymbol{x})) o_k^{(1)}(\boldsymbol{x}). \tag{3.150}$$

$\boldsymbol{w}_k^{(1)}$: We take the partial derivative of the error with respect to $\boldsymbol{w}_k^{(1)}$ (see Section A.9.4 for the derivation of (3.152)):

$$\frac{\partial e}{\partial \boldsymbol{w}_k^{1)}} = \frac{\partial e}{\partial o^{(2)}(\boldsymbol{x})} \frac{\partial o^{(2)}(\boldsymbol{x})}{\partial i^{(2)}(\boldsymbol{x})} \frac{\partial i^{(2)}(\boldsymbol{x})}{\partial o_k^{(1)}(\boldsymbol{x})} \frac{\partial o_k^{(1)}(\boldsymbol{x})}{\partial i_k^{(1)}(\boldsymbol{x})} \frac{\partial i_k^{(1)}(\boldsymbol{x})}{\partial \boldsymbol{w}_k^{(1)}} \tag{3.151}$$

$$= (o^{(2)}(\boldsymbol{x}) - y) o^{(2)}(\boldsymbol{x})(1 - o^{(2)}(\boldsymbol{x})) w_k^{(2)} o^{(1)}(\boldsymbol{x})(1 - o^{(1)}(\boldsymbol{x}))\boldsymbol{x}. \tag{3.152}$$

Similar to $\boldsymbol{w}^{(2)}$, considering all entire instances, the update rule of $\boldsymbol{w}_k^{(1)}$ for the gradient descent is given as follows:

$$\boldsymbol{w}_k^{(1):new} \leftarrow \boldsymbol{w}_k^{(1):old} + \eta \frac{\partial e}{\partial \boldsymbol{w}_k^{(1)}} \tag{3.153}$$

$$= \boldsymbol{w}_k^{(1):old} + \eta \sum_{\boldsymbol{x} \in \boldsymbol{X}} (o^{(2)}(\boldsymbol{x}) - y) o^{(2)}(\boldsymbol{x})(1 - o^{(2)}(\boldsymbol{x})) w_k^{(2)} o^{(1)}(\boldsymbol{x})(1 - o^{(1)}(\boldsymbol{x}))\boldsymbol{x}. \tag{3.154}$$

We summarize the above parameter estimation algorithm into a pseudocode, which is shown in **Algorithm 3.15**.

Time Complexity

First, each part of (3.150) can be computed by the number of features, i.e. M, and also multiplying them is also by the number of features, i.e. M. Thus, the

---

Algorithm 3.15: Backpropagation algorithm for training three-layered neural network of Fig. 3.21(a).

Function `Backpropagation_for_three-layered_neural_networks(`$\boldsymbol{X}$, $\boldsymbol{y}$, K`)`

Data: training data: $\boldsymbol{X}$, training data labels: $\boldsymbol{y}$, #intermediate layers: K (# logistic regression)

Result: trained parameters

Assign initial values to parameters, $\boldsymbol{w}_k^{(1)} (k = 1, \ldots, K)$ and $\boldsymbol{w}^{(2)}$.;

repeat

  Feedforward: compute output $o^{(2)}(\boldsymbol{x})$ of the output layer, due to (3.138)–(3.143).;

  Backpropagation: update parameters $\boldsymbol{w}_k^{(1)} (k = 1, \ldots, K)$ and $\boldsymbol{w}^{(2)}$, due to (3.154) and (3.150).;

until *convergence*;

Output the trained parameters $\boldsymbol{w}$.;

---

update rule of (3.150) can be computed by $O(NM)$. Second, each part of (3.154) can be computed by the number of features, i.e. M, and multiplication of (3.154) also needs only M. Then (3.154) needs only $O(NM)$. Entirely the complexity at each iteration is $O(NM)$. Our network is with only three layers, but if the number of layers is larger, the number of layers would also affect the time complexity.

Layered neural network has the parameter of weights between layers, meaning that two types of parameters for three layers. The output layer of Step 2 uses logistic regression. Thus the gradient of (3.150), i.e. the update rule of $\boldsymbol{w}^{(2)}$, is reasonably equivalent to the gradient of the update rule (3.127), which was parameter estimate for single logistic regression. The difference is that the last of the gradient in (3.150) is the output of the intermediate layer, while single logistic regression is like that with only two layers (no intermediate layer), and so the corresponding part is the output of the input layer, i.e. $\boldsymbol{x}$.

Deep Learning with a Larger Number of Intermediate Layers

Fig. 3.22 shows one possibility we increase the number of layers of the above three-layered neural network. In this case, similar to the above three-layered neural network, an intermediate layer computes the linearly weighted sum over the outputs of the preceding intermediate layer, gives that value to a sigmoid function as the input and transfers the output of the sigmoid function to the next intermediate layer. This procedure is repeated from the first intermediate layer to the last intermediate layer to the output layer.

Also, again, similar to the above three layered neural network, the number of nodes in one intermediate layer corresponds to the number of logistic regression between the intermediate layer closer to the input layer and the current intermediate layer. Then again the output layer has only one node, i.e. only one logistic regression between the last intermediate layer and the output layer. This type of

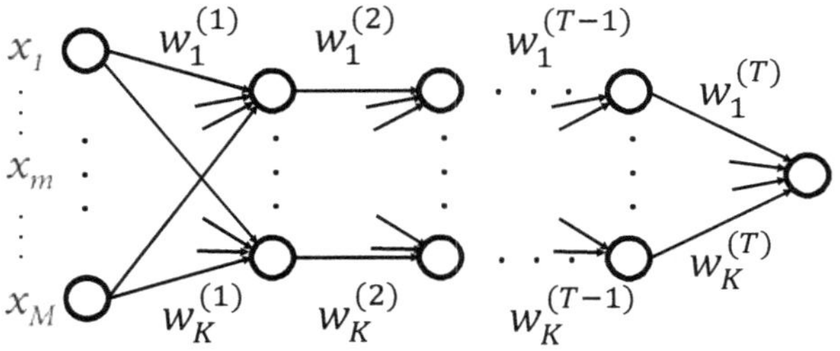

Figure 3.22: Deep learning: neural network with a larger number of intermediate layers.

neural network with a large number of layers is called *deep learning*[11] [57].

Parameter Estimation of Deep Learning

We have showed two-step parameter estimation in the three-layered neural network. The gradient was obtained by taking the partial derivative of the error with respect to the parameter. In fact this gradient has some pattern.

Below we write the gradient of $w_k^{(2)}$ and $\boldsymbol{w}_k^{(1)}$, i.e. (3.148) and (3.152) again:

$$w_k^{(2)} \quad : \quad \frac{\partial e}{\partial w_k^{(2)}} = (o^{(2)}(\boldsymbol{x}) - y)o^{(2)}(\boldsymbol{x})(1 - o^{(2)}(\boldsymbol{x}))o_k^{(1)}(\boldsymbol{x}). \tag{3.155}$$

$$\boldsymbol{w}_k^{(1)} \quad : \quad \frac{\partial e}{\partial \boldsymbol{w}_k^{(1)}} = (o^{(2)}(\boldsymbol{x}) - y)o^{(2)}(\boldsymbol{x})(1 - o^{(2)}(\boldsymbol{x}))w_k^{(2)}o^{(1)}(\boldsymbol{x})(1 - o^{(1)}(\boldsymbol{x}))\boldsymbol{x}. \tag{3.156}$$

The above two gradients share the same first part. Also each part of the gradients is a contribution of each layer, being back-propagated from the output layer to the input layer. We write this for each of the five parts for $\boldsymbol{w}_k^{(1)}$:

1. $(o^{(2)}(\boldsymbol{x}) - y)$: The first $(o^{(2)}(\boldsymbol{x}) - y)$ is from the output at the output layer.

2. $o^{(2)}(\boldsymbol{x})(1 - o^{(2)}(\boldsymbol{x}))$: The next is from the sigmoid function of at the output layer.

3. **$o_k^{(1)}(\boldsymbol{x})$ of $w_k^{(2)}$ and $w_k^{(2)}$ of $\boldsymbol{w}_k^{(1)}$** : Since the input of the output layer is linear combination of $w_k^{(2)}$ and $o_k^{(1)}(\boldsymbol{x})$, $w_k^{(2)}$ is $o_k^{(1)}(\boldsymbol{x})$, i.e. the output of the intermediate layer, and $\boldsymbol{w}_k^{(1)}$ is $w_k^{(2)}$. The last part is important, because this indicates that

[11] Note that practically there are a huge variety of types in deep learning. Here we just show a simple, straight-forward extention from the three-layered neural network.

4. $o^{(1)}(\boldsymbol{x})(1 - o^{(1)}(\boldsymbol{x}))$ **of** $\boldsymbol{w}_k^{(1)}$: This and below are only for $\boldsymbol{w}_k^{(1)}$. Step 1 goes through the intermediate layer, and so $o^{(1)}(\boldsymbol{x})(1 - o^{(1)}(\boldsymbol{x}))$ is from the sigmoid function. This is the same as the above 2, meaning that this part must be repeated, if the intermediate layer is increased.

5. $\boldsymbol{x}$ **of** $\boldsymbol{w}_k^{(1)}$: Finally the input of the intermediate layer is the linear combination of $\boldsymbol{w}_k^{(1)}$ and input $\boldsymbol{x}$. Thus, $\boldsymbol{x}$ is multiplied.

From this observation, we have four points to consider:

1. The above first point shows that this $(o(\boldsymbol{x}) - y)$ is needed.
2. The above second and fourth points are the same and this $o^{(t)}(\boldsymbol{x})(1-o^{(t)}(\boldsymbol{x}))$ must be repeated for each intermediate layer t.
3. The third point indicates that $w_k^{(t)}$ must be repeated for the t-th intermediate layer, while $o_k^{(t)}(\boldsymbol{x})$ would be added only one time at t.
4. The fifth point indicates if the intermediate layer is the closest one to the input layer, the above $o_k^{(t)}(\boldsymbol{x})$ should be replaced with just the input, i.e. $\boldsymbol{x}$.

Thus we can further summarize these four points below, as patterns to be considered, to estimate parameters of deep learning. The network we consider here has $T+1$ layers, where the first layer is the input layer and the $T+1$ layer is the output layer. The gradient of the t-th intermediate layer should be from the following four parts:

1. From the output of the output layer:
   $o^{(T)}(\boldsymbol{x}) - y$.
2. From the sigmoid function at the $t'(\in \{t+1, \ldots, T\})$-th intermediate layer:
   $o^{(t')}(1 - o^{(t')})$.
3. From the input of the $t'(\in \{t+1, \ldots, T\})$-th intermediate layer:
   $\boldsymbol{w}_k^{(t')}$.
4. From the input of the previous intermediate layer:
   $o^{(t-1)}(\boldsymbol{x})$, however $\boldsymbol{x}$ if $t = 2$.

Thus all in all, the gradient of deep learning with respect to $\boldsymbol{w}_k^{(t)}$ can be summarized into the following:

$$\frac{\partial e}{\partial \boldsymbol{w}_k^{(t)}} = (o^{(T)}(\boldsymbol{x}) - y) \prod_{t'=t}^{T} o^{(t')}(1 - o^{(t')}) \boldsymbol{w}_k^{(t')} \boldsymbol{x} \qquad \text{if} \qquad t = 2. \tag{3.157}$$

$$\frac{\partial e}{\partial \boldsymbol{w}_k^{(t)}} = (o^{(T)}(\boldsymbol{x}) - y) \prod_{t'=t}^{T} o^{(t')}(1 - o^{(t')}) \boldsymbol{w}_k^{(t')} o^{(t-1)}(\boldsymbol{x}); \qquad \text{otherwise.} \tag{3.158}$$

Time Complexity

The computation of the gradient can be implemented by the loop of $t = T, \ldots, 1$, where the computation at t has to use those already computed in the previous t. For each pair of the two neighboring intermediate layers, we need compute K logistic regression, as we did in the three-layered neural network. This means that for the neural network with $T+1$ layers, we need compute $(K(T-1)+1)$ logistic regression. That is, the computational complexity reaches $O(TNM)$, where T is the number of layers.

Notes on Deep Learning

Overfitting training data: Ensemble logistic regression allows to deal with multiple segmented subspaces. Neural network with a larger number of layers will allow this more, while the possibility of overfitting training data would be also higher.

Comprehensible results: One advantage of logistic regression is that regression coefficients can directly tell the important features, because of keeping the linear regression property. However ensemble of logistic regression loses this information, in the sense that too much information are given by a lot of logistic regression. In particular, this will be more for neural network with a larger number of layers.

Summary of extension: Logistic regression was expanded in the direction of ensemble learning as follows:

(decision tree →) linear regression → logistic regression (naive Bayes classifier) → layered neural network, deep learning

---

Comparison of Layered Neural Network with Finite Mixture Model

In Section 3.2.5, we already showed that logistic regression can be a special case of naive Bayes classifier. Logistic regression can be understood in a naive-Bayes-classifier way by focusing on its linear function and replacing each weighted independent variable in the linear function with a probability. Mathematically, logistic regression is

$$p_i(y=1|\boldsymbol{x}) \quad = \quad \sigma(\sum_m w_{i,m}^{(1)} \cdot x_m + \epsilon), \tag{3.159}$$

and the inside of each m can be replaced with a conditional probability, resulting in

$$p_i(y=1|\boldsymbol{x}) \quad = \quad p_i(y=1) \prod_m p_i(x_m|y=1). \tag{3.160}$$

Original data	1	2	3	4	5
Bootstrap sample 1	2	3	3	5	5
Bootstrap sample 2	1	2	2	3	4
Bootstrap sample 3	1	1	2	4	5
Bootstrap sample 4	1	2	3	3	5
Bootstrap sample 5	1	4	4	4	5

Figure 3.23: Bootstrapping.

Similarly three-layered neural network is

$$p(y=1|\boldsymbol{x}) \quad = \quad \sigma(\sum_i w_i^{(2)} \cdot \sigma(\sum_m w_{i,m}^{(1)} \cdot x_m + \epsilon)), \tag{3.161}$$

which can be written in the following way:

$$p(y=1|\boldsymbol{x}) \quad = \quad \prod_i p_i(y=1) \prod_m p_i(x_m|y=1). \tag{3.162}$$

Taking the logarithm of the both sides of (3.162):

$$\log p(y=1|\boldsymbol{x}) \quad = \quad \sum_i \log p_i(y=1) \prod_m p_i(x_m|y=1). \tag{3.163}$$

This shows that layered neural network is a "finite mixture model", which has naive Bayes classifier as a component and $p_i(y = 1)$ as a weight of the i-th component. In fact as mentioned in 3.1.4, the structure of a finite mixture model is the weighted majority (sum), which is the same as that of layered neural network, which takes the weighted majority (sum) over the outputs of the nodes in the intermediate layer. This would be one type of ensemble learning.

In the next section and later on, we explain ensemble learning with an arbitrary hypothesis as a component.

3.2.8 Ensemble Learning via Sampling

The objective of ensemble learning is to improve the predictive performance, thinking about the complex data space, such as XOR. When we consider an arbitrary classifier, one approach for the objective is not to change the classifier but to change the data. In fact we can change the data distribution and then improve the performance of classifiers trained by using the data with diverse distributions.

Bagging

Given the (original) data, bootstrapping is to repeat sampling with replacement (see Section A.1) randomly a certain number of times, for example, the data size. Fig. 3.23 shows a schematic example of repeating bootstrapping five times over the original data: $\{1, 2, 3, 4, 5\}$, showing that a set by sampling with replacement

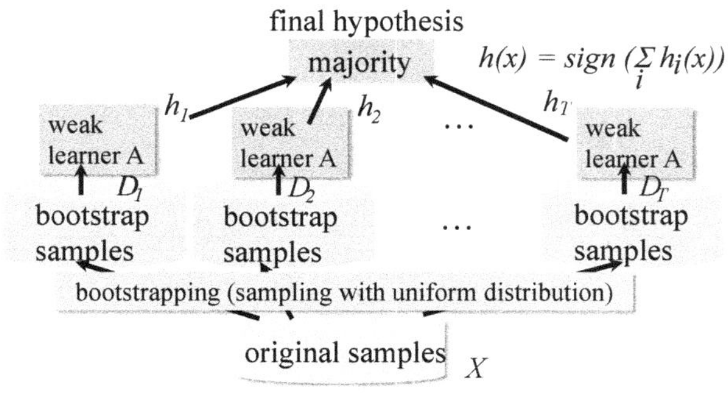

Figure 3.24: Bagging.

Algorithm 3.16: Bagging.

Function `Bagging(`$\boldsymbol{X}$, $\boldsymbol{y}$, T`)`

 Data: training data: $\boldsymbol{X}$, training data labels: $\boldsymbol{y}$, #trees to be generated: T

 Result: final hypothesis:h

 for $t \leftarrow 1$ **to** T **do**

  Obtain $\boldsymbol{D}_t$ by bootstrapping $\boldsymbol{X}$.;

  Learn hypothesis h_t from $\boldsymbol{D}_t$.;

 Output final hypothesis $h : h(\boldsymbol{x}) = \text{sign}(\sum_{t=1}^{T} h_t(\boldsymbol{x}))$.;

---

shows different from the original set. For example, the same element appears more than once or an element never appears in the sampled set. Theoretically it is proved that around only two thirds of the original elements appear in the sampled set by using sampling with replacement. This value, two thirds, is relatively low, meaning that different sets can be obtained for training a hypothesis, resulting that we might have a different classifier.

Suppose that a component classifier is decision tree. We run bootstrapping over a given original dataset to generate a new dataset, which would have a different number of each instance from the original. Thus recursive partitioning of decision tree over each dataset generated in this manner would be different from each other. We repeat generating decision trees and combine the generated decision trees, by which the boundary between positives and negatives would be smoother than one decision tree, which was a totally stepwise function (see Fig. 3.14). This smoother boundary would work well for improving the predictive performance.

Fig. 3.24 shows this procedure schematically. This simple method is called *bootstrap aggregating* or *bagging*. **Algorithm 3.16** shows a pseudocode of bagging. Bagging uses uniform weights over components to integrate, and a key point is that generating a hypothesis is independent of generating another hypothesis. Also

Algorithm 3.17: Random forest.

Function `Random_forest(`$\boldsymbol{X}$, $\boldsymbol{y}$, T`)`
  Data: training data: $\boldsymbol{X}$, training data labels: $\boldsymbol{y}$, #decision trees: T, #features to choose: $M' << M$
  Result: final hypothesis: h
  for $t \leftarrow 1$ **to** T **do**
    Do bootstrap sampling over $\boldsymbol{X}$ and also do random sampling M' features to generate data $\boldsymbol{D}_t$.;
    Learn hypothesis h_t from $\boldsymbol{D}_t$.;
  Output the final hypothesis h: $h(\boldsymbol{x}) = \text{sign}(\sum_{t=1}^{T} h_t(\boldsymbol{x}))$.;

any classifier (or hypothesis) can be used in the framework of bagging, instead of decision tree we have considered.

Time Complexity

The complexity of the learning algorithm is very simple. The method needs T components, and so $T\times$ "the complexity of generating the component classifier".

Random Forests

A similar approach to Bagging is *random forests* [14], in which not only instances but also features are randomly sampled, while the number of features in each dataset is kept at a smaller number than the given dataset (by sampling without replacement). Component classifiers are decision tree. We can say that multiple decision trees generated are independent of each other, since each dataset is generated by sampling with replacement. Interestingly in random forests, pruning is not done for decision tree (meaning that each decision tree is forced to overfitting training data obtained by bootstrapping). **Algorithm 3.17** shows a pseudocode of random forests.

Time Complexity

The time complexity of random forests is, similar to bagging, $T\times$ "complexity of each component", where T is the number of components.

3.2.9 Ensemble Learning with Three Hypotheses

Indeed ensemble learning is useful, and we might achieve a better performance by a more strategic sampling approach than random sampling. For example, a simple idea is to sample incorrectly predicted instances more than correctly predicted instances.

Simple Procedure

A simple question is if we can make the error rate smaller by combining hypotheses trained with more instances incorrectly predicted.

We can start with the minimum number of classifiers, i.e. three classifiers: Given dataset D, we do sampling to have three datasets (D_1, D_2, D_3) and generate three hypotheses by these three datasets and check if combining the three hypotheses can beat the performance of any one of the three hypotheses.

Then we would take the following strategy to obtain three datasets: D_1, D_2 and D_3 and then three hypotheses (h_1, h_2 and h_3) [75]. One assumption here is that the performance of any predictor is better than random guessing. In other words, the percentage of incorrectly predicted instances by a predictor should be always less than a half.

1. D_1: the first hypothesis h_1 is trained from the original data D, i.e. $D_1 = D$.

2. D_2: D_2 is generated by incorporating the instances incorrectly predicted by h_1. That is, we generate D_2, repeating the following procedure: every time when we sample an instance, we flip a coin, and if the head of a coin comes, we sample an instance correctly predicted by h_1; otherwise we sample an instance incorrectly predicted by h_1.

   The idea is to keep the instances correctly and incorrectly predicted by h_1 "fifty-fifty", suppose that the coin we use has no bias.

   Our assumption was that any predictor can be better than random sampling (the percentage of incorrectly predicted instance is always less than 50%), and so this strategy incorporates incorrectly predicted instances more than just randomly sampling from D.

   Then h_2 is trained from D_2.

3. D_3: we have several possibilities on how we can use the prediction results by h_1 and h_2, to generate D_3. However, to make the prediction procedure simple, we collect instances to which the two predictions by h_1 and h_2 are inconsistent, to generate D_3. Then we train h_3 from D_3.

Now we can build a very simple ensemble prediction hypothesis, h_{ens}, using the above hypotheses: h_1, h_2 and h_3:

Ensemble hypothesis h_{ens}:

1. **Given instance $\boldsymbol{x}$, we apply $h_1(\boldsymbol{x})$ and $h_2(\boldsymbol{x})$.**
2. **If the two predictions are consistent, we adopt this prediction.**
3. **Otherwise we adopt the prediction by h_3, i.e. $h_3(\boldsymbol{x})$.**

Algorithm 3.18 shows a pseudocode of this prediction algorithm.

Algorithm 3.18: Ensemble prediction by using three hypotheses: h_{ens}.

```
Function Ensemble_prediction:   h_ens(x)
    Data: unknown instance: x
    Result: prediction result
    if h1(x) == h2(x) then h1(x);
    else h3(x);
```

Table 3.4: 2×2 contingency table.

		h_2	
		correct ($h_2(\boldsymbol{x}) = y_{\boldsymbol{x}}$)	incorrect ($h_2(\boldsymbol{x}) \neq y_{\boldsymbol{x}}$)
h_1	correct ($h_1(\boldsymbol{x}) = y_{\boldsymbol{x}}$)	p_{11}	p_{10}
	incorrect ($h_1(\boldsymbol{x}) \neq y_{\boldsymbol{x}}$)	p_{01}	p_{00}

Performance Analysis: Prediction can be Improved?

We now consider the error by the ensemble prediction hypothesis h_{ens}. Let ϵ_t be the error rate by h_t for D_t. Note that ϵ_t is the error rate for training data, and we expect that the error rate for test data would be larger than training data, meaning $\epsilon_t \leq \epsilon \leq 1/2$ for some ϵ.

We can write 2×2 contingency table on the results by h_1 and h_2 which is shown in Table 3.4.

Note that any $\boldsymbol{x}$ is fallen into one of the four cells in Table 3.4, in terms of the correct or incorrect predictions by two hypotheses h_1 and h_2 Then we write the probability that $\boldsymbol{x}$ falls into one of the four cells as p_{11}, p_{10}, p_{01} and p_{00} as follows:

p_{11}: $h_1(\boldsymbol{x}) = h_2(\boldsymbol{x}) = y_{\boldsymbol{x}}$.

p_{10}: $h_2(\boldsymbol{x}) \neq h_1(\boldsymbol{x}) = y_{\boldsymbol{x}}$.

p_{01}: $h_1(\boldsymbol{x}) \neq h_2(\boldsymbol{x}) = y_{\boldsymbol{x}}$.

p_{00}: $h_2(\boldsymbol{x}) = h_1(\boldsymbol{x}) \neq y_{\boldsymbol{x}}$.

We now think about ϵ_t $(t = 1, 2)$.

First, since D_1 for hypothesis h_1 is simply D,

The probability that x is correctly predicted by h_1 :

$$1 - \epsilon_1 \quad = \quad p_{11} + p_{10}. \tag{3.164}$$

The probability that x is incorrectly predicted by h_1 :

$$\epsilon_1 \quad = \quad p_{01} + p_{00}. \tag{3.165}$$

Second, for h_2, first by flipping a coin, we have two cases: correctly predicted by h_1 (which is $p_{11} + p_{10}$) and incorrectly by h_1 (which is $p_{01} + p_{00}$) with equal probabilities. Then when we focus on instances correctly predicted by h_2, there are also two cases: one is the instances correctly predicted by h_1 (p_{11}) already and the other is those incorrectly by h_1 (p_{01}). Below we summarize this observation as follows:

The probability that x is correctly predicted by h_2 :

$$1 - \epsilon_2 = \frac{1}{2}\frac{p_{11}}{p_{11}+p_{10}} + \frac{1}{2}\frac{p_{01}}{p_{01}+p_{00}} \quad (3.166)$$

$$= \frac{1}{2}\frac{p_{11}}{(1-\epsilon_1)} + \frac{1}{2}\frac{p_{01}}{\epsilon_1}. \quad (3.167)$$

Also when we focus on instances incorrectly predicted by h_2, there are also two cases, one is the instances correctly predicted by h_1 (p_{10}) and the other is those incorrectly by h_1 (p_{00}). Then,

The probability that x is incorrectly predicted by h_2 :

$$\epsilon_2 = \frac{1}{2}\frac{p_{10}}{p_{11}+p_{10}} + \frac{1}{2}\frac{p_{00}}{p_{01}+p_{00}} \quad (3.168)$$

$$= \frac{1}{2}\frac{p_{10}}{(1-\epsilon_1)} + \frac{1}{2}\frac{p_{00}}{\epsilon_1}. \quad (3.169)$$

We here can see some p_{**} can be represented by other p_{**}.
For example, from (3.165),

$$p_{00} = \epsilon_1 - p_{01}. \quad (3.170)$$

Also by multiplying both sides of (3.169) by $2(1-\epsilon_1)$ and using (3.170),

$$p_{10} = (2\epsilon_2 - 1)(1-\epsilon_1) + \frac{p_{01}(1-\epsilon_1)}{\epsilon_1}. \quad (3.171)$$

We now consider ϵ_{ens} of ensemble learning h_{ens}. There are only two cases that h_{ens} predicts an instance incorrectly:

1. **Both h_1 and h_2 incorrect** Predictions by hypothesis h_1 and hypothesis h_2 were consistent and then the prediction was incorrect.

2. **Either h_1 or h_2 incorrect** Predictions by hypothesis h_1 and hypothesis h_2 were inconsistent and prediction by hypothesis h_3 is incorrect.

The probability of the above 1 is p_{00}, and that of the above 2 is $\epsilon_3(p_{10} + p_{01})$. Thus the error by ensemble learning is given as follows:

$$\epsilon_{ens} = p_{00} + \epsilon_3(p_{10} + p_{01}). \quad (3.172)$$

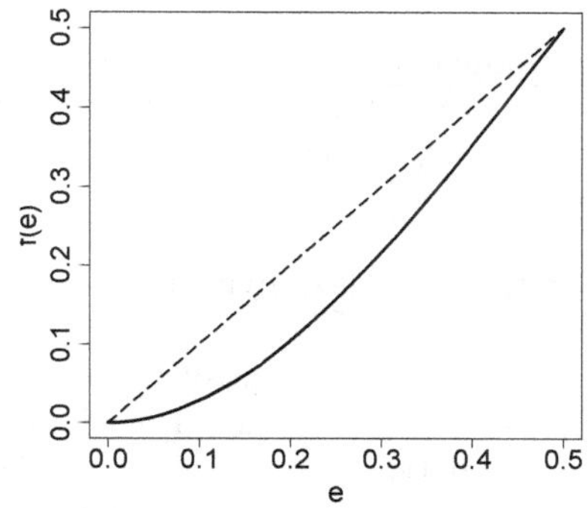

Figure 3.25: $\epsilon_{ens} = 3\epsilon^2 - 2\epsilon^3$ (solid line) and $f(\epsilon) = \epsilon$ (dotted line).

We modify (3.172) by using $\epsilon_t \leq \epsilon \leq 1/2$, (3.170) and (3.171) as follows:

$$\begin{aligned}
\epsilon_{ens} &= p_{00} + \epsilon_3(p_{10} + p_{01}) && (3.173)\\
&\leq p_{00} + \epsilon(p_{10} + p_{01}) && (3.174)\\
&= \epsilon(2\epsilon_2 - 1)(1 - \epsilon_1) + \epsilon_1 + \frac{p_{01}(\epsilon - \epsilon_1)}{\epsilon_1} && (3.175)\\
&\leq \epsilon(2\epsilon_2 - 1)(1 - \epsilon_1) + \epsilon_1 && (3.176)\\
&\leq \epsilon(2\epsilon - 1)(1 - \epsilon) + \epsilon && (3.177)\\
&= 3\epsilon^2 - 2\epsilon^3. && (3.178)
\end{aligned}$$

This $f(\epsilon)$ can be regarded as the worst error of ensemble learning. Fig. 3.25 shows $f(\epsilon) = 3\epsilon^2 - 2\epsilon^3$ as the solid line for ϵ of $[0, 1/2]$, also with $f(\epsilon) = \epsilon$ as the dotted line. Thus this figure indicates the worst error ϵ_{ens} of ensemble learning hypothesis h_{ens} always achieves a lower value than the error by one hypothesis. In summary, even the simple ensemble learning hypothesis, h_{ens}, over three hypotheses can improve the worst case error.

3.2.10 Ensemble Learning: AdaBoost

We have shown that a simple ensemble learning algorithm of three hypotheses improves the error of any of the three hypotheses, while this algorithm was specialized into the "three" hypotheses. Thus the next question is that if we repeat learning a hypothesis sequentially for ensemble learning, how should we sample incorrectly predicted instances at each iteration?

Multiplicative Sampling Weights and Predictive Hypotheses Weights

For simplicity, we will use only the last hypothesis to decide the way of sampling instances. That is, we use the t-th hypothesis and its prediction results to decide the $t + 1$-th method of sampling instances. In other words, we do not use any

Table 3.5: Product of true label and predicted label.

Correct prediction	$\rightarrow$	$y_{\boldsymbol{x}} \cdot h(\boldsymbol{x}) > 0$	$\rightarrow$	$0 < \exp(-y_{\boldsymbol{x}} \cdot h(\boldsymbol{x})) < 1$
Incorrect prediction	$\rightarrow$	$y_{\boldsymbol{x}} \cdot h(\boldsymbol{x}) < 0$	$\rightarrow$	$\exp(-y_{\boldsymbol{x}} \cdot h(\boldsymbol{x})) > 1$

hypothesis obtained older than the last hypothesis for deciding the manner of sampling.

Suppose that binary labels take +1 (positive) or -1 (negative), i.e. $y_{\boldsymbol{x}} = \{+1, -1\}$ for $\boldsymbol{x}$. Thus given an instance $\boldsymbol{x}$, if the output $h(\boldsymbol{x})$ of the trained hypothesis is positive, the prediction is positive; otherwise the prediction is negative. Thus this indicates that if the product of the true label and the predicted label, i.e. $y_{\boldsymbol{x}} h(\boldsymbol{x})$, is positive, prediction is correct; otherwise prediction is incorrect. Furthermore the absolute value of $y_{\boldsymbol{x}} h(\boldsymbol{x})$ indicates the significance of the correctness or incorrectness. Thus this product must be useful for sampling incorrectly predicted instances more. More in detail, $y_{\boldsymbol{x}} h(\boldsymbol{x})$ becomes large when the prediction is correct, and so we take its negative and use the negative value as the shoulder of some value, which is convenient, since the value is compared with not zero but 1. That is, as shown in Table 3.5, $e^{-y_{\boldsymbol{x}} \cdot h(\boldsymbol{x})}$ becomes less than 1 if the prediction is correct; otherwise this becomes larger than 1. Also again the absolute value is larger, the more significant the correctness/incorrectness of the prediction is. This would be useful for sampling incorrectly predicted instances more.

The next question is what would be the best way to predict the label of an instance from multiple hypotheses? In fact using only part of the trained hypotheses, like the last hypothesis, would not be ensemble learning. Thus we should use all hypotheses trained, while these hypotheses might include those which may cause large errors. Then we set up weights over hypotheses to integrate them as the ensemble hypothesis, and these weights should be estimated to be optimum. That is, we introduce weight w_t to, given instance $\boldsymbol{x}$, compute the final hypothesis $h_B(\boldsymbol{x})$ after repeating training T times, as follows:

$$h_B(\boldsymbol{x}) = \mathrm{sign}(\sum_{t=1}^{T} w_t h_t(\boldsymbol{x})). \tag{3.179}$$

AdaBoost Algorithm

We summarize our thought on sampling and also prediction:
letting $D_t(\boldsymbol{x})$ ($\sum_{\boldsymbol{x}} D_t(\boldsymbol{x}) = 1$) be the probability of sampling $\boldsymbol{x}$ at the t-th iteration, we can give the probability at the $t+1$-th iteration by using $y_{\boldsymbol{x}} h_t(\boldsymbol{x})$ and w_t, as follows:

$$D_{t+1}(\boldsymbol{x}) = \frac{D_t(\boldsymbol{x}) \exp(-w_t y_{\boldsymbol{x}} h_t(\boldsymbol{x}))}{Z_t}, \tag{3.180}$$

where Z_t is just for normalization: $Z_t = \sum_{\boldsymbol{x}} D_t(\boldsymbol{x}) \exp(-w_t y_{\boldsymbol{x}} h_t(\boldsymbol{x}))$.

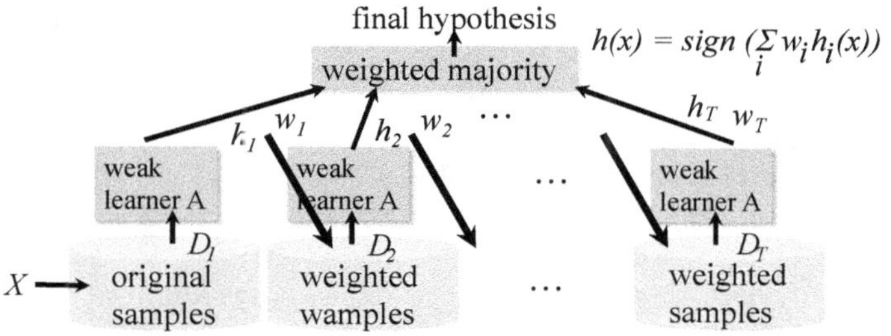

Figure 3.26: Boosting.

Algorithm 3.19: AdaBoost.

Function `Ada_boost(`$\boldsymbol{X}$, $\boldsymbol{y}$, T`)`
    Data: training data $\boldsymbol{X}$, training data labels: $\boldsymbol{y}$, #repetitions: T
    Result: final hypothesis: h_B
    Initialization: $D_1(\boldsymbol{x}) = 1/N$.;
    for $t \leftarrow 1$ **to** T **do**
        Learn hypothesis: h_t from D_t.;
        Update w_t according to (3.193).;
        Update D_{t+1} according to (3.180).;
    Output the final hypothesis: $h_B(\boldsymbol{x}) = \text{sign}(\sum_{t=1}^{T} w_t h_t(\boldsymbol{x}))$.;

Once again the idea behind this distribution is two fold: 1) sampling: as we saw in the above, $e^{-y_{\boldsymbol{x}} \cdot h(\boldsymbol{x})}$ becomes larger than 1 if the prediction for $\boldsymbol{x}$ was incorrect, and so $\boldsymbol{x}$ will be sampled more, and vice versa. 2) prediction: also each hypothesis, for example, the t-th hypothesis is weighted by w_t, according to the predictive performance of the t-th hypothesis.

This gives the procedure of an ensemble learning method, called AdaBoost [77], and Fig. 3.26 shows this procedure schematically. **Algorithm 3.19** shows a pseudocode of AdaBoost.

Time Complexity

The complexity at each iteration is not heavy. The w_t needs summing up over all instances, resulting in $O(N)$. Also the update of $D_{t+1}(\boldsymbol{x})$ is done over all instances, also resulting in $O(N)$. Thus learning a component, i.e. hypothesis learning h, would be heavier than ensemble learning.

Training Error Analysis to Estimate Weights

Similar to Table 3.5, we can have the following for the final hypothesis h_B:

1. If the final hypothesis is incorrect ($h_B(\boldsymbol{x}) \neq y_{\boldsymbol{x}}$),

$$y_{\boldsymbol{x}} \sum_t w_t h_t(\boldsymbol{x}) \leq 0 \quad \rightarrow \quad \exp(-y_{\boldsymbol{x}} \sum_t w_t h_t(\boldsymbol{x})) \geq 1. \tag{3.181}$$

2. If the final hypothesis is correct ($h_B(\boldsymbol{x}) = y_{\boldsymbol{x}}$),

$$y_{\boldsymbol{x}} \sum_t w_t h_t(\boldsymbol{x}) > 0 \quad \rightarrow \quad 0 < \exp(-y_{\boldsymbol{x}} \sum_t w_t h_t(\boldsymbol{x})) < 1. \tag{3.182}$$

Then, suppose that [ ] is a function which returns 1 if the inside of the bracket is correct; otherwise this function returns zero. We can have the following:

$$[h_B(\boldsymbol{x}) \neq y_{\boldsymbol{x}}] \leq \exp(-y_{\boldsymbol{x}} \sum_t w_t h_t(\boldsymbol{x})). \tag{3.183}$$

Also (3.180) can be transformed as follows:

$$\begin{aligned}
D_{t+1}(\boldsymbol{x}) &= \frac{D_t(\boldsymbol{x}) \exp(-w_t y_{\boldsymbol{x}} h_t(\boldsymbol{x}))}{Z_t} && (3.184)\\
&= \frac{D_{t-1}(\boldsymbol{x}) \exp(-w_t y_{\boldsymbol{x}} h_t(\boldsymbol{x})) \exp(-w_{t-1} y_{\boldsymbol{x}} h_{t-1}(\boldsymbol{x}))}{Z_t Z_{t-1}} && (3.185)\\
&= \frac{D_1(\boldsymbol{x}) \exp(-w_t y_{\boldsymbol{x}} h_t(\boldsymbol{x})) \cdots \exp(-w_1 y_{\boldsymbol{x}} h_1(\boldsymbol{x}))}{Z_t \cdots Z_1} && (3.186)\\
&= \frac{D_1(\boldsymbol{x}) \exp(-y_{\boldsymbol{x}} \sum_t w_t h_t(\boldsymbol{x}))}{\prod_t Z_t} && (3.187)\\
&= \frac{\exp(-y_{\boldsymbol{x}} \sum_t w_t h_t(\boldsymbol{x}))}{N \prod_t Z_t}. && (3.188)
\end{aligned}$$

Then, for all $\boldsymbol{x}$,

$$\sum_{\boldsymbol{x}} \frac{\exp(-y_{\boldsymbol{x}} \sum_t w_t h_t(\boldsymbol{x}))}{N \prod_t Z_t} = \sum_{\boldsymbol{x}} D_{t+1}(\boldsymbol{x}) = 1. \tag{3.189}$$

Thus, we have the following upper bound on the training error:

Proposition 3.2 (Upper bound of training error by AdaBoost)**.**

$$\frac{1}{N} |\{\boldsymbol{x} | h_B(\boldsymbol{x}) \neq y_{\boldsymbol{x}}\}| \leq \prod_{t=1}^{T} Z_t. \tag{3.190}$$

Proof. **Proposition 3.2** can be directly obtained by (3.183) and (3.189). □

This proposition indicates that the training error can be given by the product of $\boldsymbol{Z}_t$ over t, meaning that to minimize the training error, we can minimize Z_t at

Table 3.6: Error functions of Boosting.

Name	loss function
AdaBoost	$\exp(-y_{\boldsymbol{x}} h(\boldsymbol{x}))$.
LogitBoost	$\ln(1 + exp(-y_{\boldsymbol{x}} h(\boldsymbol{x})))$.

each t, independently.

$$Z_t = \sum_{\boldsymbol{x}} D_t(\boldsymbol{x}) \exp(-w_t y_{\boldsymbol{x}} h_t(\boldsymbol{x})) \quad (3.191)$$

$$\leq \sum_{\boldsymbol{x}} D_t(\boldsymbol{x}) (\frac{1 + y_{\boldsymbol{x}} h_t(\boldsymbol{x})}{2} \exp(-w_t) + \frac{1 - y_{\boldsymbol{x}} h_t(\boldsymbol{x})}{2} \exp(w_t)). \quad (3.192)$$

We then obtain the w_t, which minimize the right-hand side, analytically. That is, we estimate w_t at each iteration t, as follows:

$$w_t = \frac{1}{2} \ln(\frac{1 + \sum_{\boldsymbol{x}} D_t(\boldsymbol{x}) y_{\boldsymbol{x}} h_t(\boldsymbol{x})}{1 - \sum_{\boldsymbol{x}} D_t(\boldsymbol{x}) y_{\boldsymbol{x}} h_t(\boldsymbol{x})}). \quad (3.193)$$

Generalization of AdaBoost

AdaBoost decides the way of sampling at the $t + 1$-th iteration, depending on the prediction results of the t-th iteration. In fact, the point of this algorithm, shown in **Algorithm 3.19**, is, as shown in Table 3.5, the correctness of prediction is checked by $\exp(-y_{\boldsymbol{x}} \cdot h(\boldsymbol{x}))$. In more detail, if the prediction is incorrect, $\exp(-y_{\boldsymbol{x}} \cdot h(\boldsymbol{x}))$ is larger than 1, by which the corresponding instance is sampled more. This means that incorrectly predicted instances are more sampled. Thus $\exp(-y_{\boldsymbol{x}} \cdot h(\boldsymbol{x}))$ is monotonically decreasing in terms of $y_{\boldsymbol{x}} \cdot h_(\boldsymbol{x})$ and reaches 1 when $y_{\boldsymbol{x}} \cdot h_(\boldsymbol{x})$ is 0. That is, this is an error function, which takes 1 at the boundary between correctness and incorrectness:

$$L(\boldsymbol{x}, y_{\boldsymbol{x}}) = \exp(-y_{\boldsymbol{x}} h(\boldsymbol{x})). \quad (3.194)$$

This objective function (error function) is called *Exponential loss*. Again the property that this loss has to satisfy that $y_{\boldsymbol{x}} \cdot h_(\boldsymbol{x})$ is decreasing monotonically and reaches 1 when $y_{\boldsymbol{x}} \cdot h_(\boldsymbol{x})$ is zero.

Note that any objective function can be used for sampling, if the function satisfies this property (however weight α_t over hypothesis also depends on the function, and so the objective function must be the one from which the weight over the hypothesis can be derived). Thus so far a variety of ensemble learning methods in this literature have been proposed, just by changing the objective (error) function. Table 3.6 raises two examples of them. Also there are already lots of discussion on the functions which satisfy the above property [32].

As well as AdaBoost, a series of ensemble learning methods, which repeat sampling incorrectly predicted instances more are called *boosting* [76]. The boosting literature have theoretically presented that even by using a method (called

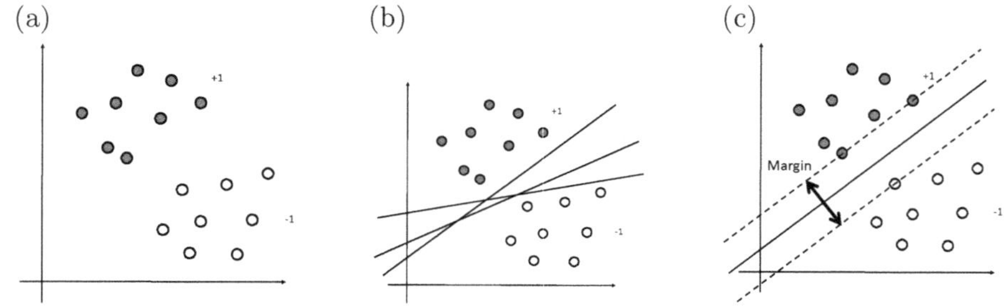

Figure 3.27: Margin: key of support vector machine.

weak learner) which is slightly better in prediction than random guessing, a high-performance predictor can be generated by ensemble learning, specifically boosting.

3.2.11 Support Vector Machine

Focus on Support Vectors

We already mentioned that there are two approaches to represent XOR: 1) ensemble learning, and 2) projecting the original data space into another data space, in which instances can be classified. So far, regarding ensemble learning, we have seen three-layered neural network, deep learning, more general ensemble learning of using arbitrary classifiers, i.e. ensemble learning via sampling and also boosting. In particular, we have seen that boosting can make even weak learners (as components) a high-predictive predictor.

In this section, we explore the other approach for XOR. We explain support vector machine (SVM), which can do classification in a data space different from the original input space. SVM is linear regression. Linear and logistic regression explained so far used all given training instances to estimate the linear function. On the other hand, SVM is motivated by the idea that we can focus on only the instances which are around a boundary between positives and negatives, instead of using all instances.

Fig. 3.27 (a) is a toy sample dataset, which should be classified into two classes (dark circles and white circles) by linear regression (linear function). As shown in (b), when we think about what linear function (i.e. a line) would be good, intuitively a good solution would be, as shown in (c), the line most distant from both positives (more precisely, positives on the boundary) and negatives (also negatives on the boundary). In this case, the instances on the boundary are called *support vectors*, which are the instances on the dotted lines in Fig. 3.27 (c). Also the linear function (linear line), which is the most distant from the support vectors, is called *hyperplane* (the line between the two dotted lines in Fig. 3.27 (c)). Furthermore, the double of the distance between the support vector and the hyperplane is called the *margin*. The idea that the hyperplane most distant from both positives and negatives is the best, is called *maximizing the margin*, which is the basis of SVM.

Hard Margin

Let $\boldsymbol{w}$ be weight coefficients of the linear function of hyperplane:

$$y = \boldsymbol{w}^\top \boldsymbol{x} + b. \tag{3.195}$$

Then to maximize the margin, y must be the following (Fig. 3.27 (a)):

$$\text{For support vectors of positives} \quad y = +1. \tag{3.196}$$
$$\text{On the hyperplane} \quad y = 0. \tag{3.197}$$
$$\text{For support vectors of negatives} \quad y = -1. \tag{3.198}$$

That is, we can write that support vectors of positives are $(\boldsymbol{x}_+, +1)$ and the closest points on hyperplane from these support vectors are $(\boldsymbol{x}_0, 0)$. Then from 3.195,

$$+1 = \boldsymbol{w}^\top \boldsymbol{x}_+ + b. \tag{3.199}$$
$$0 = \boldsymbol{w}^\top \boldsymbol{x}_0 + b. \tag{3.200}$$

From the difference of these two,

$$+1 = \boldsymbol{w}^\top (\boldsymbol{x}_+ - \boldsymbol{x}_0). \tag{3.201}$$

Note that $||\boldsymbol{x}_+ - \boldsymbol{x}_0||$ is exactly the half of the distance of the margin, and also exactly $\frac{1}{||\boldsymbol{w}||}$. Similarly the distance between the support vectors of negatives and the hyperplane is $\frac{1}{||\boldsymbol{w}||}$. That is, maximizing the margin means maximizing $\frac{1}{||\boldsymbol{w}||}$. This further means minimizing $||\boldsymbol{w}||$ or $||\boldsymbol{w}||^2$ (for simpler derivation hereafter) as the objective function.

The above thought just focused on the boundary between positives and negatives, while we need to incorporate the information of all instances, which must be separated correctly by the $\boldsymbol{w}$ optimized in the above objective function. So the following two must be satisfied for all instances:

$$\boldsymbol{w}^\top \boldsymbol{x} + b \geq +1 \quad \text{for} \quad y \geq 1. \tag{3.202}$$
$$\boldsymbol{w}^\top \boldsymbol{x} + b < -1 \quad \text{for} \quad y < -1. \tag{3.203}$$

This can be written in the following way:

$$y_i(\boldsymbol{w}^\top \boldsymbol{x}_i + b) \geq 1 \quad \text{for all} \quad i. \tag{3.204}$$

This can be the constraints for the objective function.

Thus all in all, to maximize the margin, we have to solve the following optimization problem:

$$\min_{\boldsymbol{w}} \frac{1}{2}||\boldsymbol{w}||^2 \quad \text{subject to} \quad \alpha_i y_i(\boldsymbol{w}^\top \boldsymbol{x}_i + b) \geq 1 \quad \text{for all} \quad i. \tag{3.205}$$

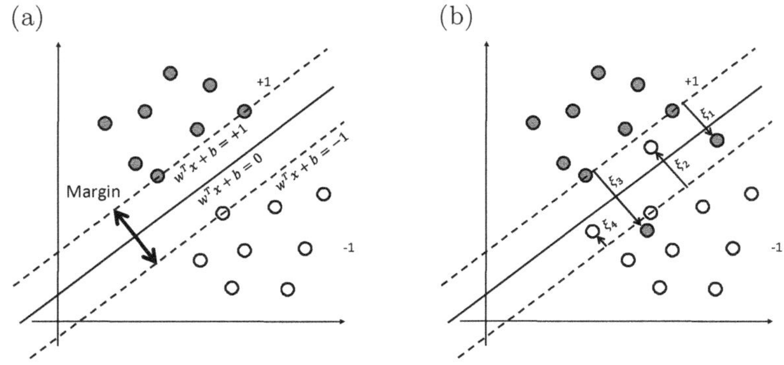

Figure 3.28: Maximizing the margin: (a) hard margin, (b) soft margin.

This is called a *primal form*. Fig. 3.28 (a) shows a schematic picture of this maximization problem (minimization of $\boldsymbol{w}$).

In this formulation, the constraints: $\alpha_i y_i(\boldsymbol{w}^\top \boldsymbol{x}_i + b) \geq 1$ can be regarded as equality conditions, and we can use the method of Lagrange multipliers. We can then define the Lagrange function as follows:

$$L(\boldsymbol{w}, b, \boldsymbol{\alpha}) = \frac{1}{2}\boldsymbol{w}^\top \boldsymbol{w} - \sum_i \alpha_i (y_i(\boldsymbol{w}^\top \boldsymbol{x}_i + b) - 1), \tag{3.206}$$

where $\alpha_i (i = 1, \ldots, N)$ are Lagrange multipliers.

We set the partial derivative of the Lagrange function with respect to each parameter equal to zero, which results in the following:

$$\frac{\partial L(\boldsymbol{w}, b, \boldsymbol{\alpha})}{\partial \boldsymbol{w}} = 0 \quad \Rightarrow \quad \boldsymbol{w} = \sum_i \alpha_i y_i \boldsymbol{x}_i. \tag{3.207}$$

$$\frac{\partial L(\boldsymbol{w}, b, \boldsymbol{\alpha})}{\partial b} = 0 \quad \Rightarrow \quad \sum_i \alpha_i y_i = 0. \tag{3.208}$$

We can have the following by using the above for (3.206) (see Section A.9.5 for derivation):

$$L(\boldsymbol{\alpha}) = \sum_i \alpha_i - \frac{1}{2}\sum_{i,j} y_i y_j \alpha_i \alpha_j \boldsymbol{x}_i^\top \boldsymbol{x}_j. \tag{3.209}$$

This is called a *dual form*.

Now our problem is given as follows:

$$\max_{\boldsymbol{\alpha}} \quad (\sum_i \alpha_i - \frac{1}{2}\sum_{i,j} y_i y_j \alpha_i \alpha_j \boldsymbol{x}_i^\top \boldsymbol{x}_j) \tag{3.210}$$

$$\text{subject to} \quad \sum_i^N y_i \alpha_i = 0. \tag{3.211}$$

$$\alpha_i \geq 0 \ (i = 1, \ldots, N). \tag{3.212}$$

This is a quadratic programming problem (also a convex optimization problem), and so we can just apply an appropriate solver to this problem to obtain the optimal α_i. We then, if necessary, can have $\boldsymbol{w}$ by using (3.207), which shows us the hyperplane.

The reason why we write "if necessary" in the above is that the primal objective of SVM is not to show the trained results but to predict the label of an unknown instance. That is, we can predict the label of a new instance $\boldsymbol{x}_{new}$ by using (3.207) as follows:

$$\hat{y} = \boldsymbol{w}^\top \boldsymbol{x}_{new} \tag{3.213}$$

$$= \sum_i \alpha_i y_i \boldsymbol{x}_i^\top \boldsymbol{x}_{new}. \tag{3.214}$$

The (3.214) shows that α_i can be directly used to predict the label $\hat{y}$ of the new instance $\boldsymbol{x}_{new}$. Note again that this prediction is possible without $\boldsymbol{w}$. That is, we do not have to compute $\boldsymbol{w}$.

Soft Margin

Fig. 3.27 shows the case that instances in one class are easily separated from those in the other class, with the good enough margin. Practically, this would be rare, and instances must not be separable so clearly. We now consider such a more practical situation, shown in Fig. 3.28 (b), where each non-separable instance has some value (ξ_i for instance i in Fig. 3.28), which reduces the margin. This setting of SVM is called *soft margin*, while the setting in Fig. 3.28 (a) is called *hard margin*. For soft margin, instead of (3.202) and (3.203), the following conditions must be satisfied for all instances:

$$\boldsymbol{w}^\top \boldsymbol{x} + b \geq +1 - \xi \quad \text{for} \quad y \geq 1. \tag{3.215}$$

$$\boldsymbol{w}^\top \boldsymbol{x} + b < -1 + \xi \quad \text{for} \quad y < -1. \tag{3.216}$$

where ξ_i $(i = 1, \ldots, N)$ can be defined for all instances with the possibility of reducing the margin, and are called *slack variables*. Also the objective function, instead of $\frac{1}{2}||\boldsymbol{w}||^2$, can be set as follows:

$$\frac{1}{2}||\boldsymbol{w}||^2 + C\sum_i^N \xi_i. \tag{3.217}$$

Then the primal form of the optimization problem can be given as follows:

$$\min_{\boldsymbol{w}} \quad \frac{1}{2}||\boldsymbol{w}||^2 + C\sum_i^N \xi_i \tag{3.218}$$

$$\text{subject to} \quad \alpha_i y_i(\boldsymbol{w}^\top \boldsymbol{x}_i + b) \geq 1 - \xi_i \quad \text{for all} \quad i. \tag{3.219}$$

$$\xi_i \geq 0 \quad \text{for all} \quad i. \tag{3.220}$$

The objective function, given by (3.218), has two terms, which looks very normal, because the second term is the error term to be minimized and the first term is the regularization term. So the optimization problem is to make balance between these two terms. The constraints are again regarded as the equality constraints, and then the Lagrange function is given as follows:

$$L(\boldsymbol{w}, b, \boldsymbol{\alpha}, \boldsymbol{\xi}) = \frac{1}{2}\boldsymbol{w}^\top \boldsymbol{w} + C\sum_i^N \xi_i - \sum_i \alpha_i(y_i(\boldsymbol{w}^\top \boldsymbol{x}_i + b) - 1 + \xi_i) - \sum_i \mu_i \xi_i, \tag{3.221}$$

where α_i $(i = 1, \ldots, N)$ and ξ_i $(i = 1, \ldots, N)$ are Lagrange multipliers.

We now have to solve (3.221), as we did for hard margin. A good news is that the dual form of soft margin is the same as that of hard margin, except that α_i has the upper bound. Thus (3.210), (3.211) and (3.212) are given as follows:

$$\max_{\boldsymbol{\alpha}} \quad (\sum_i \alpha_i - \frac{1}{2}\sum_{i,j} y_i y_j \alpha_i \alpha_j \boldsymbol{x}_i^\top \boldsymbol{x}_j) \tag{3.222}$$

$$\text{subject to} \quad \sum_i^N y_i \alpha_i = 0. \tag{3.223}$$

$$C \geq \alpha_i \geq 0 \ (i = 1, \ldots, N). \tag{3.224}$$

We solve this optimization problem, in the same manner for hard margin, to obtain $\boldsymbol{\alpha}$. Then, similar to hard margin, we can predict the label of a new instance $\boldsymbol{x}_{new}$ just by using $\boldsymbol{\alpha}$:

$$\hat{y} = \boldsymbol{w}^\top \boldsymbol{x}_{new} + b \tag{3.225}$$

$$= \sum_i \alpha_i y_i \boldsymbol{x}_i^\top \boldsymbol{x}_{new} + b. \tag{3.226}$$

Again we do not have to compute $\boldsymbol{w}$ to estimate the label y of any unknown instance.

Replacing the Inner Product with a Kernel Function

We have derived the dual form to obtain the linear regression coefficients of SVM, both for hard margin and soft margin. This derivation has a clear purpose, or we have derived the dual form for a certain purpose. In fact, as you can see from (3.222), (3.223) and (3.224), this optimization problem (of dual form) does not need feature values of input data any more, except the inner product in (3.222). Thus, we can say that this optimization problem can be solved just if we already know the inner products over given instances. In other words, we do not need vectors of instances, and instead what we need is the inner product values between instances in training data. An inner product of two instances is a similarity between the two instances. Importantly, we can replace the inner product with a similarity function in the inner product space (strictly speaking *reproducing kernel Hilbert space* (RKHS), while we just call that the inner product space for simplicity), where the similarity function in the inner product space (or RKHS) is called the *kernel function* (see Section A.5 for more details):

$$\boldsymbol{x}_i^{\mathsf{T}}\boldsymbol{x}_j \quad \Rightarrow \quad \boldsymbol{K}(\boldsymbol{x}_i, \boldsymbol{x}_j). \tag{3.227}$$

Then any kernel function $K(\boldsymbol{x}_i, \boldsymbol{x}_j)$ can be used, instead of the inner product $\boldsymbol{x}_i^{\mathsf{T}}\boldsymbol{x}_j$.

$$\max_{\boldsymbol{\alpha}} \quad (\sum_i \alpha_i - \frac{1}{2}\sum_{i,j} y_i y_j \alpha_i \alpha_j \boldsymbol{K}(\boldsymbol{x}_i, \boldsymbol{x}_j)) \tag{3.228}$$

$$\text{subject to} \quad \sum_i^N y_i \alpha_i = 0. \tag{3.229}$$

$$C \geq \alpha_i \geq 0 \ (i = 1, \ldots, N). \tag{3.230}$$

Also in prediction again we can replace the inner product in (3.226) with the kernel function between an unknown instance $\boldsymbol{x}_{new}$ and the training instances, to predict the label of the new instance:

$$\boldsymbol{x}_i^{\mathsf{T}}\boldsymbol{x}_{new} \quad \Rightarrow \quad \boldsymbol{K}(\boldsymbol{x}_i, \boldsymbol{x}_{new}). \tag{3.231}$$

and

$$\hat{y} \quad = \quad \sum_i \alpha_i y_i \boldsymbol{K}(\boldsymbol{x}_i, \boldsymbol{x}_{new}). \tag{3.232}$$

Thus we do not need the original vectors, and instead we just need the value/output of the kernel function between training instances in learning and also between the new, unknown instance and training instances in prediction.

Note that the kernel function has merits in that we can design kernel functions using prior knowledge on the given data so that the input instances should be separable more easily. This is equivalent to, as shown in Fig. 3.29, projecting the input instances into a high-dimensional hyperspace, in which the training instances can be more accurately classified than the original space, by which a higher predictive performance is achieved.

Kernel functions can be introduced into not only SVM in classification, but also methods in other problem settings. We will call the kernel function-based methods in other settings entirely *kernel learning*, which are explained more in Section 3.4.

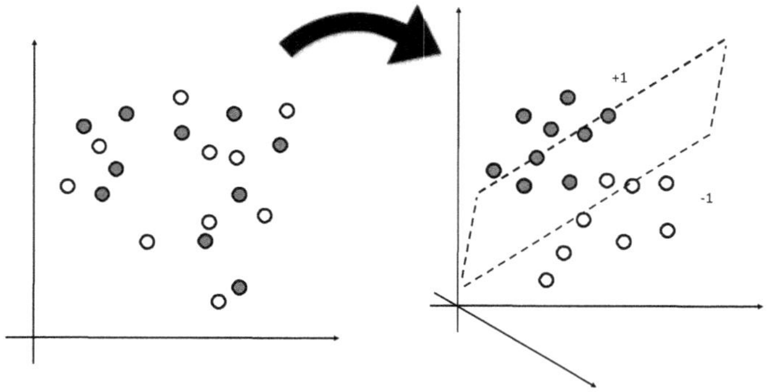

Figure 3.29: Projection to a high-dimensional hyperspace and classification in the space.

Efficient Prediction

In prediction of SVM, we need to compute/know all similarities/kernels between a new test instance and all training instances. This sounds like K-nearest neighbors or case based reasoning, which also check the labels of all closest training instances, resulting in high computational costs for prediction, making KNN and CBR unscalable.

However the trained $\boldsymbol{\alpha}$, obtained by optimizing the dual form, are very sparse, because α_i are non-zero only for support vectors and zero for all other instances. Thus after training, in prediction of a new, unknown instance, we do not have to check all training instance. Instead, in (3.232), we can consider only support vectors. That is, letting $\mathcal{S}_v$ be the set of support vectors, the prediction for unknown instance $\boldsymbol{x}_{new}$, which is given as (3.232) already, is newly given as follows:

$$\hat{y} = \sum_{i|i \in \mathcal{S}_v} \alpha_i y_i \boldsymbol{K}(\boldsymbol{x}_i, \boldsymbol{x}_{new}). \tag{3.233}$$

Thus again in prediction we do not have to check all instances, but we can estimate the label of $\boldsymbol{x}_{new}$ just only by support vectors.

Algorithm 3.20 shows a pseudocode of the algorithm of learning SVM and predicting an unknown test instance $\boldsymbol{x}_{new}$ by using $\alpha_i (i \in \mathcal{S}_v)$ and the given kernel function.

Time Complexity

The time complexity of SVM learning is the cost of obtaining $\boldsymbol{\alpha}$, which can be bounded by the complexity of a solver for the quadratic programming problem. The time complexity for prediction depends on the kernel function. However, for example, the simplest linear kernel function needs computing the inner product over all training instances (under the worst case that all instances are support vectors), resulting in $O(NM)$.

Algorithm 3.20: Support vector machine (learning and prediction).

Function Support_vector_machine($\boldsymbol{K}$, $\boldsymbol{y}$, $\boldsymbol{K}_{new}$)
Data: kernel between training instances: $\boldsymbol{K}$, Labels for training instances: $\boldsymbol{y}$, Kernel between training and testing instances: $\boldsymbol{K}_{new}$
Result: predicted labels for testing instances
Learning: Compute $\boldsymbol{\alpha}$ from $\boldsymbol{K}$ and $\boldsymbol{y}$ by solving (3.228)-(3.230).;
Prediction: Predict unknown labels from $\boldsymbol{\alpha}$, $\boldsymbol{y}$ and $\boldsymbol{K}_{new}$ according to (3.233).;

3.3 Feature Learning

Machine learning can be classified into unsupervised and supervised learning basically. However before working on either of the above learning, input data might have redundant features, which should be reduced or summarized into a smaller size of features. Also maybe just extracting important features might be useful for understanding the given data. In fact in both of the above two situations, given data might be huge. Thus working on only features, *feature learning*, which is a problem setting peculiar to vectors, is a relatively large research area in machine learning of vectors.

We note that feature learning are irreversible, in the sense that the original set of features cannot be recovered from the output of these methods. Feature learning has the following two objectives:

1. **Make input vectors (features) learnable:** Maybe the size of given features is too large or too redundant, and then the given data cannot be an input of a machine learning method. Then like a preprocessing step over input data before applying the machine learning method, feature learning changes the feature set to allow machine learning methods to run over the processed data easily.

2. **Make input vectors (features) interpretable (comprehensible):** Given features might be complex and incomprehensible. Thus feature learning changes the feature set to make the points of the data more comprehensible and interpretable.

Feature learning can be categorized into the following three approaches, to achieve basically both (or sometimes either) of the above two objectives[12].

1. **Feature selection:** We select only significant features out of the original given features, and unselected features are discarded. This is called *feature selection.*

   Feature selection makes the input data smaller, allowing a machine learning method to apply to the data feasible, even if it is infeasible for the same

[12] We note that this book focuses on basic feature learning rather than non-linear feature learning, such as non-linear dimensionality reduction.

machine learning method to apply to the original large dataset. Also a smaller number of features are more easily understandable. Thus feature selection has the possibility of achieving the above two objectives.

2. **Dimensionality reduction:** To reveal the properties embedded in given data, we can output a small number of "newly made" features out of the original data. This is called *dimensionality reduction*, where all original features are discarded. Thus dimensionality reduction also makes learning feasible. Also a small number of features may be useful for making results understandable, while this point depends on methods. Thus at least the above first objective is addressed by dimensionality reduction.

   Dimensionality reduction usually does not use labels, i.e. unsupervised learning and summarizing data. There are many eminent dimensionality reduction methods, while we explain principal component analysis (PCA) and Laplacian eigenmaps [9], because PCA is the most classical and typical dimensionality reduction method in machine learning, and Laplacian eigenmeps is a method which is strongly connected to graph learning. Other well-known dimensionality reduction methods include multidimensional scaling (MDS) [12] and locally linear embedding (LLE) [74].

3. **Sparse learning:** In regular machine learning methods, we can change the formulation for optimization to make the trained results sparse and more visible. Sparse learning makes only significant parameter values are kept at non-zero and others are at zero. Sparse learning makes the learning results visible, understandable and comprehensible, and at the same time a fewer number of features are shown, making further learning over the results easier. Thus sparse learning allows to realize the above both objectives.

3.3.1 Feature Selection

If we do not use labels, i.e. unsupervised learning, a simple approach to select important features in training data is first to do clustering data over features (not instances) and then select features which are closest to cluster centers. In other words, unsupervised learning is summarizing data, and if we have data summary as features, which are already the features to be selected. Thus in unsupervised learning, we already have a solution for feature selection or unsupervised learning itself is feature selection.

Thus below we consider feature selection in the context of supervised learning, i.e. using the label of given data.

Applying Decision Stump Iteratively?

In supervised learning, intuitively important features should be correlated with the label (Fig. 3.30 (a)). Selecting the most relevant feature to the label has been already done by decision stump (Section 3.2.2). Thus we can just repeat the two steps: 1) applying decision stump to the data to find the most relevant feature and 2) moving this feature from the data to a list named "feature selection", until the

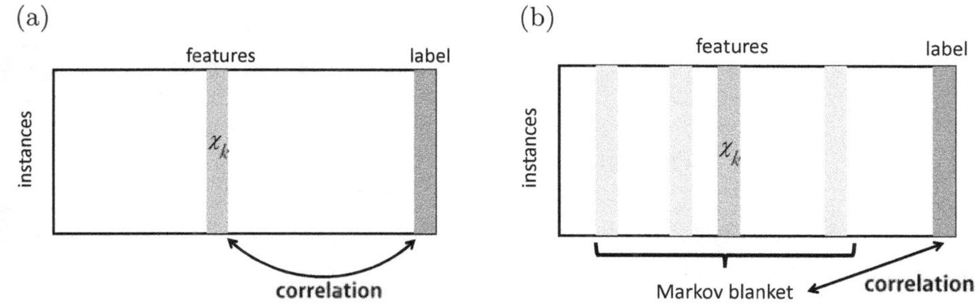

Figure 3.30: Feature selection: (a) select only features correlated with the label (b) Markov blanket is a set of features similar to χ_k, and then χ_k can be removed if (3.234) is satisfied.

size, say K, of this list reaches the number of features we want to select. This list gives us the K most relevant features to the label, and as shown in Fig. 3.30 (a), can be efficiently computed by just checking the pair of two vectors. However the serious problem of features in this list is they must be highly similar to each other, regarding the label vector, resulting in just a very redundant list. Thus it would be not enough to just consider the relevance to the label.

Markov Blanket

We then focus on the fact that if we have features which are both relevant to the label vector, they are redundant and can be reduced. In more detail, we can borrow the idea, called *Markov blanket*, which can be defined below:

Definition 3.1 (Markov Blanket). *Markov blanket $\mathcal{M}_k$ of feature χ_k is a set of features which are correlated with feature χ_k.*

Markov blanket is useful, because of the following:

Proposition 3.3. *If χ_k and its Markov blanket $\mathcal{M}_k$ satisfy* (3.234), *we can remove feature χ_k, since this feature is redundant.*

$$p(\boldsymbol{y}|\mathcal{M}_k, \chi_k) = p(\boldsymbol{y}|\mathcal{M}_k). \tag{3.234}$$

That is, if the distribution of the label $\boldsymbol{y}$ under $\mathcal{M}_k$ is irrelevant to feature χ_k, we can remove χ_k, since it is unnecessary. When we generate Markov blanket for each feature, say χ_k, we cannot consider all features except χ_k. Thus we need to place some upper bound on the number of features to be included in Markov blanket. Even so, the above process needs to check the relevance between multiple features and the label vector, causing high computational costs.

Algorithm

Again (3.234) needs high computational cost. Thus we first compare each feature and the label vector and select the most relevant M_1 features out of the M input

Algorithm 3.21: Feature selection by using Markov blanket.

```
Function Feature_selection_with_Markov_blanket(X, y, M1, M2, M)
    Data: training data: X, training data labels: y, #features to be
          selected in step 1: M1, #features to be selected in step 2: M2,
          Maximum Markov blanket size: |M|
    Result: M2 features most relevant to the label
    Step 1: Select M1 features most relevant to the label.;
    Step 2-1: Compute Markov blanket with the size of |M| for each of M1
    features.;
    Step 2-2: Repeat removing features satisfying (3.234) until #features is
    M2.;
    Output remaining features.;
```

features (Step 1). Then these features are redundant and reduced to M_2 features by using Markov blanket (Step 2). **Algorithm 3.21** shows a pseudocode of this algorithm. In this algorithm, various metrics can be candidates for measuring the correlation in Step 1, such as Pearson correlation, t-test, Area under the ROC curve (AUC), etc. In Step 2, there would be not so many features which can satisfy (3.234), and so we need to make approximation such that if the both sides are very close, we regard them that they are equal to [98, 62].

Time Complexity

Step 1 is, as shown in Fig. 3.30 (a), comparison between all features and the label vector, resulting in the time complexity of $O(NM)$. Step 2 needs to consider at least pair of two features with the label vector, resulting in at least $O(NM^2)$.

3.3.2 Principal Component Analysis

Background

We consider to generate a new feature (vector) by a linear combination of the current features so that this new feature should explain the data visually well. Let w_i be the weight over input feature (vector) $\boldsymbol{\chi}_i$.

$$w_1\boldsymbol{\chi}_1 + w_2\boldsymbol{\chi}_2 + \cdots + w_M\boldsymbol{\chi}_M = \boldsymbol{X}\boldsymbol{w}. \tag{3.235}$$

That is, we use numerical vector $\boldsymbol{w}$, which has the same length as that of the number of features in input data $\boldsymbol{X}$, and project each instance of $\boldsymbol{X}$ into just one value by $\boldsymbol{X}\boldsymbol{w}$ (so the entire instances are one vector). Fig. 3.31 shows one example of this projection, where the projected vector is on an one-dimensional axis. To explain the given data visually well, these points on one-dimensional axis should have a broad distribution, because it would be hard to see if these points are very close together. This means that the variance of this distribution should be as large as possible. Again Fig. 3.31 shows the projection by $\boldsymbol{w}$ into

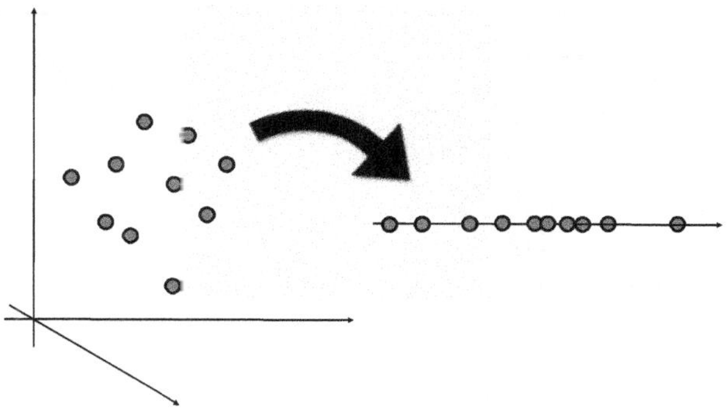

Figure 3.31: Concept of principal component analysis (PCA).

the one-dimension in the right-hand side, where the point should be distributed as broadly (widely) as possible, for visual purpose.

Algorithm

Suppose that $\boldsymbol{X}\boldsymbol{w}$ is normalized (the means is zero) already, the variance of this vector can be computed by the inner product of the same variance (see Section A.2.1):

$$(\boldsymbol{X}\boldsymbol{w})^\top \boldsymbol{X}\boldsymbol{w} = \boldsymbol{w}^\top \boldsymbol{X}^\top \boldsymbol{X}\boldsymbol{w}. \tag{3.236}$$

This is the objective function to be maximized. At the same time, we place constraint on $\boldsymbol{w}$ (of the L^2 norm) to reduce the flexibility of $\boldsymbol{w}$, and also to make $\boldsymbol{w}$ weights by which we can analyze what features are selected by the weights. Thus the optimization problem can be formulated as follows:

$$\max_{\boldsymbol{w}} \boldsymbol{w}^\top \boldsymbol{X}^\top \boldsymbol{X}\boldsymbol{w} \quad \text{subject to} \quad \boldsymbol{w}^\top \boldsymbol{w} = \text{Constant}. \tag{3.237}$$

Because of the equality constraints, Lagrange function can be defined as follows:

$$L(\boldsymbol{w}) = \boldsymbol{w}^\top \boldsymbol{X}^\top \boldsymbol{X}\boldsymbol{w} - \lambda(\boldsymbol{w}^\top \boldsymbol{w} - \text{Constant}). \tag{3.238}$$

We set the gradient of the Lagrangian with respect to $\boldsymbol{w}$ equal to zero, as follows:

$$\frac{dL(\boldsymbol{w})}{d\boldsymbol{w}} = 0 \qquad \Rightarrow \qquad \boldsymbol{X}^\top \boldsymbol{X}\boldsymbol{w} = \lambda \boldsymbol{w}. \tag{3.239}$$

The right-hand side of (3.239) is exactly the eigenvalue problem of $\boldsymbol{X}^\top \boldsymbol{X}$. Thus we solve this eigenvalue problem to have eigenvectors. These eigenvectors are called *principal components.* The principal components are ordered by eigenvectors, which are further ordered according to the amount of variance. Thus the first principal component shows the projected instances most broadly distributed on the one-dimensional axis.

Algorithm 3.22 shows a pseudocode of the algorithm of generating principal components, which is called *principal component analysis* (PCA).

Algorithm 3.22: Principal component analysis.

Function `Principal_component_analysis(`$\boldsymbol{X}$, K`)`
  Data: training data: $\boldsymbol{X}$, #principal components: K
  Result: K principal components
  Preprocessing: normalize $\boldsymbol{X}$.;
  Compute $\boldsymbol{V} = \boldsymbol{X}^\top \boldsymbol{X}$.;
  Run `Eigen_decomposition(`$\boldsymbol{V}, K$`)` and output K eigenvectors as K principal components.;

Time Complexity

The complexity of the algorithm follows that of solving eigenvalue problem. The complexity is $O(M^3)$ for $(M \times M)$-matrix.

Notes on Principal Component Analysis

Singular value decomposition: The $\boldsymbol{X}^\top \boldsymbol{X}$ is a symmetric matrix, and the same eigenvectors can be obtained by using singular value decomposition.

Covariance matrix: The $\boldsymbol{X}^\top \boldsymbol{X}$ is a covariance matrix (A.3.1), and so the principal components are the eigenvectors of the covariance matrix [102].

Rayleigh quotient: The optimization problem setting (3.237) and Lagrange function (3.293) show that solving PCA is given as the following Rayleigh quotient, which can be solved as the eigenvalue problem as well as other typical machine learning problems:

$$R(\boldsymbol{X}^\top \boldsymbol{X}, \boldsymbol{w}) = \frac{\boldsymbol{w}^\top \boldsymbol{X}^\top \boldsymbol{X} \boldsymbol{w}}{\boldsymbol{w}^\top \boldsymbol{w}}. \quad (3.240)$$

3.3.3 Laplacian Eigenmaps

We introduce another approach of dimensionality reduction, which is called *Laplacian eigenmaps* (or *spectral embedding*). A brief summary of this method is, to represent the input features by only a few features, keeping not only the distance between each pair of instances but also the distance among multiple instances.

Algorithm, based on Spectral Learning

We just introduce the procedure following [9].

1. We generate a graph (adjacency matrix) from input data $\boldsymbol{X}$.

   Each node corresponds to a feature. If two features are similar to each other, the corresponding two nodes are connected by an edge. There are two possible heuristics on the way to connect two nodes by an edge [9]:

(a) If the distance (similarity) of two features is smaller (larger) than ϵ, the corresponding two nodes are linked by an edge:

$$||\chi_i - \chi_j||^2 < \epsilon. \tag{3.241}$$

(b) If for both features, one feature is one of the top K most similar features of the other feature, the corresponding node is linked by an edge (this graph is called *K-nearest neighbor graph*).

2. We attach weights over graph nodes, by using either of the following two heuristics:

   (a) For $\boldsymbol{W}_{ij}$ of feature pair i and j, we use the following heat kernel:

$$\boldsymbol{W}_{ij} = \exp(-\frac{||x_i - x_j||^2}{\tau}). \tag{3.242}$$

   This is a heuristic for generating a graph. τ is a parameter adjusting the strength of the weight[13].

   (b) We can just use a binary (zero or one, indicating if the edge exists for a node pair). We do not have to consider the hyperparameter.

3. From the graph weight matrix, i.e. $\boldsymbol{W}_{ij}$ for the weight of nodes i and j (this matrix, $\boldsymbol{W}$ is the adjacency matrix of the graph), we compute diagonal matrix $\boldsymbol{D}$ by $\boldsymbol{D}_{ii} = \sum_j \boldsymbol{W}_{ij}$. Let $\boldsymbol{L} = \boldsymbol{D} - \boldsymbol{W}$, we can solve the following generalized eigenvalue problem to have new features:

$$\boldsymbol{L}\boldsymbol{v} = \lambda \boldsymbol{D}\boldsymbol{v}. \tag{3.243}$$

   $\boldsymbol{L}$ is a graph Laplacian (see Section A.7.1) and so a nonsingular matrix. Thus by multiplying the inverse of $\boldsymbol{L}$ from the left-hand side of (3.243), the generalized eigenvalue problem can be an eigenvalue problem. Then by solving this problem, we obtain the top K eigenvectors as new features.

Algorithm 3.23 shows a pseudocode of the above procedure, where we adopt the first heuristics for both Steps 1 and 2.

Time Complexity

The time complexity of the algorithm follows that of solving the eigenvalue problem. Graph Laplacian $(N \times N)$-matrix, and a straight-forward solution needs the complexity of $O(N^3)$.

[13] In Laplacian eigenmaps, τ is originally given as a hyperparameter. However τ can be determined from given data [50, 51]. See Section 9.3.1 for in detail of how we can fix τ from data.

Algorithm 3.23: Laplacian eigenmaps.

Function `Laplacian_eigenmaps(`$\boldsymbol{X}$, K, ϵ, τ`)`
  Data: training data: $\boldsymbol{X}$, #features to be generated: K, ϵ, τ
  Result: K features
  Generate a graph by linking two features as an edge, which satisfies (3.241).;
  Weigh edges, according to (3.242).;
  Solve eigenvalue problem (3.243) to obtain the top K eigenvectors.;
  Output the resultant eigenvectors.;

Notes on Laplacian Eigenmaps

Background We did not explain the way that the optimization problem (3.243) was derived and what the objective of this formulation is. In fact, obtaining eigenvalues by solving (3.243) means that originally the optimization problem was formulated as follows:

$$\min_{\boldsymbol{v}} \quad \boldsymbol{v}^\top \boldsymbol{L}\boldsymbol{v} \quad \text{subject to} \quad \boldsymbol{v}^\top \boldsymbol{D}\boldsymbol{v} = \text{Constant}. \tag{3.244}$$

This formulation is similar to the constrained K-means clustering, i.e. (3.18), where the objective function $\boldsymbol{Z}^\top \boldsymbol{E}\boldsymbol{Z}$ shows the smoothness over the variance $\boldsymbol{E}$. Similarly the objective function $\boldsymbol{v}^\top \boldsymbol{L}\boldsymbol{v}$ indicates that $\boldsymbol{v}$ should be smooth over the graph (Laplacian). Also in (3.18), the constraint can be arbitrary given by $\boldsymbol{D}$, such as the equal size of clusters. On the other hand, in (3.244), each diagonal of $\boldsymbol{D}$ is the edge degree (# of edges) of each node. Thus $\boldsymbol{v}^\top \boldsymbol{D}\boldsymbol{v}$ = Constant, means that $\boldsymbol{v}$ should be consistent with the degree (# edges) distribution, rather than $\boldsymbol{v}^\top \boldsymbol{v}$ = Constant, meaning L^2 norm, simply implying that $\boldsymbol{v}$ can be weights. Thus entirely the optimization problem in (3.244) makes $\boldsymbol{v}$ consistent with the generated graph, indicating that generating the graph is extremely important. In other words, generating the graph without losing any information is the key to achieve a better performance. Again we note that hyperparameter τ can be reasonably estimated from given data (see Section 9.3.1).

Rayleigh quotient This problem is also a generalized Rayleigh quotient, which can be a Rayleigh quotient, because graph Laplacian is a nonsingular matrix and has its inverse matrix. Thus Laplacian eigenmaps is also a typical problem setting of machine learning, and the optimization problem can be solved by an eigenvalue problem.

3.3.4 Sparse Learning

Several supervised learning methods are already feature learning themselves. For example, decision stump selects a feature most relevant to the label. Decision tree uses decision stump over partitioned data recursively. Thus decision stump and

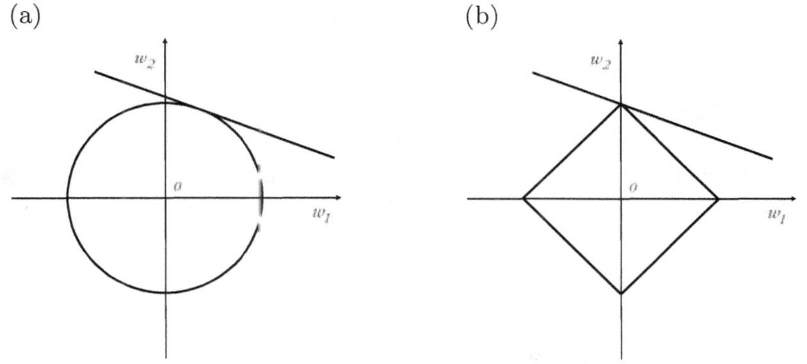

Figure 3.32: (a) L^2 norm and (b) L^1 norm in linear regression.

decision tree are simply feature learning in the framework of supervised learning. Similarly, linear regression and logistic regression, as shown in (3.94), have weights (for all features) to be learned and the learned weights indicate important features.

The parameter estimation of ridge regression is the optimization problem with respect to regression coefficients $\boldsymbol{w}$, being formulated with the regularization term, i.e. L^2 norm (see Section A.6) to avoid overfitting training data:

$$\hat{\boldsymbol{w}} = \arg\min_{\boldsymbol{w}}(||\boldsymbol{y} - \boldsymbol{X}\boldsymbol{w}||_2^2 + \lambda||\boldsymbol{w}||_2^2). \quad (3.245)$$

Then $\boldsymbol{w}$ is estimated, following

$$\hat{\boldsymbol{w}} = 2(\boldsymbol{X}^\top\boldsymbol{X} + \lambda\boldsymbol{I})^{-1}\boldsymbol{X}^\top\boldsymbol{y}. \quad (3.246)$$

When we have only two features, and consider the possible space of regression coefficients (w_1,w_2), the range of (w_1,w_2) for L^2 norm would be the circle, shown in Fig. 3.32 (a), meaning no bias on two features. The solution would be points (w_1,w_2), which are the intersection of this circle and the linear line of regression. Thus again we note that no bias on the circle of Fig. 3.32 (a) and all points on this circle can be the solution without any bias. On the other hand, in order to specify what features should be used more clearly, we can use the L^p norm with p of lower than 2, for example L^1 norm, instead of L^2 norm. Again when we consider two features (w_1,w_2), i.e. the two-dimensional space, as shown in Fig. 3.32 (b), the L^1 norm is not a circle but a square, i.e. a set of four linear lines. The optimal solution is a point (intersection) between this square and a linear line of regression. However both are linear lines, and so the optimum intersection is likely to be one of the four end points of the square of the L^1 norm, as in Fig. 3.32 (b). Note that in the L_1 norm the optimum solution is biased, while no bias in the L^2 norm. At the end of the points, either of w_1 or w_2 is zero, meaning that the L_1 norm is likely to choose either of the two features. More generally, the end points indicate that regularization coefficients are likely to have only one feature with non-zero and all other features with zero. This indicates that only the feature with non-zero is important. This way of controlling the regularization term allows learning to implement that only a small number of features non-zero is called *sparse learning*. Sparse learning allows to present important features more easily and in general discovering knowledge also easier. Sparse learning is used in a variety of applications including life sciences [18]. Particularly linear regression with the L^1 norm is called *Least Absolute Shrinkage and Selection Operator* (LASSO), formulated as follows:

$$\hat{\boldsymbol{\beta}} = \arg\min_{\boldsymbol{w}}(||\boldsymbol{y} - \boldsymbol{X}\boldsymbol{w}||_2^2 + \lambda||\boldsymbol{w}||_1). \tag{3.247}$$

Also sometimes the L^1 norm reduces the number of non-zero features to extremely small, say only one, because as mentioned in Fig. 3.32 (b), in general only one feature is non-zero and others are zero. In order to reduce the number of non-zeros more mildly, combining the L^1 norm and the L^2 norm is useful. This formulation of linear regression is called *elastic net*:

$$\hat{\boldsymbol{w}} = \arg\min_{\boldsymbol{w}}(||\boldsymbol{y} - \boldsymbol{X}\boldsymbol{w}||_2^2 + \lambda(\alpha||\boldsymbol{w}||_2^2 + (1-\alpha)||\boldsymbol{w}||_1), \tag{3.248}$$

where α is a regularization coefficient, i.e. hyperparameter, taking the balance between the two norms.

3.4 Kernel Learning

We already saw that kernel functions are useful for supervised learning or classification, in learning of support vector machine (SVM) in Section 3.2.11. We call this idea of projecting instances of the input space into a hyperspace of kernels, as in Fig. 3.29, *kernel trick*, and the entire machine learning methods using kernel trick *kernel learning*, for any machine learning paradigm, such as supervised, unsupervised learning and whatever else. Kernel learning needs only kernel function and not need instances of vectors any more.

Here are several points of kernel function:

Basic input and output are two instances and the similarity between the two input instances, respectively.

Kernel function over the entire data is a symmetric matrix, where each element is a similarity between the two corresponding instances, and needs to satisfy the positive semidefinite property.

Kernel design is very flexible, allowing to incorporate prior knowledge so that the similarity of two instances should be consistent with the conditions of application. This property of kernel functions makes the predictive performance of kernel learning higher (Fig. 3.29).

See Section A.5 for kernels for vectors.

Below we describe kernel learning methods under three problem settings: clustering (kernel K-means), regression (kernel ridge regression) and feature learning (kernel principal component analysis), for which we already explained methods without kernels.

3.4.1 Kernel K-means

Background: Objective Function by Projection to Kernel Space

The objective function of K-means, i.e. (3.8), is given as follows, by using cluster assignment matrix $\boldsymbol{Z}$:

$$J = \sum_{k=1}^{K}\sum_{i=1}^{N} \boldsymbol{Z}_{ik}||\boldsymbol{x}_i - \boldsymbol{\mu}_k||_2^2. \tag{3.249}$$

Now we project each given instance $\boldsymbol{x}$ into hyperspace:

$$\boldsymbol{x}_i \Rightarrow \phi(\boldsymbol{x}_i). \tag{3.250}$$

$$\boldsymbol{\mu}_k \Rightarrow \frac{1}{|C_k|}\sum_i \boldsymbol{Z}_{ik}\phi(\boldsymbol{x}_i). \tag{3.251}$$

Note that $\boldsymbol{Z}_{ik}$ is one if instance i is in cluster k; otherwise zero. Thus $\sum_i \boldsymbol{Z}_{ik}\phi(\boldsymbol{x}_i)$ means that $\phi(\boldsymbol{x}_i)$ are summed up over all i if instance i is in cluster k. Then this is divided by the size of cluster, $|C_k|$, resulting in the mean.

Then the objective function in the kernel space is given as follows:

$$J = \sum_{k=1}^{K}\sum_{i=1}^{N} \boldsymbol{Z}_{ik}||\phi(\boldsymbol{x}_i) - \frac{1}{|C_k|}\sum_j^N \boldsymbol{Z}_{jk}\phi(\boldsymbol{x}_j)||_2^2. \tag{3.252}$$

Kernel K-means is to solve this problem of minimizing J with respect to $\boldsymbol{Z}$ under the constraint of $\boldsymbol{Z}$.

We expand the objective function and introduce kernel function $\boldsymbol{K}(\boldsymbol{x}_i, \boldsymbol{x}_j) = \phi(\boldsymbol{x}_i)\phi(\boldsymbol{x}_j)$. Then we can have the following (see Section A.9.6 for derivation).

$$J = -\sum_{k=1}^{K}\frac{1}{|C_k|}\sum_{i=1}^{N}\boldsymbol{Z}_{ik}\sum_{j=1}^{N}\boldsymbol{Z}_{jk}\boldsymbol{K}(\boldsymbol{x}_i, \boldsymbol{x}_j). \tag{3.253}$$

Two instances i and j are connected by cluster k. Here to clarify the structure of Kernel K-means, we remove coefficients unused, and finally obtained the objective function by removing all unnecessarily as follows:

$$J = \sum_{k=1}^{K}\sum_{j=1}^{N}\sum_{i=1}^{N}\boldsymbol{Z}_{ik}\boldsymbol{Z}_{jk}\boldsymbol{K}(\boldsymbol{x}_i, \boldsymbol{x}_j) \tag{3.254}$$

$$= \mathrm{trace}(\boldsymbol{Z}^\mathsf{T}\boldsymbol{K}\boldsymbol{Z}). \tag{3.255}$$

Algorithm 3.24: Kernel K-means.

1 **Function** `Kernel_`K`-means(`$\boldsymbol{K}$, K, K_1`)`
  Data: kernel: $\boldsymbol{K}$, #clusters: K, #eigenvectors: K_1
  Result: K clusters
2 Obtain $\tilde{\boldsymbol{K}}$ from $\boldsymbol{K}$ by (3.259).;
3 Run `Eigen_decomposition(`$\tilde{\boldsymbol{K}}$, K_1`)` to generate $\boldsymbol{X}$ with K_1 eigenvectors.;
4 Run a clustering algorithm (say K`-means(`$\boldsymbol{X}$, K`)`) over $\boldsymbol{X}$.;
5 Output the resultant K clusters.;

Here $\boldsymbol{K}$ is a matrix, where each element is a kernel function. $\boldsymbol{Z}$ had the following constraint by using $\boldsymbol{D}$ in (3.5):

$$\boldsymbol{Z}^\top \boldsymbol{Z} = \boldsymbol{D}. \tag{3.256}$$

The objective function of the original K-means is given in (3.10). Thus the difference is that $\boldsymbol{E}$ given by (3.11) is just replaced with kernel $\boldsymbol{K}$. Also K-means could be formulated as the problem with the constraint on cluster sizes, and similarly kernel K-means can be formulated as the following maximization problem:

$$\max_{\boldsymbol{Z}} \quad \text{trace}(\boldsymbol{Z}^\top \boldsymbol{K} \boldsymbol{Z}) \quad \text{subject to} \quad \boldsymbol{Z}^\top \boldsymbol{Z} = \boldsymbol{D}. \tag{3.257}$$

Similar to the constrained K-means algorithm, for example, on $\boldsymbol{D}$, we can place the constraint that cluster sizes are equal to each other in (3.5):

$$|C_1| = |C_2| = \cdots = |C_K|. \tag{3.258}$$

Algorithm

Thus the problem of (3.257) can also be solved by using the Lagrange function and setting the derivative of the Lagrange function with respect to $\boldsymbol{Z}$ equal to zero, again resulting in a generalized eigenvalue problem. This procedure is totally the same as the constrained K-means algorithm, and using the following $\tilde{\boldsymbol{K}}$, the problem can be solved by the eigenvalue problem of $\tilde{\boldsymbol{K}}$.

$$\tilde{\boldsymbol{K}} = \boldsymbol{D}^{-1} \boldsymbol{K}. \tag{3.259}$$

Algorithm 3.24 shows a pseudocode of the kernel K-means algorithm. In this algorithm, the input is kernel function $\boldsymbol{K}$, which is normalized into $\tilde{\boldsymbol{K}}$ (line 2) by (3.259), and an eigenvalue decomposition algorithm is run over $\tilde{\boldsymbol{K}}$, to have the first K_1 eigenvectors (line 3), over which the K-means algorithm is run to have clusters (line 4). Again this is the same as the constrained K-means algorithm.

Notes on Kernel K-means

Difference from K-means: The original K-means problem was, as shown in (3.19), a minimization problem. In fact $\boldsymbol{E}$, which is sandwiched between two cluster assignment matrices $\boldsymbol{Z}$, indicates, as shown in (3.11), the distance between instance $\boldsymbol{x}_i$ and instance $\boldsymbol{x}_j$. Then learning is to minimize $\boldsymbol{Z}^\top \boldsymbol{E}\boldsymbol{Z}$. On the other hand, the objective function with kernel function $\boldsymbol{K}$ between two $\boldsymbol{Z}$ must be maximized, since $\boldsymbol{K}$ shows the similarity between instances. In summary the difference is that $\boldsymbol{E}$ is replaced with $\boldsymbol{K}$, and the minimization problem was changed into the maximization problem.

Rayleigh quotient: This problem is also represented by Rayleigh quotient.

Connection to spectral clustering: We describe the similarity between spectral clustering (for graphs) and kernel K-means [25] in Section 8.1.

3.4.2 Kernel Ridge Regression

Background

Support vector machine (SVM) is a linear regression, if we do not think about the kernel space. SVM focuses on support vectors around boundaries between positives and negatives, by which prediction is made efficient and effective rather than using all instances. Thus we do not have to consider ridge regression, which uses all instances. However to show an example of developing a kernel learning method, we introduce a method of using kernel functions with ridge regression, resulting in kernel ridge regression.

First we follow the derivation of SVM.

Derivation Following Support Vector Machine

The minimization problem of ridge regression was formulated in (3.100), as follows:

$$\min_{\boldsymbol{w}}(\frac{1}{2}||\boldsymbol{y} - \boldsymbol{X}\boldsymbol{w}||_2^2 + \frac{\lambda}{2}||\boldsymbol{w}||_2^2). \tag{3.260}$$

We describe this by vectors, instead of matrices:

$$\min_{\boldsymbol{w}}(\frac{1}{2}\sum_i (y_i - \boldsymbol{w}^\top \boldsymbol{x}_i) + \frac{\lambda}{2}\boldsymbol{w}^\top \boldsymbol{w}). \tag{3.261}$$

Let $\xi_i = y_i - \boldsymbol{w}^\top \boldsymbol{x}_i$ and the optimization problem can be formulated as follows:

$$\min_{\boldsymbol{u}} \quad \frac{1}{2}\sum_i \xi_i^2 + \frac{\lambda}{2}\boldsymbol{w}^\top \boldsymbol{w} \tag{3.262}$$

$$\text{subject to} \quad \xi_i = y_i - \boldsymbol{w}^\top \boldsymbol{x}_i \quad \text{for all} \quad i. \tag{3.263}$$

Then Lagrange function is given as follows:

$$L(\boldsymbol{w}, \boldsymbol{\xi}) = \frac{1}{2}\sum_i \xi_i^2 + \frac{\lambda}{2}\boldsymbol{w}^\top\boldsymbol{w} + \sum_i \alpha_i(y_i - \boldsymbol{w}^\top\boldsymbol{x}_i - \xi_i). \quad (3.264)$$

By setting the partial derivative of the Lagrange function equal to zero, we can have the following:

$$\frac{\partial L(\boldsymbol{w}, \boldsymbol{\xi})}{\partial \boldsymbol{w}} = 0 \quad \Rightarrow \quad \boldsymbol{w} = \frac{1}{\lambda}\sum_i \alpha_i \boldsymbol{x_i}. \quad (3.265)$$

$$\frac{\partial L(\boldsymbol{w}, \boldsymbol{\xi})}{\partial \xi_i} = 0 \quad \Rightarrow \quad \xi_i = \alpha_i. \quad (3.266)$$

Substituting the above for the corresponding part of (3.264), we can have the following (see Section A.9.7 for derivation):

$$L(\boldsymbol{\alpha}) = -\frac{1}{2}\sum_i \alpha_i^2 - \frac{1}{2\lambda}\sum_{i,j}\alpha_i\alpha_j\boldsymbol{x}_i^\top\boldsymbol{x}_j + \sum_i \alpha_i y_i \quad (3.267)$$

$$= -\frac{1}{2}\boldsymbol{\alpha}^\top\boldsymbol{\alpha} - \frac{1}{2\lambda}\boldsymbol{\alpha}^\top\boldsymbol{X}\boldsymbol{X}^\top\boldsymbol{\alpha} + \boldsymbol{\alpha}^\top\boldsymbol{y}. \quad (3.268)$$

We maximize this Lagrange function. Letting $\frac{dL(\boldsymbol{\alpha})}{d\boldsymbol{\alpha}} = 0$,

$$\boldsymbol{\alpha} = 2(\boldsymbol{I} + \frac{1}{\lambda}\boldsymbol{X}\boldsymbol{X}^\top)^{-1}\boldsymbol{y}. \quad (3.269)$$

$\boldsymbol{X}\boldsymbol{X}^\top$ is called a *variance-covariance matrix* or a covariance matrix (see Section A.3.1) of $\boldsymbol{x}_i(i = 1, \ldots, N)$, and can be replaced with kernel function $\boldsymbol{K}$ in kernel learning:

$$\boldsymbol{\alpha} = \frac{2}{\lambda}(\boldsymbol{K} + \lambda\boldsymbol{I})^{-1}\boldsymbol{y}. \quad (3.270)$$

When we predict the label of a new instance, (3.265) says

$$\hat{w} = \boldsymbol{X}^\top\boldsymbol{\alpha}. \quad (3.271)$$

Thus given $\boldsymbol{X}_{new}$ of multiple new instances, we can estimate their labels:

$$\hat{\boldsymbol{y}} \Leftarrow \boldsymbol{X}_{new}\boldsymbol{w} \quad (3.272)$$

$$= \boldsymbol{X}_{new}\boldsymbol{X}^\top\boldsymbol{\alpha} \quad (3.273)$$

$$= \boldsymbol{K}_{n\ vs.\ t}\boldsymbol{\alpha} \quad (3.274)$$

$$= \frac{2}{\lambda}\boldsymbol{K}_{n\ vs.\ t}(\boldsymbol{K} + \lambda\boldsymbol{I})^{-1}\boldsymbol{y}, \quad (3.275)$$

where $\boldsymbol{K}_{n\ vs.\ t}$ $(= \boldsymbol{X}_{new}\boldsymbol{X}^\top)$ is kernel function, which can be generated from the covariance matrix between training instances and test instances. Note that $\boldsymbol{K}$ is on kernel function of training instances. Thus, similar to SVM, we do not need $\boldsymbol{w}$. We can use (3.265) and the label of test, unknown instance $\boldsymbol{x}_{new}$ can be estimated by using $\boldsymbol{\alpha}$ and the kernel function between between training instances and test instances.

More Direct Derivation

We have seen the derivation following SVM. The purpose of this derivation was to represent the prediction rule by using only the inner product:$\boldsymbol{x}_i^\top \boldsymbol{x}_j$ by vector representation or covariance matrix: $\boldsymbol{X}\boldsymbol{X}^\top$ by matrix representation.

For kernel ridge regression, we have another derivation, which can achieve this purpose more directly. Originally the objective function to be minimized is as follows:

$$J(\boldsymbol{w}) = \frac{1}{2}||\boldsymbol{y} - \boldsymbol{X}\boldsymbol{w}||_2^2 + \frac{\lambda}{2}||\boldsymbol{w}||_2^2 \quad (3.276)$$

$$= \frac{1}{2}(\boldsymbol{y} - \boldsymbol{X}\boldsymbol{w})^\top(\boldsymbol{y} - \boldsymbol{X}\boldsymbol{w}) + \frac{\lambda}{2}\boldsymbol{w}^\top\boldsymbol{w}. \quad (3.277)$$

We here assume that

$$\boldsymbol{w} = \boldsymbol{X}^\top\boldsymbol{\beta}. \quad (3.278)$$

That is, for linear coefficients $\boldsymbol{w}$, each value $\boldsymbol{w}_k$ can be represented by a linear combination of the values of the corresponding instances $\boldsymbol{X}_{ik}$:

$$\boldsymbol{w}_k = \sum_{i=1}^{N} \beta_i \boldsymbol{X}_{ik} \quad \text{for} \quad \text{all} \quad k. \quad (3.279)$$

The objective of this assumption is to derive the covariance matrix more directly. We use (3.278) for (3.277), resulting in:

$$J(\boldsymbol{\beta}) = \frac{1}{2}(\boldsymbol{y}^\top\boldsymbol{y} - 2\boldsymbol{\beta}^\top\boldsymbol{X}\boldsymbol{X}^\top\boldsymbol{y} + \boldsymbol{\beta}^\top\boldsymbol{X}\boldsymbol{X}^\top\boldsymbol{X}\boldsymbol{X}^\top\boldsymbol{\beta}) + \frac{\lambda}{2}\boldsymbol{\beta}^\top\boldsymbol{X}\boldsymbol{X}^\top\boldsymbol{\beta}. \quad (3.280)$$

By $\frac{dJ(\boldsymbol{\beta})}{d\boldsymbol{\beta}} = 0$,

$$-\boldsymbol{y} + \boldsymbol{X}\boldsymbol{X}^\top\boldsymbol{\beta} + \frac{\lambda}{2}\boldsymbol{\beta} = 0. \quad (3.281)$$

We write covariance matrix $\boldsymbol{X}\boldsymbol{X}^\top$ by kernel function $\boldsymbol{K}$, resulting in

$$\boldsymbol{\beta} = (\boldsymbol{K} + \frac{\lambda}{2}\boldsymbol{I})^{-1}\boldsymbol{y}. \quad (3.282)$$

Thus (except the minor details) we can have (3.282), which is the same as (3.270), derived following the derivation of SVM.

Also the regression coefficients of the original ridge regression were analytically computed, as follows:

$$\hat{\boldsymbol{w}} = (\boldsymbol{X}^\top\boldsymbol{X} + \lambda\boldsymbol{I}_M)^{-1}\boldsymbol{X}^\top\boldsymbol{y} \quad (3.283)$$

On the other hand, from (3.278) and (3.282),

$$\hat{\boldsymbol{w}} = \boldsymbol{X}^\top\boldsymbol{\beta} \quad (3.284)$$

$$= \boldsymbol{X}^\top(\boldsymbol{K} + \frac{\lambda}{2}\boldsymbol{I})^{-1}\boldsymbol{y}. \quad (3.285)$$

Algorithm 3.25: Kernel ridge regression (learning and prediction).

```
1 Function Kernel_ridge_regression(K, y, K_{n vs. t})
    Data: kernel: K (between training instances), training instance labels:
          y, kernel: K_{n vs. t} (between training and test instances)
    Result: estimated labels of test instances
2   Learning: Compute β from K and y by (3.270).;
3   Prediction: Estimate ŷ from β and K_{n vs. t} by (3.288).;
```

These two manners of estimating $\hat{\boldsymbol{w}}$, i.e. (3.283) and (3.285), are very similar.

Suppose that multiple test instances $\boldsymbol{X}_{new}$ (rows are instances and columns are features) are given. We consider how to estimate the labels of instances in $\boldsymbol{X}_{new}$. For this, we first estimate $\boldsymbol{\beta}$ by (3.282), using training instances. Then we estimate labels by (3.278):

$$
\begin{aligned}
\hat{\boldsymbol{y}} &= \boldsymbol{X}_{new}\boldsymbol{w} && (3.286)\\
&= \boldsymbol{X}_{new}\boldsymbol{X}^{\top}\boldsymbol{\beta} && (3.287)\\
&= \boldsymbol{K}_{n\ vs.\ t}\boldsymbol{\beta} && (3.288)\\
&= \boldsymbol{K}_{n\ vs.\ t}(\boldsymbol{K} + \frac{\lambda}{2}\boldsymbol{I})^{-1}\boldsymbol{y}, && (3.289)
\end{aligned}
$$

where $\boldsymbol{K}_{n\ vs.\ t}$ is a kernel function between test instances and training instances, while $\boldsymbol{K}$ is the kernel function between training instances themselves.

Algorithm

Algorithm 3.25 shows a pseudocode of the learning and prediction algorithms that we have shown for kernel ridge regression.

Time Complexity

The biggest complexity for learning is computing the inverse matrix in (3.270) for computing α. $\boldsymbol{K}$ is $(N \times N)$-matrix, and a straight-forward inverse matrix computation needs $O(N^3E)$. On the other hand, for prediction, in (3.289), we can compute $(\boldsymbol{K} + \lambda\boldsymbol{I})^{-1}\boldsymbol{y}$ beforehand and keep the result as a vector, then the computation for prediction is the product of $\boldsymbol{K}_{nvw.t}$ and vector y, resulting in complexity of $O(N_{new}N)$, where N_{new} is the number of test instances.

Notes on Kernel Ridge Regression

Difference from SVM: SVM focused on support vectors only, and then $\alpha_i (i = 1, \ldots, N)$ are sparse in the sense that α_i are non-zero for support vectors only and zero for others. This sparseness allows the efficiency of SVM to compute only non-zero α_i, i.e. support vectors. On the other hand, kernel ridge regression has no benefits like that, and predicts the label of a test instance by using all training instances, as we can see from, for example, (3.288).

Linear coefficients not needed for prediction: We can see from (3.289) that the labels of test instances can be estimated from the following three:

1. $\boldsymbol{K}$: Kernel function among training instances.
2. $\boldsymbol{y}$: Labels of training instances.
3. $\boldsymbol{K}_{n\ vs.\ t}$: Kernel function between test instances and training instances.

This is the same as SVM.

3.4.3 Kernel Principal Component Analysis

We follow the more direct derivation that we performed for kernel ridge regression, to derive kernel principal component analysis (kernel PCA).

Derivation

The optimization problem of principal component analysis (PCA) was formulated as follows:

$$\max_{\boldsymbol{w}} \boldsymbol{w}^\top \boldsymbol{X}^\top \boldsymbol{X} \boldsymbol{w} \quad \text{subject to} \quad \boldsymbol{w}^\top \boldsymbol{w} = \text{Const.} \tag{3.290}$$

We here assume that

$$\boldsymbol{w} = \boldsymbol{X}^\top \boldsymbol{\beta}. \tag{3.291}$$

and substitute this for (3.292), and the problem can be formulated as follows:

$$\max_{\boldsymbol{\beta}} \boldsymbol{\beta}^\top \boldsymbol{X}\boldsymbol{X}^\top \boldsymbol{X}\boldsymbol{X}^\top \boldsymbol{\beta} \quad \text{subject to} \quad \boldsymbol{\beta}^\top \boldsymbol{X}\boldsymbol{X}^\top \boldsymbol{\beta} = \text{Const.} \tag{3.292}$$

Due to the equality condition, we can define the Lagrange function as follows:

$$L(\boldsymbol{\beta}) = \boldsymbol{\beta}^\top \boldsymbol{X}\boldsymbol{X}^\top \boldsymbol{X}\boldsymbol{X}^\top \boldsymbol{\beta} - \lambda \boldsymbol{\beta}^\top \boldsymbol{X}\boldsymbol{X}^\top \boldsymbol{\beta}. \tag{3.293}$$

We can set the derivative of the Lagrange function with respect to $\boldsymbol{\beta}$ equal to zero:

$$\frac{dL(\boldsymbol{\beta})}{d\boldsymbol{\beta}} = 0 \qquad \Rightarrow \qquad \boldsymbol{X}\boldsymbol{X}^\top \boldsymbol{\beta} = \lambda \boldsymbol{\beta}. \tag{3.294}$$

We replace variance-covariance matrix $\boldsymbol{X}\boldsymbol{X}^\top$ with kernel function $\boldsymbol{K}$:

$$\boldsymbol{X}\boldsymbol{X}^\top \qquad \Rightarrow \qquad \boldsymbol{K}. \tag{3.295}$$

Thus simply, $\boldsymbol{\beta}$ can be estimated from the eigenvalue decomposition of the kernel function:

$$\boldsymbol{K}\boldsymbol{\beta} = \lambda \boldsymbol{\beta}. \tag{3.296}$$

Algorithm 3.26: Kernel principal component analysis.

1 **Function** `Kernel_principal_component_analysis(`$\boldsymbol{K}$, K`)`
  Data: kernel: $\boldsymbol{K}$, #principal components: K
  Result: K principal components
2 Initialization: Normalize $\boldsymbol{K}$ and obtain $\tilde{\boldsymbol{K}}$.;
3 Run `Eigen_decomposition(`$\tilde{\boldsymbol{K}}, K$`)` to have K eigenvectors: $\boldsymbol{\beta}_1, \ldots, \boldsymbol{\beta}_K$.;
4 **for** $k \leftarrow 1$ **to** K **do**
5   k-th principal component $= \tilde{\boldsymbol{K}}\boldsymbol{\beta}_k$.;
6 Output all resultant K principal components.;

Originally in PCA, instances $\boldsymbol{X}$ are projected into one dimensional vector by using $\boldsymbol{w}$:

$$\boldsymbol{X}\boldsymbol{w} = \sum_k w_k \chi_k. \tag{3.297}$$

We then use (3.291) and (3.297), and projection into one dimensional vector can be done by using $\boldsymbol{\beta}$:

$$\begin{aligned} \boldsymbol{X}\boldsymbol{w} &= \boldsymbol{X}\boldsymbol{X}^\top\boldsymbol{\beta} & (3.298) \\ &= \boldsymbol{K}\boldsymbol{\beta}. & (3.299) \end{aligned}$$

Algorithm

Algorithm 3.26 shows a pseudocode of the algorithm to run kernel PCA. In this algorithm, the input is the kernel function among training instances. First the kernel function is normalized (line 2). Eigenvalue decomposition is run over the normalized kernel function to have K eigenvectors (line 3). We can have kernel principal component vectors from the obtained eigenvectors sequentially (lines 4-5).

Time Complexity

Complexity follows the eigenvalue problem to obtain eigenvectors of the kernel function.

Notes on Kernel PCA

$\boldsymbol{X}^\top\boldsymbol{X} \Rightarrow$ kernel function: PCA runs eigenvalue decomposition over $\boldsymbol{X}^\top\boldsymbol{X}$ to have principal components. Interestingly, Kernel PCA uses the kernel function, instead of $\boldsymbol{X}^\top\boldsymbol{X}$.

Rayleigh quotient: Note that as you can see from (3.292) easily, kernel PCA also becomes the problem of Rayleigh quotient.

3.4.4 Summary of Kernel Learning

We have seen the derivation of various algorithms of kernel learning, such as SVM, kernel K-means, kernel ridge regression, kernel principal component analysis. The advantage of kernel learning is we can use prior knowledge to design the kernel function, which can be consistent with the ideal data space of applications. By using this property, we can achieve efficient and effective learning within the framework of kernel learning. In particular, regarding predictive performance, kernel learning, such as SVM, achieves quite high performance in cross-validation, a performance evaluation manner in supervised learning (see Chapter 10).

However we need to point out several problems peculiar to kernel learning.

Incomprehensible results: The original motivation of machine learning was to acquire rules or hypotheses. The objective of sparse learning was also to realize the exploitation by using only a few number of features. However kernel learning is unable to present the obtained knowledge explicitly. We can say that kernel learning achieves high predictive performance, while exploitation ability was lost on behalf of high performance in kernel learning.

Missing data: Kernel functions need to satisfy the positive semidefinite property, while this constraint does not allow missing values in the matrix. That is, kernel learning has a merit of not using vectors but also kernel functions, i.e. only similarity between instances. However this function cannot be unknown, and any element of the kernel function must have some value. In fact real data have definitely missing parts, which are inevitable. Also in machine learning, there exist methods which can be trained from data with missing values, such as decision trees, matrix factorization, etc. However kernel learning does not allow that.

In Chapter 8, we will explain a variety of machine learning approaches for the case that instances are nodes in a graph. For graphs, we can see that edges indicate similarities between nodes connected by those edges, resulting in that a graph can be an adjacency matrix. Thus kernel functions might be thought as a graph or an adjacency matrix. However a decisive difference between graph learning and kernel learning is that in graph learning, if similarities between nodes are not known, we can just think them as no edges and then no problems happen in learning after that. However, this is not possible in kernel learning. We have to make all pairs of given instances have similarities in kernel learning (one way is to assign zero to missing values, while we are not sure if this is fine enough). Thus, graph learning allows more flexible learning than kernel learning.

Scalability: Kernel function is a $(N \times N)$-square matrix, where N is the number of instances. This means that this matrix becomes huge as the number of instances goes up. Furthermore, as mentioned in the above, no missing values are allowed. Thus clearly a serious problem lies in data scalability for kernel learning. Kernel learning was matured, and performance advantages are advertised, while at the same time, the limitation of kernel learning is

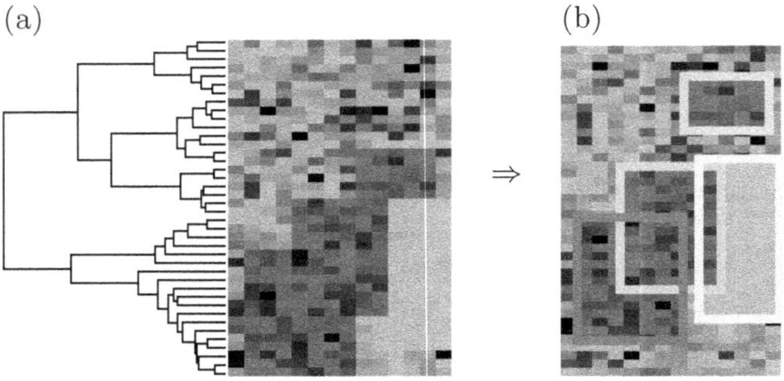

Figure 3.33: (a) Hierarchical clustering over expression data (genes × experimental conditions) and (b) biclusters.

getting somewhat exposed: particularly we are in the era of big data, the current data size making creating kernels for the entire data simply very hard.

3.5 Applications to Life Sciences

3.5.1 Unsupervised Learning: Biclustering

In life sciences, labels have to be obtained through experiments, while experiments need time and cost, making hard to have a good number of labels. In fact the discovery of life sciences are "positive examples", implying the difficult of obtaining labels. Thus unsupervised learning is well used in life sciences. Particularly visibly understandable machine learning methods, such as hierarchical clustering and principal component analysis (PCA), are often run over the data.

Clustering over Gene Expression

The most typical example is clustering gene expression data. The data matrix is (genes × experimental conditions), where elements show the expression of the corresponding genes and the corresponding experimental conditions [81]. This matrix is usually used in a reversible way, in the sense that both genes and experimental conditions can be instances or features.

Fig. 3.33 (a) shows that results of running hierarchical clustering over the row side of the matrix, where instances of the row side are ordered, according to the clustering results (dendrogram). Interestingly, from (a), we can see that even neighboring instances can be similar not entirely but rather partially, indicating that we can do clustering on the column side also. Fig. 3.33 (b) shows the results by running hierarchical cluster over both sides, i.e. rows and columns (dendrograms are not shown). The resultant groupings, so-called *biclustering*, are shown by rectangles. A key advantage of biclustering is, as shown in (b), to capture the

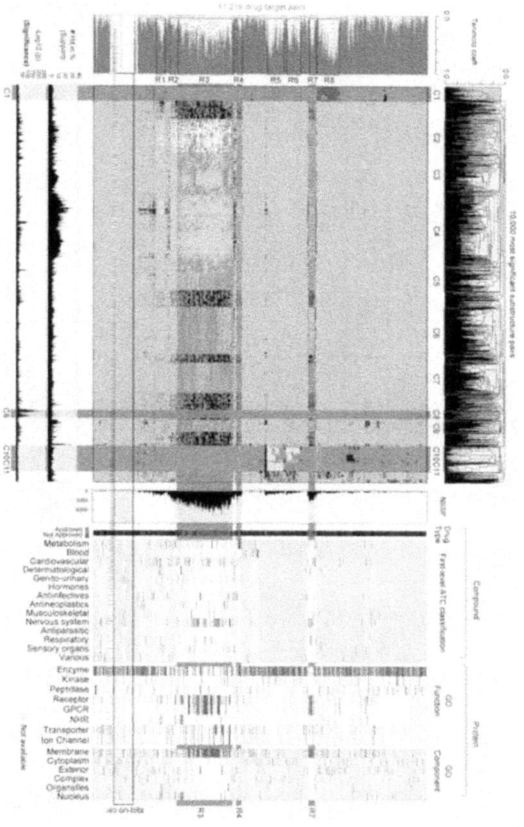

Figure 3.34: Biclustering over drug-target interactions with their substructure features.

relationships between genes and experimental conditions as one group (bicluster) or one single set with both certain genes and conditions together. This would lead to some biological findings. For example, in (b), one rectangle shows a gene set which share a series of experimental conditions and their expression values which should be the same values. This implies that these genes are working together under particular conditions, leading to further biological experiments and analysis.

Clustering over Drug-Target Interactions: Data

We now further show biological knowledge discovery from data through biclusters obtained by using hierarchical clustering. Fig. 3.34 shows an example obtained by biclustering over drug-target (protein) interactions using both features of drugs and targets [89]. Instances (columns in Fig. 3.34) are 11,219 drug-target interactions (pairs; downloaded from DrugBank [97]). Drugs are chemical compounds,

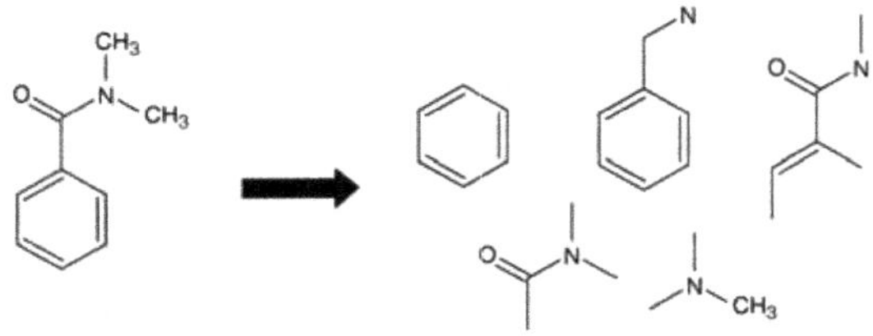

Figure 3.35: Molecular graph and substructures (subgraphs).

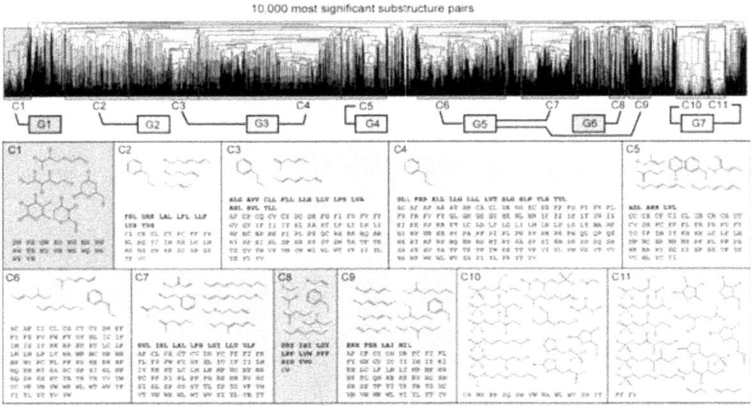

Figure 3.36: Feature side substructure information.

and targets are proteins. Drugs (chemical compounds) have chemical structures, which can be regarded as molecular graphs. Fig. 3.35 shows examples of molecular graphs. Note that each molecular graph has multiple *subgraphs*. On the other hand, target proteins are amino acid sequences, generated from 20 types of amino acids (20 letters), meaning that one protein is a string with multiple *substrings*. We generate features by counting how many times each pair of a subgraph and a substring (maximum length is three) appears in the 11,219 instances and selecting the top 10,000 in terms of frequencies (more specifically statistical significance). These features are rows in Fig. 3.34). Thus the matrix of Fig. 3.34 has 11,219 instances (columns) with 10,000 features (rows).

Clustering over Drug-Target Interactions: Results

We used hierarchical clustering for biclustering, showing the both dendrogram at the top (instances) and right-hand side (features). Using these dendrograms, instances (columns) can be divided into R1 to R8, while features (rows) can be divided into C1 to C11. From the figure, we can see that there are several

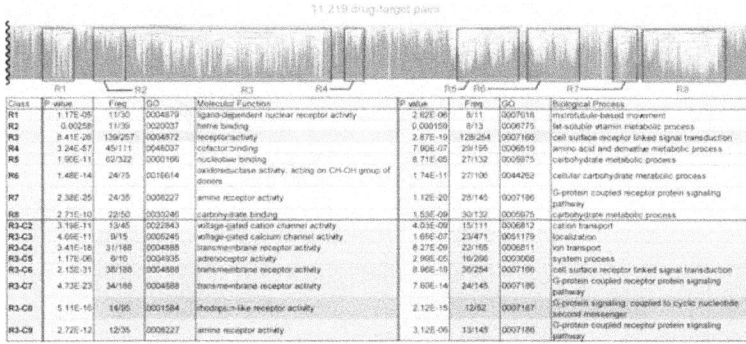

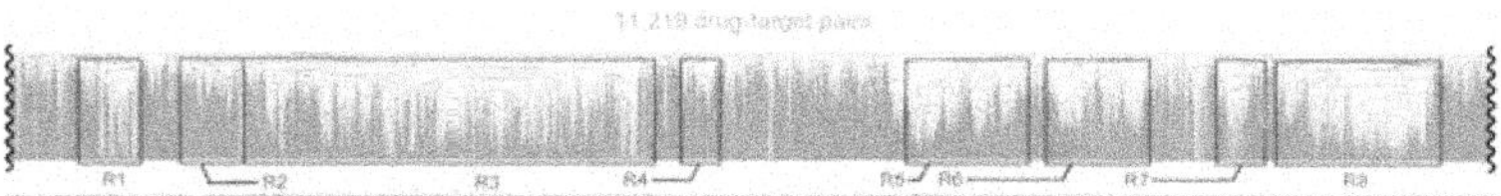

Class	P value	Freq	GO	Molecular Function	P value	Freq	GO	Biological Process
R1	1.17E-05	11/30	0004879	ligand-dependent nuclear receptor activity	2.82E-06	8/11	0007018	microtubule-based movement
R2	0.00258	11/39	0020037	heme binding	0.000159	8/13	0006775	fat-soluble vitamin metabolic process
R3	8.41E-26	139/257	0004872	receptor activity	2.87E-19	128/254	0007166	cell surface receptor linked signal transduction
R4	3.24E-57	45/111	0048037	cofactor binding	7.90E-07	29/195	0006519	amino acid and derivative metabolic process
R5	1.90E-11	62/322	0000166	nucleotide binding	8.71E-05	27/132	0005975	carbohydrate metabolic process
R6	1.48E-14	24/75	0016614	oxidoreductase activity, acting on CH-OH group of donors	1.74E-11	27/106	0044262	cellular carbohydrate metabolic process
R7	2.38E-25	24/35	0008227	amine receptor activity	1.12E-20	28/145	0007186	G-protein coupled receptor protein signaling pathway
R8	2.71E-10	22/50	0030246	carbohydrate binding	1.53E-09	30/132	0005975	carbohydrate metabolic process
R3-C2	3.19E-11	13/45	0022843	voltage-gated cation channel activity	4.03E-09	15/111	0006812	cation transport
R3-C3	4.66E-11	9/15	0005245	voltage-gated calcium channel activity	1.65E-07	23/471	0051179	localization
R3-C4	3.41E-18	31/188	0004888	transmembrane receptor activity	6.27E-09	23/165	0006811	ion transport
R3-C5	1.17E-06	6/10	0004935	adrenoceptor activity	2.99E-05	16/296	0003008	system process
R3-C6	2.13E-31	38/188	0004888	transmembrane receptor activity	8.96E-18	36/254	0007166	cell surface receptor linked signal transduction
R3-C7	4.73E-23	34/188	0004888	transmembrane receptor activity	7.60E-14	24/145	0007186	G-protein coupled receptor protein signaling pathway
R3-C8	5.11E-16	14/95	0001584	rhodopsin-like receptor activity	2.12E-15	12/62	0007187	G-protein signaling, coupled to cyclic nucleotide second messenger
R3-C9	2.72E-12	12/35	0008227	amine receptor activity	3.12E-06	13/145	0007186	G-protein coupled receptor protein signaling pathway

Figure 3.37: Instance side information.

correlations between these groups of columns and those of rows, as follows:

R3 and C2–C9: Most of feature (row) groups have biclusters with R3 of instances (column).

R5–6 and C10: R5 and R6 have biclusters with C10.

R5 and C11: R5 and C11 have bicluster, which is shown just below the above two biclusters.

R8 and C1: R8 has only one bicluster with C1, and also C1 has only one bicluster with R8. This bicluster is very unique, which is isolated from other biclusters.

Other combinations: As well as R3, other groups of instances, such as R1, R2, R4 and R7 have rather small biclusters with a lot of groups (C2–C9) of features.

Fig. 3.36 shows main feature values in the groups (C1–C11) of the feature (row) side. In each box of a group, such as C1, the upper and lower sides show the subgraphs and substrings, respectively. From this figure, we can see that feature values in C2 to C9 are rather similar each other, comparing with other groups, C1, C10 and C11. In other words, particularly regarding subgraphs, C1 has different subgraphs from those in C2 to C9, and similarly C10 and C11 are different from C2 to C9. Interestingly, features in C10 and those in C11 are rather similar, and in fact, these groups (C10 and C11) have both biclusters with R5.

Fig. 3.37 shows gene functions, which are most relevant to each group of instances (columns). Note that instances are drug-target interactions, which are pairs of drugs and targets (genes), while here we focus on genes only. This relevance is measured by the idea, so-called *gene set enrichment analysis (GSEA)* [86], which is a problem of finding the most relevant functions to the input gene set[14].

[14]There exist a lot of software implementing GSEA, and gene functions are obtained from Gene Ontology (GO) [90], a database of gene functions and working process of genes.

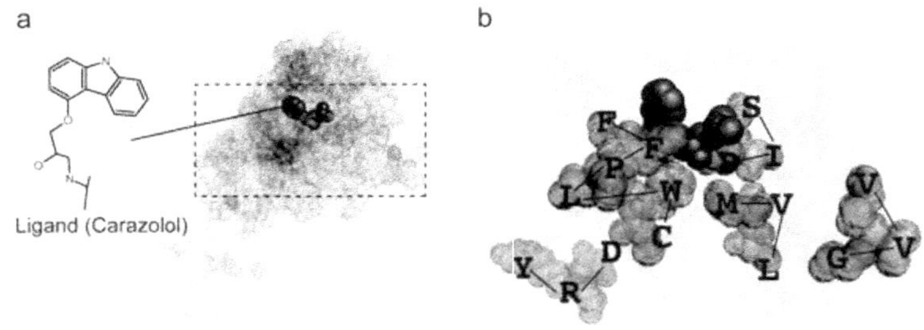

Figure 3.38: 3D image of ligand docking to G-protein coupled receptor (GPCR).

We can first see that from the column of 'Molecular Function', all eight functions of R1 to R8 are different, indicating that R1 to R8, i.e. groups obtained by hierarchical clustering, are significant.

Then again from the clustering of the column side in Fig. 3.34, clearly R3 is the biggest group occupying a lot of instances, giving a bias in the instance distribution. We then focus on R3, and checks each bicluster between R3 and one of C2 to C9 by using each gene function assigned by a tool of GSEA. The gene function most relevant to each bicluster is shown in the lower half of Fig. 3.37. In fact in the literature of drug design, target proteins of more than 30% of all drugs are a group of proteins called *G-protein coupled recepter* (GPCR) [41]. From the lower half of Fig. 3.37, R3–C8 is with 'rhodopsin-like receptor activity', where this 'rhodopsin-like receptor' is the biggest group of the GPCR family. Thus we now focus on the group, C8.

The lower part (substrings) of C8 of Fig. 3.36 shows eight substrings with the length of three, as follows:

DRY, ISI, LGY, LPF, LVM, PFF, SID, VVG

GPCR is a membrane protein, which is hard to be crystallized, and thus so far the three-dimensional (3D) structure of GPCR has not been experimentally determined so much, regardless of the importance of this protein. Fig. 3.38 (a) shows the 3D structure of GPCR with a ligand[15], called *Carazolol*. Fig. 3.38 (b) shows the enlargement of the region surrounded by dotted lines in (a). In (b), in the region surrounding the ligand, letters (amino acids) corresponding to the eight substrings with three letters are shown. From this figure, out of the eight substrings, the following six substrings appear in the region close to the ligand:

DRY, LPF, LVM, PFF, SID, VVG

[15] Ligands are chemical compounds, which can bind to proteins. So ligands are a subclass of chemical compounds, and at the same time a superclass of drugs.

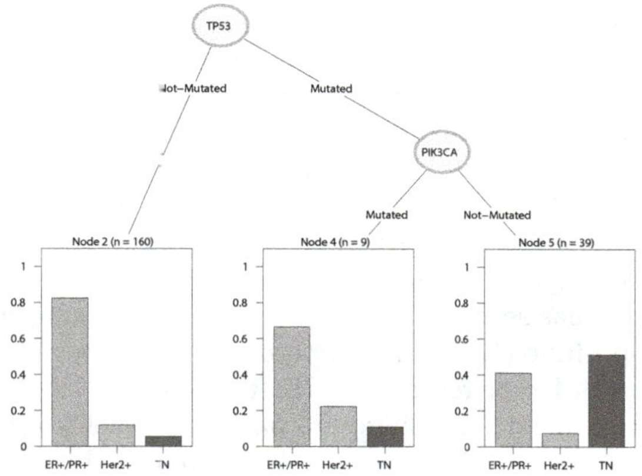

Figure 3.39: Decision tree to classify breast cancer patients with genes of single nucleotide polymorphisms (SNPs).

Thus, we can summarize this result as follows: we focus on the most significant bicluster of R3-C8, which seems on GPCR, because of the gene function, rhodopsin-like receptor activity, assigned to this bicluster. Then we checked the most representative features in this bicluster, particularly substrings, which indeed show the protein sequences surrounding the ligand bound to GPCR, as shown in Fig. 3.38 (b). This result reveals that the (bi)cluster automatically found shows the most significant group of drug-target interactions, and also in the important interactions, the most significantly conserved region was again found automatically. Then the region was indeed clearly close to the ligand bound to GPCR, without any knowledge on the protein 3D structure

3.5.2 Supervised Learning

As mentioned earlier, supervised learning needs labels, and so the situation we can have class labels are limited in life sciences. In fact there are only two types of data: high-throughput data, such as gene expression, and gene (or protein) sequences.

Classification for High-throughput Data

The most typical high-throughput data is gene expression, which is a matrix of (genes × experimental conditions). Classification task is for example, experimental conditions can be binary, like disease patients and healthy people, and then genes are features. This is a typical classification problem to classify people into two cases, by using features, i.e. gene expression.

Another high-throughput data is genome sequences themselves. In this case, again individuals are instances to classify, and features are the positions of nucleotides variable over individuals, though most of positions keep the same nucleotide over people. In fact for example, *single nucleotide polymorphisms (SNPs)* are already found at more than one hundred thousand positions in the human genome. Then in this case, the positions with variable nucleotides are already fixed, and so the data can be a matrix of (individuals × a fixed size of nucleotide positions). Then we have two classes of individuals like disease patients and healthy people again. This is a binary classification problem, which explains the patients by nucleotide sequences variants.

We here explain a classification problem of breast cancer patients, as an example. Breast cancer has several types already known. Among them, we focus on Estrogen receptor-positive (ER+), Progesterone receptor-positive (PR+) and human epidermal growth factor receptor 2 (HER2)-positive (HER2+). These types are connected to treatment, and there are patients, which cannot be fallen into any of the above three categories (these patients are called triple negative (TN)). Thus if we can assign such a TN patient to one of the three types, that would be useful to assign an appropriate treatment to the TN patient.

We then generate a dataset of (patients × SNPs) with three labels: (ER+/PR+, HER2+, TN), and set up a classification problem with three classes [103]. Fig. 3.39 shows the result obtained by running decision tree over this dataset. From the decision tree we obtained in Fig. 3.39, where the classification rule (dendrogram) is rather simple, using only two internal nodes, which can be summarized into the following:

$$\text{TP53} \neq \text{mutated} \quad \rightarrow \quad \text{ER+PR+}. \tag{3.300}$$

$$\text{TP53} = \text{mutated } \& \text{ PIK3CA} \neq \text{mutated} \quad \rightarrow \quad \text{Her2+ or ER+PR+}. \tag{3.301}$$

$$\text{TP53} = \text{mutated } \& \text{ PIK3CA} = \text{mutated} \quad \rightarrow \quad \text{TN or ER+PR+}. \tag{3.302}$$

Both TP53 and PIK3CA are gene names, and for example, 'TP53 = mutated' indicates that there are mutated positions in TP53. In Fig. 3.39, each leaf has the distribution of patients in each class (type). In fact the rules cannot be classified so perfectly, but this distribution implies some relations exist between gene mutations and types, which would be further analyzed by increasing the data size.

Sequence classification

Originally bioinformatics started with sequence analysis. Thus it has long time well-accepted and performed to transform sequences into vectors (in a matrix), generate a classifier from sequences according to labels, and finally apply the classifier to the new sequence to assign a new label to the sequence.

We can classify the current problems by their problem settings:

I. Entire sequence with one label: Each given (entire) sequence has one label, like if one peptide (short protein) is bound to some protein or not, and each peptide has one label.

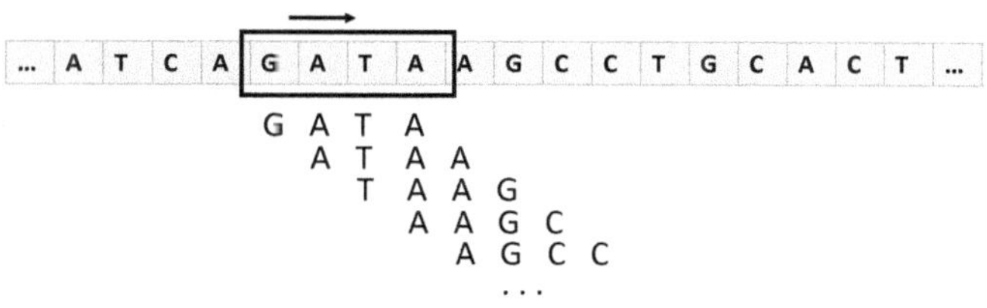

Figure 3.40: Generating instances with a fixed size by sliding a window over a sequence.

1. **Protein localization prediction [45]:** Each protein is fixed to exist in a cell, like in nucleus, in membrane, etc. Such a place name is a label, and this label is assigned to each sequence. Thus each sequence has a label, to be learned and be predicted.
2. **MHC binding peptide prediction [105]:** The immune system in our body starts with the recognition of an antigen (peptide) by a protein called Major Histocompatibility Complex (MHC). This recognition is initiated by the binding of a peptide to MHC. Thus the question is if a given peptide is bound to MHC, which is a binary classification problem.

II. Each position with one label: This setting is that the focus is on one position of the given sequence, and the label is assigned to the position and not to the sequence.

3. **Protein secondary structure prediction [27]:** Prediction is done by position-wise, and so each instance is a sequence with a fixed number of letters, and prediction is to estimate the secondary structure (α-helix, β-sheer and others) at the center position of the given sequence.
4. **Calpain substrate cleavage site prediction [28]:** This problem setting is almost the same as the above. Given sequences have a fixed size of substrates, where the center position is the cleavage site, to be learned and also be predicted. The difference from the above is the classes of this problem is binary.

Original sequences have a variable size of sequences, and so to make a certain size of sequences, turning into vectors and then eventually a matrix of instances (sequences) $\times$ features (positions of sequences). This procedure follows the two steps below:

1. Cut out subsequences

   Fig. 3.40 shows a schematic picture of cutting out subsequences from a given sequence. The idea is to first set up a fixed length window, and slide this window to output the subsequence in the window. Let T and W be the

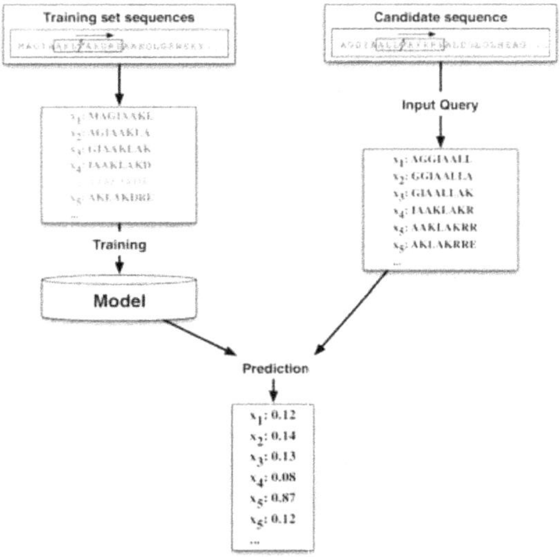

Figure 3.41: Entire procedure of learning and prediction of sequence classification (calpain substrate cleavage site prediction).

length of a given sequence and that of the window, we can have $T - W + 1$ instances from a given dataset.

2. Assign labels

   We then have simply two choices on assigning the label to each instance, according to the above I and II. The first one (I) is simple, all subsequences cut out have the same label, while the second one (II) is that depending on the locations of the center positions, labels are different.

Fig. 3.41 shows a typical procedure of supervised learning for sequence classification [28]. As shown in the left-hand side of the figure, subsequences are cut out from given sequences, to generate training data with labels. Then the classifier (model in the figure) is learned from the training data. On the other hand, as shown in the right-hand side of the figure, subsequences are also cut out (so that the length of subsequences should be the same as those of training subsequences) from the test sequence to be predicted, and each subsequence is predicted by the model already trained.

This procedure is a classical manner in bioinformatics to transform sequences into vectors to use machine learning methods for vectors. However, originally data are sequences, and machine learning methods for sequences would be more appropriate than using those for vectors. Chapter 5 introduces various machine learning methods for sequences.

Chapter 4

Learning Sets

The data with vectors as instances had the following two characteristics:

Fixed size: The size (i.e. the number of features) of each vector is fixed at some constant.

Ordered features: The features have their own order, being kept through the data.

On the other hand, the data with sets as instances have the following characteristics:

Flexible size: The size (number of features) of each set is changeable and inconstant.

Unordered features: Features (elements) are not ordered in a set.

We may think both discrete and continuous values for elements, while we focus on discrete values in this chapter, except for kernel learning. Below we raise some examples of sets with discrete elements:

Market basket: A set of goods (items) in the basket a customer has.

Accessed web sites: A set of web sites a user accessed in one web surfing, ignoring the order of the accessed timing.

Words in natural language processing: A set of words which appear in one sentence, paragraph or document. This is called a *bag of words* (BOW).

Overexpressed genes: In a gene expression matrix of (genes $\times$ patients), a set of genes for each patient, where expression values are larger (or smaller) than some cut-off value.

In this chapter, we will explain frequent pattern mining, probabilistic models and kernel learning, finally followed by applications to life science.

4.1 Frequent Pattern Mining

Frequent pattern mining does not have any model, and in this sense, is different from machine learning. Generally speaking, however, the objective of frequent pattern mining is knowledge discovery, being consistent with that of machine learning or more generally data science.

4.1.1 Terms, Concepts and Problem Setting

First we define terms of frequent pattern mining (for sets, *frequent itemset mining*). Some of these terms are used not only for sets but also for sequence, trees and graphs in frequent pattern mining.

item: One element of a set is called an *item*. Then a set of items is called *itemset*. Originally the application of frequent patter mining was market basket, and then we may call an element a *goods*, while items are more mildly used than goods, in the sense that items include not only those to be purchased.

pattern:$\mathcal{P}$**:** A combination of items is called a pattern. If an instance is a set, a pattern is a subset[1]. For sets, an instance is an itemset and a pattern is also an itemset. If a pattern has K items, this pattern is called K-pattern.

support: Given multiple instances (sets) as training data, the number of patterns appeared in the set of instances in training data is called *support*. We write the support of pattern $\mathcal{P}$ as support($\mathcal{P}$).

minimum support (min_sup): A cut-off value to decide if a pattern is frequent or not, is called the *minimum support* or *min_sup* for short. This is set against the support, like that the min_sup is 10.

frequent pattern: We call a pattern with the support no less than the minimum support (min_sup) a *frequent pattern*. That is, the set of frequent patterns $\mathcal{F}$ satisfies the following:

$$\mathcal{F} = \{\mathcal{P} | \text{support}(\mathcal{P}) \geq \text{min\_sup}\}. \tag{4.1}$$

Note that min_sup is a hyperparameter, i.e. an input by a user. As you can see from the above equation, a smaller min_sup makes the number of frequent patterns increase more. On the other hand, a larger min_sup reduces frequent patterns more. That is, the min_sup can control the number of frequent patters and also searching space, resulting in computation time. In particular, in this chapter, a pattern is an itemset, and as a synonym, we use a *frequent itemset*. We write a frequent itemset with K items, *K-frequent pattern*.

We write the problem setting below:

[1] Similarly for other data types, if an instance is a sequence, tree or a graph, a pattern is a subsequence, a subtree or a subgraph, respectively.

Table 4.1: An example of itemsets.

itemset 1	A B C D E
itemset 2	A B D E
itemset 3	A C D E
itemset 4	B C D
itemset 5	B D E

Definition 4.1 (Frequent Itemset (Pattern) Mining). *Frequent pattern mining on sets or frequent itemset mining is to enumerate all itemsets (combinations of items), each with the support larger than or equal to the minimum support*[2].

Frequent pattern (itemset) mining for sets is the founded concept of data mining. Particularly this problem was derived from market basket analysis, to understand the goods/items likely to be in the same basket, through a large number of market baskets in real stores or in E-commerce.

More concretely, bread and butter would not be a surprising combination, while beers and diapers might be. The application idea is to find this type of combination, inspiring the idea of promoting sales by selling goods concurrently purchased but nobody finds so far.

4.1.2 Downward Closure Property

Enumerating all possible patterns of items is an NP-hard problem, and so we need some idea to efficiently explore the vast searching space. Then the first proposition we will use is the so-called *Downward Closure Property*:

Proposition 4.1 (Downward Closure Property). *If the support of a pattern is smaller than the minimum support, any pattern including this pattern cannot be frequent.*

We should start with an example. Table 4.1 shows a toy dataset of five itemsets (or sets of items). In this table, for example, pattern of A and D, i.e. AD, appears three times, meaning that the support of this pattern is three (Both A and D appear in itemsets 1 to 3). If B is added to AD, the support of pattern ABD is two (appear in itemsets 1 and 3). Also AD with C, i.e. ACD, is with the support of two. Finally the support of ADE is three. Thus if some item is added to AD, the support cannot be larger than three. For example, if the minimum support is four, AD is not a frequent pattern, and any pattern including AD cannot be frequent.

In summary, when you add some item to pattern $\mathcal{P}$, the support is at most the same. Thus the support of $\mathcal{P}$ is lower than the minimum support, or $\mathcal{P}$ is not frequent, any pattern including $\mathcal{P}$ is not frequent.

[2]The original objective of frequent pattern mining is not necessarily only on "frequency", while for simplicity we consider only frequent patterns

4.1.3 Two Types of Approaches

Methods for frequent pattern mining for sets can be summarized into the following two approaches:

Candidate generation: starts with the smallest pattern, repeating adding one item to frequent patterns with the size of K to generate frequent patterns with the size of $K + 1$. The database must be scanned each time when the size of patterns is increased, resulting in that the times of scanning the entire database is too large and (said to be) inefficient.

Representative method: Apriori **Algorithm 4.1**

Pattern growth: particularly FP-growth, scans the database only twice to enumerate all frequent patterns. The contribution of FP-growth is to develop the FP-tree, where all frequent patterns can be enumerated from the FP-tree (described later) without scanning the database. Thus once FP-tree is generated (by scanning the database twice), the algorithm does not have to scan the database.

Representative method: FP-growth **Algorithm 4.2**

4.1.4 Apriori Algorithm

Apriori [2] is a typical method of so-called *candidate generation*, and straightforwardly based on **Proposition 4.1**. Briefly Apriori starts with the smallest frequent patterns to increase the size of frequent patterns until no frequent patterns.

Below we write a bit more detailed procedure of the algorithm. First the given database is scanned to count the support and enumerate all K-frequent patterns ($K = 1$). Then the following two steps are repeated alternately:

1. Add one item to all K-frequent patterns to generate all $K + 1$-frequent patterns.

2. Scan the database to compute the support of all $K + 1$-frequent pattern candidates and generate all $K + 1$-frequent patterns.
   $K \leftarrow K + 1$.

Algorithm 4.1 shows a pseudocode of the above procedure.

We explain Apriori by using toy examples, given by Fig. 4.1. For example, we assume that the minimum support is three. First we check 1-patterns and found all items are 1-frequent patterns. Next we generate frequent patterns with the size of two, and scan the database to find that 2-frequent patterns are AD and AE. Adding one item to AD and AE, we find that ADE is frequent, and no frequent patterns with a larger size.

A drawback of Apriori is that Apriori has to scan the database each time the new frequent pattern candidates are generated. In practice, as mentioned earlier, an itemset is a market basket, and the set of itemsets is the database of all past purchase records, and scanning over that many times is clearly not easy.

Algorithm 4.1: Apriori: a candidate generation algorithm for mining frequent patterns.

```
Function Apriori(𝒟, Min_Sup)
    Data: input itemsets: 𝒟, minimum support: Min_Sup
    Result: all possible frequent patterns
    Initialization: Scan all instances and all frequent patterns with the size
    of K(= 1).;
    repeat
        Generate frequent pattern candidates with the size of K + 1 by
        adding one item to frequent patterns with the size of K.;
        Scan 𝒟 and keep patterns with the support larger than Min_Sup as
        frequent patterns with the size of K + 1.;
        K + +;
    until no frequent patterns any more;
```

4.1.5 FP-growth Algorithm

From a viewpoint that scanning the entire itemset database many times is inefficient, the *FP-growth* algorithm proposes the algorithm in which the database is scanned only twice. FP-growth [36] is a typical algorithm in a category called pattern generation. The FP-growth algorithm has the following four steps: 1) Preprocessing, 2) Generating FP-tree, 3) Generating conditional pattern base (CPB) and 4) Generating conditional frequent pattern (CFP).

1) Preprocessing: We scan the database to count the support of all items. We then sort items according to their supports, keep only frequent items and also in each itemset, items are sorted according to the order of items.

We explain this step by using the toy example of Table 4.1. First, we scan the database, i.e. all itemsets, to count the support of each item for all items, as shown in Table 4.2 (a). Then the items are sorted, according to the support size (Table 4.2 (b)). In (b), we keep frequent items only. That is, if we set the minimum support at three, we keep 1-frequent patterns with the support larger than or equal to 3, only. Finally we sort items in each itemset, according to the order of items in (b), as shown in Table 4.2 (c).

In this step, we check all itemsets once, meaning that the database is scanned once.

2) Generating FP-tree: Using Table 4.2 (c), we generate a tree, called *FP-tree*: FP-tree has null node at the root, and is generated by using itemsets in their order. We explain this step also by using the example of Table 4.2 (c)).

Itemset 1: From itemset 1 (in Table (c)), every time new item appears, a new node is added to the FP-tree, resulting in a tree in Fig. 4.1 (a) Note that the tree from the first itemset is always a sequence (tree

Table 4.2: (a) counted supports, (b) items ordered by their supports and (c) items in each itemset ordered according to (b).

(a)

item	support
A	3
B	4
C	3
D	5
E	4

(b)

item by support	support
D	5
B	4
E	4
A	3
C	3

(c)

	items
itemset 1	D B E A C
itemset 2	D B E A
itemset 3	D E A C
itemset 4	D B C
itemset 5	D B E

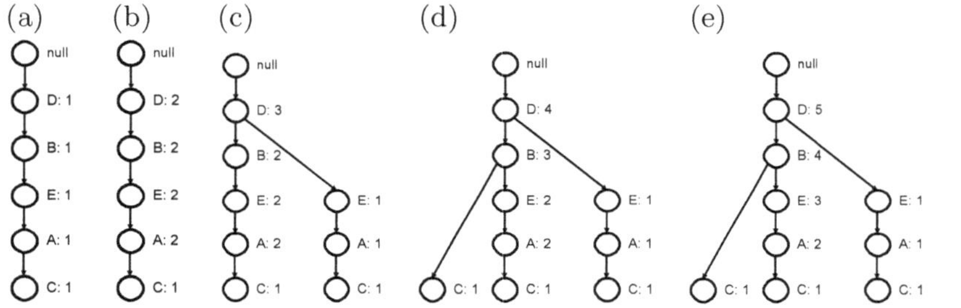

Figure 4.1: Example of generating FP-tree using itemsets 1 (a) to 5 (e).

without any branches). Also note that not only labels but also the number of appearances are added to each node.

Itemset 2: Itemset 2 is just a part of Itemset 1. So we start with the root (null) and go through in the same path as itemset 1, incrementing the number of appearances by one from D to A (Fig. 4.1 (b)).

Itemset 3: Itemset 3 (DEA) is the same as the current path of DBEA by D and different from DBEA after D. Thus a new path is made with a branch at node D, as shown in 4.1 (c). We increment the number of appearances at node D by one. In the new path, we assign the corresponding item to every newly generated node with the number of appearances at 1.

Itemset 4: Itemset 4 (DBC) is the same as DBEAC by B and generates another path (only node C) afer B, with a branch at B, as shown in Fig. 4.1 (d). Also we increment the number of appearances at each node of DB by one.

Itemset 5: Itemset 5 (DBE) is just part of DBEAC, and so we increment the number of appearances at the corresponding node by one (Fig. 4.1 (e)).

Table 4.3: Mining frequent patters (CPB: conditional pattern bases, CFP: conditional frequent patterns).

item	CPB	CFP	frequent patterns
C	BD:1, AEBD:1, AED:1	D:3	CD:3
A	EBD:2, ED:1	D:3, E:3, ED:3	AD:3, AE:3, AED:3
E	BD:3, D:1	D:4, B:3, BD:3	ED:4, EB:3, EBD:3
B	D:4	D:4	BD:4
D	-		

From the above procedure, we complete the FP-tree (Fig. 4.1 (e)) by using all itemsets in Table 4.1 (c), one by one. Then to complete the FP-tree, all itemsets (i.e. database) are scanned once, meaning the second database scan. In the next step and later on, we use only the FP-tree to enumerate all frequent patterns, and never contact the database. Thus this algorithm scans the database only twice. The key point of this algorithm is to find that all frequent patterns can be enumerated by using the FP-tree only.

3) Generating conditional pattern base (CPB):
The *conditional pattern base* (CPB) is, in the FP-tree, all paths from each item (node) to the root. Also not only items in each path but the number of the starting item (node) are copied. The path to be enumerated starts with the nodes in the most depth layer (i.e. items with the lowest support). We explain this step by using the example of Fig. 4.1 (e).

In this example, the item in the most depth (distant from the root) layer is C, and so we start with item C. There are three paths from item C to the root (null). They are, from the left, DC:1, AEBD:1 and AED:1, omitting C. Note that the numbers are those assigned to the nodes of item C. They are written in the row of C and column of CPB in Table 4.3.

In the next layer, the item is A, which appears at two places. From them, we have two paths: EBD:2 and ED:1, omitting A.

We repeat this manipulation from the bottom layer to the root, and have the column of CPB in Table 4.3.

4) Generating conditional frequent pattern (CFP): The final step is to generate *conditional frequent pattern* (CFP) from CPB for each item. We explain this again by using the toy example in Table 4.3. In the column CPB of Table 4.3, we check the rows with items, from the (item C) row to the (item B) row, one by one.

Item C: BD, AEBD and AED all appear once. We regard these three as three appearing itemsets, and we enumerate all frequent patterns from these three itemsets. More concretely, we can follow the process of the FP-growth algorithm so far (but now we skip explaining this, since we

Algorithm 4.2: FP-growth.

Function FP-growth($\mathcal{D}$, Min_Sup)
  Data: input itemsets: $\mathcal{D}$, minimum support: Min_Sup
  Result: all possible frequent patterns
  foreach *itemset* $I \in \mathcal{D}$ **do**
    Increment the support of each item in I.;
  Generate a list of items according to their support.;
  Sort items in each itemset according to their support.;
  foreach *itemset* $I \in \mathcal{D}$ **do**
    Add I to FP-tree.;
  Generate conditional pattern bases (CPB) from FP-tree.;
  Generate conditional frequent pattern (CFP) from CPB.;
  Enumerate all frequent patterns from CFP.;

have explained this algorithm already. Instead we just manually check frequent patterns). In fact from the three itemsets, the patterns which satisfy the minimum support of three are just D only. Thus CFP of item C is only D:3.

Item A: EBD appears twice and ED appears once. So the patterns which can satisfy the minimum support of three are D:3, E:3 and DE:3.

Item E: BD appears three times and D appears one. So the patterns with the support no less than three are D:4, B:3 and BD:3.

Item B: D appears four times, and directly CFP is D:4.

By the above, CFP is complete. We now present all frequent patterns, from CFP. This step is very simple, and for each item row in Table 4.3, we just add the corresponding item to each pattern of CFP. Then the column of frequent patterns in Table 4.3 shows all K-frequent patterns ($K > 1$) enumerated.

Algorithm 4.2 shows a pseudocode of the above four step procedure. Note again that this algorithm scans the entire database (all given itemsets) only twice (in **Algorithm 4.2**, the database scan corresponds to **foreach**, and you can see **foreach** appears only twice in the algorithm).

4.2 Probabilistic Model

4.2.1 Topic Model

In natural language processing, if we use documents (sentences, paragraph, etc.) as instances, each instance has an arbitrary number of words. Among the possible instances, the smallest instance would be a sentence. One sentence is a sequence of words, while in natural language processing, the order of words is often ignored and also words are regarded as discrete values. Thus one instance becomes a set of discrete values, which is equivalent to the idea of an itemset (sets) and items (elements) for a sentence and words, resulting in learning sets, the main concept of this chapter. This way of understanding sentences is called *bag of words*, in natural language processing.

We can think about a generative model, which generates elements in a given set (instance). A simple and reasonable assumption is that elements (and also instances) have some groups as a latent variable behind observable variables, and elements are generated according to the latent variable. For natural languages, sentences (instances) have several *genres*, such as science, engineering, politics, sports, etc., and words are generated differently, according to the genres. The model based on this idea is called a *topic model*, where the topic corresponds to the above genre.

The topic model is originally developed for natural language processing, and an instance and an element are always an sentence and a word, respectively. For generality, we write instances (sets) instead of sentences, and elements instead of words. For simplicity, we do not assume that a set has the same element more than once, meaning that all elements in one set are unique. This simply means that we do not count the number of appearances of words in one sentence, implying the count is always one even if the word appears many times.

The topic model has historically two approaches: pLSA (probabilistic latent semantic analysis) [43] and LDA (latent Dirichlet allocation) [11]. The latter is an extension of the former. In fact LDA is in the framework of Bayesian statistical learning, which is a research field developing in a high speed. In this sense, it might be a bit too earlier to introduce LDA here in the sense that another better approach from Bayesian statistical learning may emerge as a topic model in the future.

pLSA: Probabilistic Structure

pLSA has the following two probability parameters:

- $p(z|d)$

  The probability that instance d is assigned to topic z.

- $p(w|z)$

  The probability that element w is generated from topic z.

The probability of generating element w from instance d should be structured through topic z, the latent variable, as follows:

$$p(w|d) = \sum_z p(w|z)p(z|d). \tag{4.2}$$

This structure is shown by a graphical model, in Fig. 4.2 (a). That is, both instances and elements should be clustered, and so the latent variable is sandwiched between them to do clustering both.

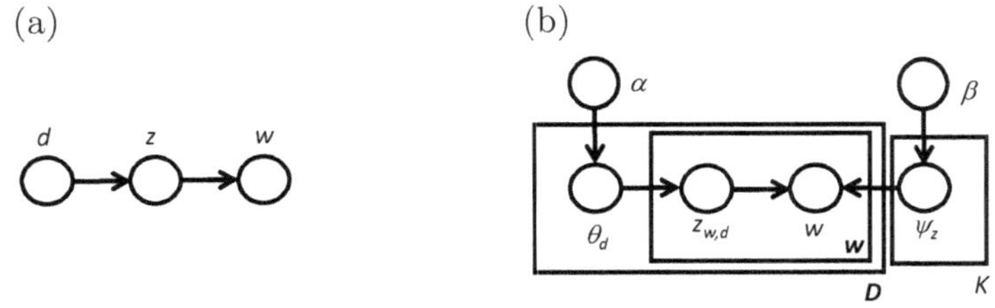

Figure 4.2: (a) pLSA and (b) LDA.

The likelihood of given instances including d under the model of pLSA is given as follows:

$$\prod_{w}\prod_{d} p(w|d)^{n(w,d)}, \tag{4.3}$$

where $n(w,d)$ is one if element w appears in instance d; otherwise zero.

pLSA: Estimating Probability Parameters

We estimate the two probability parameters by the maximum likelihood criterion, i.e. maximizing the likelihood, given by (4.3). In reality we use an Expectation and Maximization (EM) algorithm, which allows to achieve a local optimum of the maximum likelihood estimation. The EM algorithm we use repeats the following two steps alternately [43]:

E-step: Compute the posterior probability of latent variable.

$$p(z|d,w) \quad \leftarrow \quad \frac{p(w|z)p(z|d)}{\sum_z p(w|z)p(z|d)}. \tag{4.4}$$

M-step: Maximize the expectation value of the log-likelihood.

$$p(w|z) \quad \leftarrow \quad \frac{\sum_d n(d,w)p(z|d,w)}{\sum_w \sum_d n(d,w)p(z|d,w)}. \tag{4.5}$$

$$p(z|d) \quad \leftarrow \quad \frac{\sum_w n(d,w)p(z|d,w)}{\sum_w n(d,w)}. \tag{4.6}$$

Algorithm 4.3 shows a pseudocode, which summarizes the above alternate procedure of estimating the probability parameters of pLSA.

LDA: Motivation and Background

LDA was proposed by the following two motivations, which are the drawbacks of the above model of pLSA.

Algorithm 4.3: EM algorithm for pLSA.

```
1 Function pLSA(D, K)
    Data: input itemsets: D = (d_1, ..., |D|), #clusters: K
    Result: learning results
2   Initialize p(z|d, w) randomly.;
3   repeat
4       M-step: compute (4.5) and (4.6).;
5       E-step: compute (4.4).;
6   until convergence;
```

Problem 1: $p(z|d)$ for test data We can conceptually see that parameter $p(z|d)$ is the distribution (or weight) of topics to which an instance is assigned. However, in reality, d is an instance, which is fine for training data, but for testing, we are not sure what d should be used for. In other words, instances are all unique, and an unknown test instance would not be any past instance in the training data. Thus we cannot use any d for this parameter in testing. This problem is caused by the model structure, in which instance itself is a variable in a parameter to be estimated.

Problem 2: Overfitting The latent variable is z, and so we can change the number of clusters, according to the data size, by which we think that appropriate clustering is done, and overfitting can be avoided. However, even if the number of clusters is the same, if the data size is increased, the number of parameters (the sum of the number of $p(w|z)$ and that of $p(z|d)$) is increased, because $p(z|d)$ has d. Thus this indicates interestingly if the data size is increased, overfitting might happen. In fact, this problem is also caused by the model structure that instances are one variable of the parameter. In general overfitting is a problem of unsupervised learning.

LDA solves the above two problems, by the following two points, which are related with each other:

Incorporating distributions with hyperparameters as parameters: We use distributions with hyperparameters for parameters. In fact parameters had instances as a variable in pLSA, while such a variable also can be by a distribution with hyperparameters in LDA. Hyperparameters control the distributions for training data and also test data, by which Problem 1 can be solved. Also hyperparameters can be like regularization terms in the optimization problem, by which overfitting in Problem 2 can be solved.

Distributions are estimated by Bayes learning: Bayes statistical learning allows to use the above probabilities with hyperparameters and also training these probabilities within the framework of Bays learning.

LDA: Probabilistic Structure

LDA assumes a distribution for each probability parameter, such as a multinomial distribution, where its parameters are estimated in training. In particular, D of LDA stands for the *Dirichlet distribution*, which has hyperparameters which can control distributions flexibly by using prior knowledge (see Section A.2.5 for these distributions).

LDA has the following variables:

θ_d The distribution of instance d.

w Element w, and instance d is a set of elements: $\boldsymbol{w}$ ($d = \boldsymbol{w}$)

$z_{w,d}$ The topic of instance d and element w.

ψ_z The distribution of elements at topic z.

Only $\boldsymbol{w}$ is an observable variable, and all other are latent variables.

LDA assumes distributions for variables (parameters), which reduces the flexibility of the model. This is similar to that adding regularization terms to the objective function reduces the range that parameters can take. Then this makes the number of possible alternatives smaller and reduces overfitting.

Again, in fact, pLSA had a problem that the trained model cannot be applied to unknown, test instance, because test instances are not in training instances. Among the above variables, variables which might have this problem are those for instances, i.e. θ_d and w. Thus LDA uses Dirichlet distributions for probabilities of generating these variables, and also define α and β as hyperparameters of the Dirichlet distributions.

Below are the probability distributions for probability parameters of LDA:

- $p(\theta_d|\alpha) \sim \text{Dirichlet}(\alpha)$.
- $p(\psi_z|\beta) \sim \text{Dirichlet}(\beta)$.
- $p(z_{w,d}|\theta_d) \sim \text{Multinomial}(\theta_d)$.
- $p(w|\psi_z) \sim \text{Multinomial}(\psi_z)$.

Using these probability parameters, the joint probabilities of variables of instance d ($= \boldsymbol{w}$) can be defined by the following probabilistic structure:

$$p(\psi_z, \theta_d, z_{w,d}, \boldsymbol{w}|\alpha, \beta) = p(\theta_d|\alpha) \prod_{w \in \boldsymbol{w}} p(w|\psi_z)p(\psi_z|\beta)p(z_{w,d}|\theta_d). \tag{4.7}$$

On the other hand, pLSA was defined as follows:

$$p(d, z, w) = p(d)p(w|z)p(z|d). \tag{4.8}$$

Thus, from the corresponding parts of the two structures, we can see that these two models share almost the same probabilistic structure. The biggest difference is that LDA assumes distributions for probabilities and introduces hyperparameters which can control distributions. In particular by introducing the Dirichlet distributions, controlling hyperparameter α and β flexibly, the model can avoid overfitting and allow to use the model for unknown, test instances.

By computing marginalized probabilities, the probability of instance d ($= \boldsymbol{w}$) is given as follows:

$$p(\boldsymbol{w}|\alpha, \beta) = \int_{\theta_d} \int_{\psi_z} p(\theta_d|\alpha) \prod_{w \in \boldsymbol{w}} \sum_z p(w|\psi_z) p(\psi_z|\beta) p(z_{w,d}|\theta_d) d\psi_z d\theta_d. \tag{4.9}$$

The probability of all instances in training data:

$$p(\mathcal{D}|\alpha, \beta) = \prod_d \int_{\theta_d} \int_{\psi_z} p(\theta_d|\alpha) \prod_{w \in \boldsymbol{w}} \sum_z p(w|\psi_z) p(\psi_z|\beta) p(z_{w,d}|\theta_d) d\psi_z d\theta_d. \tag{4.10}$$

LDA: Generative Process

Fig. 4.2 (b) shows a graphical model of LDA. The generative process of this generative model can be given as follows:

1. For $k = 1, \ldots, K$

   $\psi_k \sim \text{Dirichlet}(\beta)$.

2. For each $d \in \mathcal{D}$

   (a) $\theta_d \sim \text{Dirichlet}(\alpha)$.

   (b) For $w \in d$

      i. $z \sim \text{Multinomial}(\theta_d)$.

      ii. $w \sim \text{Multinomial}(\psi_z)$.

LDA: Estimating Probability Distributions

For estimating parameters, we need to estimate the posterior distribution of the latent variable $z_{w,d}$:

$$p(z_{w,d}|\boldsymbol{w}) = \frac{p(z_{w,d}, \boldsymbol{w})}{p(\boldsymbol{w})}. \tag{4.11}$$

In (4.11), the numerator of the right-hand side is the marginalized probability of (4.7), and the denominator of the right-hand side is also the marginalized probability (over $z_{w,d}$ as well), which is shown in (4.9).

In particular, for the marginalized probability of the denominator of the right-hand side, the latent variable has a very large number of combinations of w and d, which makes computing all combinations for marginalization impossible. Thus, the probability structure of (4.9) is approximated, keeping the graphical model in Fig. 4.2(b). For this purpose, we have the following two possibilities for parameter estimation:

1. **Sampling (for example, Gibbs sampling)**
2. **Variational method**

LDA: Estimating Probability Distributions by Gibbs Sampling

We introduce the solution by Gibbs sampling [34]. In Gibbs sampling, we focus on one latent variable and compute the conditional probability (posterior probability) of the latent variable under the condition that all other variables are given. So we focus on $\boldsymbol{z}$ out of the three latent variables.

Below are notations to explain Gibbs sampling:

- $z_w = k$: The topic of element w is topic k.
- $\boldsymbol{z}_{-w}$: The topic for all elements except element w.
- $\boldsymbol{w}_{-w}$: The set of elements except element w

We first introduce a general Bayes theorem between latent variable z and observable variable x:

$$p(z|x) \propto p(x|z)p(z). \tag{4.12}$$

Thus our objective here is to generate a similar qualitative equation:

$$\begin{aligned} p(z_w = k|\boldsymbol{z}_{-w}, \boldsymbol{w}) &\propto p(z_w = k, \boldsymbol{z}_{-w}, \boldsymbol{w}) & (4.13)\\ &= p(w|z_w = k, \boldsymbol{z}_{-w}, \boldsymbol{w}_{-w})p(z_w = k, \boldsymbol{z}_{-w}, \boldsymbol{w}_{-w}) & (4.14)\\ &= p(w|z_w = k, \boldsymbol{z}_{-w}, \boldsymbol{w}_{-w})p(z_w = k|\boldsymbol{z}_{-w}, \boldsymbol{w}_{-w})p(\boldsymbol{z}_{-w}, \boldsymbol{w}_{-w}) & (4.15)\\ &\propto p(w|z_w = k, \boldsymbol{z}_{-w}, \boldsymbol{w}_{-w})p(z_w = k|\boldsymbol{z}_{-w}, \boldsymbol{w}_{-w}). & (4.16) \end{aligned}$$

since z_w and $\boldsymbol{w}_{-w}$ are not related with each other. Thus,

$$p(z_w = k|\boldsymbol{z}_{-w}, \boldsymbol{w}) = p(w|z_w = k, \boldsymbol{z}_{-w}, \boldsymbol{w}_{-w})p(z_w = k|\boldsymbol{z}_{-w}), \tag{4.17}$$

where we note that $p(w|z_w = k, \boldsymbol{z}_{-w}, \boldsymbol{w}_{-w})$ and $p(z_w = k|\boldsymbol{z}_{-w})$ correspond to $p(x|z)$ and $p(z)$ in (4.12), respectively.

Thus now we need to estimate these two probabilities, $p(w|z_w = k, \boldsymbol{z}_{-w}, \boldsymbol{w}_{-w})$ and $p(z_w = k|\boldsymbol{z}_{-w})$. We explain the derivation for each probability, as follows.

$p(w|z_w = k, \boldsymbol{z}_{-w}, \boldsymbol{w}_{-w})$:

Since w is controlled by ψ, the first probability of the right-hand side of (4.17) can be written as follows:

$$p(w|z_w = k, \boldsymbol{z}_{-w}, \boldsymbol{w}_{-w}) = \int p(w|z_w = k, \psi_{z_w=k})p(\psi_{z_w=k}|\boldsymbol{z}_{-w}\boldsymbol{w}_{-w})d\psi_{z_w=k}, \tag{4.18}$$

where each probability of the right-hand side of (4.18) can be given as follows:

$$p(w|z_w = k, \psi_{z_w=k}) = \psi_{z_w=k}. \tag{4.19}$$
$$p(\psi_{z_w=k}|\boldsymbol{z}_{-w}\boldsymbol{w}_{-w}) \propto p(\boldsymbol{w}_{-w}|\psi_{z_{w,d}=k}, \boldsymbol{z}_{-w})p(\psi_{z_w=k}). \tag{4.20}$$

We already decided to use the Dirichlet distribution:

$$p(\psi_{z_w=k}) = \text{Dirichlet}(\beta). \tag{4.21}$$

This is a conjugate prior distribution[3] of $p(\boldsymbol{w}_{-w}|\psi_{z_w=k}, \boldsymbol{z}_{-w})$:

$$p(\psi_{z_w=k}|\boldsymbol{z}_{-w}\boldsymbol{w}_{-w}) = \text{Dirichlet}(\beta + N_{-w,z_w=k}), \tag{4.22}$$

where $N_{-w,z_w=k}$ is the number of words, assigned to topic k, except w. Thus, we can write the expectation value of the Dirichlet distribution:

$$p(w|z_w = k, \boldsymbol{z}_{-w}, \boldsymbol{w}_w) = \int \psi_{z_w=k}\text{Dirichlet}(\beta + N_{-w,z_w=k})d\psi_{z_w=k} \tag{4.23}$$
$$= \frac{N_{-w,z_w=k} + \beta}{\sum_w (N_{-w,z_w=k} + \beta)}. \tag{4.24}$$

$p(z_w = k|\boldsymbol{z}_{-w})$:

For d including element w,

$$p(z_w = k|\boldsymbol{z}_{-w}) = \int p(z_w = k|\theta_d)p(\theta_d|\boldsymbol{z}_{-w})d\theta_d. \tag{4.25}$$

The two terms in the right-hand side of the above equation can be given as follows:

$$p(z_w = k|\theta_d) = \theta_d. \tag{4.26}$$
$$p(\theta_d|\boldsymbol{z}_{-w}) \propto p(\boldsymbol{z}_{-w}|\theta_d)p(\theta_d), \tag{4.27}$$

where

$$p(\theta_d) = \text{Dirichlet}(\alpha). \tag{4.28}$$

This is a conjugate prior distribution of $p(\boldsymbol{z}_{-w}|\theta_d)$:

$$p(\theta_d|\boldsymbol{z}_{-w}) \propto \text{Dirichlet}(\alpha + N_{-w,w\in d,z_w=k}), \tag{4.29}$$

where $N_{-w,w\in d,z_w=k}$ is the number of instances (sets), which are assigned to topic k and do not have element w.

[3]See Section A.8.10

Algorithm 4.4: Gibbs sampling for latent Dirichlet allocation.

Function `Gibbs_sampling_for_latent_Dirichlet_allocation(`$\mathcal{D}$, T`)`
  Data: input itemsets: $\mathcal{D}$, #iterations: T
  Result: element assignment matrix: $\boldsymbol{z}$, $N_{d,k}, N_{w,k}$
  Generate element set $\boldsymbol{w}$ from $\mathcal{D}$ (record the sets from which each element was).;
  Initialization: assign random values to $N_{w,k}(w = 1, \ldots, |\boldsymbol{w}|)$.;
  for $t \leftarrow 1$ **to** T **do**
    foreach $w \in \boldsymbol{w}$ **do**
      $k \leftarrow z_w; N_{d|w\in d,k} --; N_{w,k} --; N_k --$.;
      for $k \leftarrow 1$ **to** K **do**
        $p_k \leftarrow p(z_w = k|\boldsymbol{z}_{-w}, \boldsymbol{w})$ (computed in (4.32)).;
      Do sampling $k' \leftarrow p_k(k = 1, \ldots, K)$.;
      $z_w \leftarrow k'; N_{d|w\in d,k'} ++; N_{w,k'} ++; N_{k'} ++;$.;

From the above,

$$p(z_w = k|\boldsymbol{z}_{-w}) = \int \theta_d \text{Dirichlet}(\alpha + N_{-w,w\in d,z_w=k}) d\theta_d \quad (4.30)$$

$$= \frac{N_{-w,w\in d,z_w=k} + \alpha}{\sum_k (N_{-w,w\in d,z_w=k} + \alpha)}. \quad (4.31)$$

The denominator of the right-hand side of (4.31) has nothing to do with any topic. Thus this is a constant determined by given data, and we can ignore this denominator.

We now put the derivations of the two probabilities together, which are given as follows:

$$p(z_w = k|\boldsymbol{z}_{-w}, \boldsymbol{w}) \propto \frac{N_{-w,w\in d,z_w=k} + \alpha}{\sum_k (N_{-w,w\in d z_w=k} + \alpha)} \frac{N_{-w,z_w=k} + \beta}{\sum_w (N_{-w,z_w=k} + \beta)}. \quad (4.32)$$

LDA: Algorithm (Gibbs Sampling)

We can implement Gibbs sampling, using (4.32). **Algorithm 4.4** shows a pseudocode of the Gibbs sampling algorithm of LDA.

Algorithm 4.4 gives the following two numbers:

1. $N_{d,k}$: The number of instances d, including element w, which is assigned to topic k.

2. $N_{w,k}$: The number of element w, which is assigned to topic k.

Table 4.4: Vectors from Table 4.1.

	A	B	C	D	E
itemset 1	1	1	1	1	1
itemset 2	1	1	0	1	1
itemset 3	1	0	1	1	1
itemset 4	0	1	1	1	0
itemset 5	0	1	0	1	1

Using these two numbers, we can estimate the latent variable, as follows:

$$\theta_{d,k} = \frac{N_{d,k} + \alpha}{\sum_k (N_{d,k} + \alpha)}. \tag{4.33}$$

$$\phi_{w,k} = \frac{N_{w,k} + \beta}{\sum_w (N_{w,k} + \beta)}. \tag{4.34}$$

4.3 Kernel Learning

The advantage of kernel learning is that as shown in Section 3.4, a lot of general machine learning problem settings, such as K-means, PCA, etc. can be already *"kernelized"*. Thus if once kernel functions are generated, we can solve many machine learning problems just by using the generated kernel, regardless of data types which are used for generating the kernel. Here we can explore what kernels can be generated for sets.

4.3.1 Kernels for Sets

If the elements of each set are discrete values, the number of discrete values for all instances is a fixed size, say M. This means that we can generate a vector with the size of M for each instance, meaning the entire dataset is a matrix of instances $\times M$ features. Each feature is a unique discrete value appearing in the original dataset, where one instance is a set. Thus we show an example: the sets of Table 4.1 are changed into vectors, as shown in Table 4.4. As such, if each element is a discrete value, each instance can be a (relatively huge) vector (and the dataset becomes a matrix). We then can compute a kernel function for all pairs of vectors. (see Section A.5 for general kernels (for vectors)).

Match Kernel: Sets with Continuous Values

As long as elements are discrete values, sets can be vectors, while we now think about continuous values for elements. For continuous values, we are unable to have a fixed number of discrete values from the entire dataset, i.e. all instances, and so we cannot make a fixed number of features. This means, we cannot take

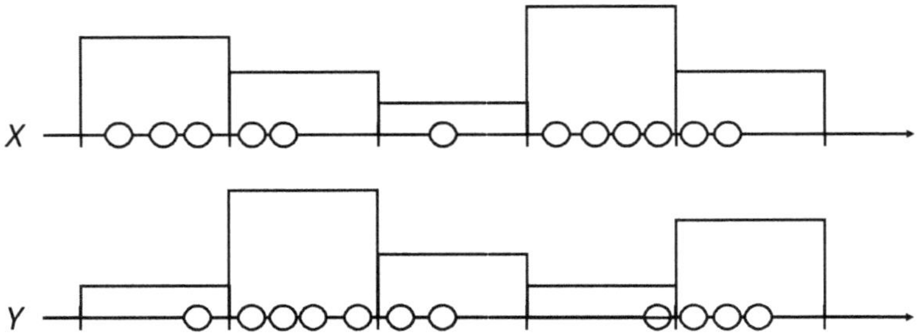

Figure 4.3: Nonparametric distributions (histograms) of $\boldsymbol{X}$ and $\boldsymbol{Y}$.

a strategy that we first generate a vector for one instance and a kernel function for pairs of vectors.

Thus for continuous values, we need to compute a kernel function between two sets of continuous values directly. This kernel function for sets of elements with continuous values is called a *match kernel*. Again the key point of match kernels is that the size of elements (continuous values) of each instance (set) is flexible. Also for generality we can make each element a d-dimensional numerical vector, where d is constant over all elements. However $d = 1$ (one element has only one value) for the setting of this chapter.

Match Kernel: Problem Setting

We now set the problem below: To compute kernel functions for a pair of instances, where two instances can have different number of elements. Thus we write these two instances $\boldsymbol{X}$ and $\boldsymbol{Y}$:

$$\boldsymbol{X} = \{\boldsymbol{x}_1, \ldots, \boldsymbol{x}_m\}, \quad (4.35)$$

$$\boldsymbol{Y} = \{\boldsymbol{y}_1, \ldots, \boldsymbol{y}_n\}, \quad (4.36)$$

where $m \leq n$, without loss of generality. As mentioned above, $\boldsymbol{x}_i$ and $\boldsymbol{y}_j$ are both d-dimensional vectors.

Match Kernel: Ideas

To solve this problem, the first, intuitive, straight-forward approach is that we think π, which is a function (permutation) that changes the order of indices of $\boldsymbol{Y}$ so that the first m elements of $\boldsymbol{Y}$ are most consistent with $\boldsymbol{X}$ in a pair-wise and element-wise manner:

$$\mathcal{M}(\boldsymbol{X}, \boldsymbol{Y}; \pi) = \{(\boldsymbol{x}_1, \boldsymbol{y}_{\pi_1}), \ldots, (\boldsymbol{x}_m, \boldsymbol{y}_{\pi_m}).\} \quad (4.37)$$

The cost of this matching $\mathcal{M}(\boldsymbol{X}, \boldsymbol{Y}; \pi)$ can be computed by, for example, the sum over all pairwise distances, as follows:

$$\mathcal{C}(\mathcal{M}(\boldsymbol{X}, \boldsymbol{Y}; \pi)) = \sum_{\boldsymbol{x}_i} ||\boldsymbol{x}_i - \boldsymbol{y}_{\pi_i}||_1. \quad (4.38)$$

Thus, this approach can be summarized into the following two steps:

1. We first find the function π^*, which minimizes the cost:

$$\pi^* \quad = \quad \arg\min_{\pi} \mathcal{C}(\mathcal{M}(\boldsymbol{X}, \boldsymbol{Y}; \pi)). \tag{4.39}$$

2. We then compute the kernel between $\boldsymbol{X}$ and $\boldsymbol{Y}$ under π^*.

However this method has the following two problems:

Combinatorial explosion: The number of combinations of elements in $\boldsymbol{X}$ and those in $\boldsymbol{Y}$ is huge, and it is not easy to find the optimal π^*.

Limited comparison: Only m elements are compared between $\boldsymbol{X}$ and $\boldsymbol{Y}$, and the remaining $n-m$ elements in $\boldsymbol{Y}$ are not considered at all. However these remaining elements should be considered for the difference between $\boldsymbol{X}$ and $\boldsymbol{Y}$.

Thus, we need a method, which can avoid the above two problems, and in this point the method we need must keep the following two points, which corresponds to the above two problems:

No combination: The method does not have to consider the combinations of elements in the two sets.

All features: The method does not use only m features, but also all other features can be considered.

Then a possible method that can address these two issues has the following three steps [48]:

Step 1: We plot the numerical values on the d-dimensional space for each instance

Step 2: We approximate the distribution of the numerical values by using some (parametric) distribution for each instance.

Step 3: Finally we compute a kernel function by using similarity between the two parametric distributions, each being derived from one instance.

Again this manner of computing a kernel function has the following two clear and serious problems.

Limited data: First of all, we need approximate probabilistic distributions, and so we need a good number of elements to estimate the distributions.

No prior knowledge on distribution: The structure of the distributions to be estimated is unknown. Thus there are no prior information on what type of (parametric) distributions should be assumed.

Thus one important solution for these two problems is to estimate *nonparametric* distributions against given numerical values. More concretely, as shown in Fig. 4.3 for each of the d dimensions, we use histograms to estimate the distribution of numerical values. We then compare the two sets of histograms to compute kernels. In other words, simply we segment the axis of one dimension into several intervals (called *bins*), and count how many elements are in each bin, resulting in a real-valued vector, where the length of the vector is the number of bins for one instance. Thus one instance is a vector, and we can compute a kernel function using these vectors for a pair of instances. For example, Fig. 4.3 shows five bins for both $\boldsymbol{X}$ and $\boldsymbol{Y}$, where the number of elements in each bin from left to right can be a vector, i.e. (3, 2, 1, 4, 2) for $\boldsymbol{X}$ and (1, 4, 2, 1, 3) for $\boldsymbol{Y}$. The above story of obtaining a kernel function is about only one of the d-dimensional space, while we can have a kernel function over all d-dimensions by the sum/product of the kernel for each dimension. This is because the product/sum of kernels is also a kernel.

The above nonparametric approach of using histograms corresponds to modifying the original numerical values to a discrete valued vector. This is called in many ways, such as *digitalization*, *quantization*, *discretization*, etc., depending on application fields.

However, this solution is still not perfect. The problem is the number of bins. In fact if the number of bins is too small, we are unable to find the overlap between two vectors easily. On the other hand, if the size of bins is too large, two vectors would be easily overlapped too much.

Pyramid Match Kernel

We now explain a kernel function, called *pyramid match kernel*, which addresses the above bin-size problem, by integrating the comparison (in overlap) between two instances at different bin sizes [33]. In fact the idea behind pyramid match kernel is to use not only one but also many different sizes of bins and integrate their results, instead of trying to find the optimal number of bins.

We here focus on $d = 1$ or just one of d-dimension. As mentioned earlier, we can compute the kernel function for d dimensions rather easily, just by using the product/sum over d one-dimensional kernel functions.

Let $H(\boldsymbol{X})$ be the histogram of instance $\boldsymbol{X}$ and $H^{(j)}(\boldsymbol{X})$ be the number of elements in the j-th bin of the histogram of $\boldsymbol{X}$. Also let R be number of bins.

We can measure the overlap between two histograms $H(\boldsymbol{X})$ and $H(\boldsymbol{Y})$, obtained for two instances $\boldsymbol{X}$ and $\boldsymbol{Y}$, respectively, as follows:

$$I(H(\boldsymbol{X}), H(\boldsymbol{Y})) \; := \; \sum_{j=1}^{R} \min(H^{(j)}(\boldsymbol{X}), H^{(j)}(\boldsymbol{Y})). \tag{4.40}$$

That is, the overlap in each bin is measured by the smaller (minimum) number of elements between two instances. This means that the overlap of each bin is captured as the *intersection*. Then the total overlap is obtained by summing up the intersection of each bin over all bins.

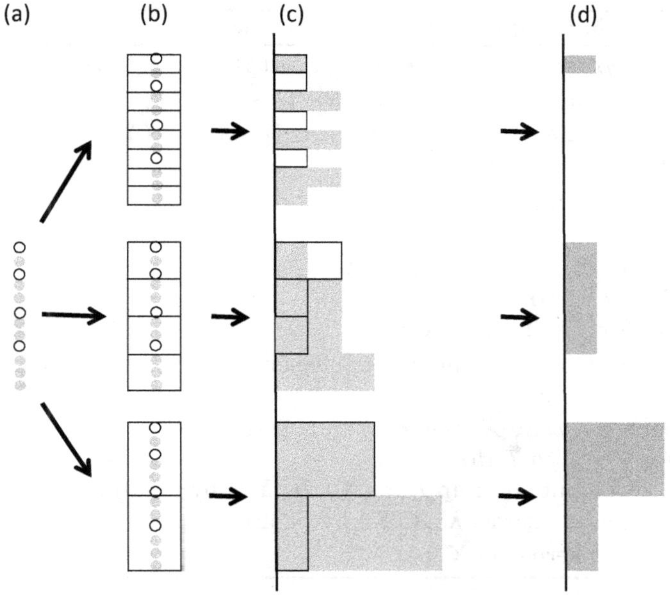

Figure 4.4: A toy example procedure of computing pyramid match kernel, where one instance is white circles and the other instance is gray circles.

Then we do not know what bin size is the best. Again the key point of pyramid match kernel is to use many different sizes of bins, instead of trying to find the optimal bin size.

Thus let L be the number of bin sizes we test. For $\boldsymbol{X}$ and $\boldsymbol{Y}$, we can consider a variety of histogram pairs with L different bin sizes:
$(H_0(\boldsymbol{X}), H_0(\boldsymbol{Y})), \ldots, (H_{L-1}(\boldsymbol{X}), H_{L-1}(\boldsymbol{Y}))$.
We then use weight w_i $(i = 0, \ldots, L-1)$ for L histograms. We define the kernel function as follows:

$$\mathcal{K}(\boldsymbol{X}, \boldsymbol{Y}) = w_{L-1} I(H_{L-1}(\boldsymbol{X}), H_{L-1}(\boldsymbol{Y})) + \sum_{i=1}^{L-1} (w_i - w_{i-1}) I(H_i(\boldsymbol{X}), H_i(\boldsymbol{Y})). \tag{4.41}$$

This kernel sums up L overlaps with weights. In practice, this can be computed more easily as follows:

$$\mathcal{K}(\boldsymbol{X}, \boldsymbol{Y}) = \sum_{i=1}^{L} w_i (I(H_i(\boldsymbol{X}), H_i(\boldsymbol{Y})) - I(H_{i-1}(\boldsymbol{X}), H_{i-1}(\boldsymbol{Y}))). \tag{4.42}$$

That is, after computing the overlaps of two histograms for L different bin sizes, we can take the weighted sum over the difference between two histograms.

Algorithm 4.5: Computing pyramid match kernel.

Function Pyramid_match_kernel_for_sets($\mathcal{D}$, Min_Sup)

  Data: training Data: $\mathcal{D}$, Bins: $\boldsymbol{B}_i^{(j)} (i = 1, \ldots, L, j = 1, \ldots, R_L)$, weights over different bins: $w_i (i = 1, \ldots, L)$

  Result: kernel over $\mathcal{D}$

  foreach *Set* $\boldsymbol{S} \in \mathcal{D}$ **do**

    Sort values in $\boldsymbol{S}$.;

  foreach *Set* $\boldsymbol{S} \in \mathcal{D}$ **do**

    for $i \leftarrow 1$ **to** L **do**

      for $j \leftarrow 1$ **to** R_L **do**

        Count the number of elements, $H_i^{(j)}$, in j-th bin of i-th bin size.;

  foreach *Set pair,* $\boldsymbol{X}$ *and* $\boldsymbol{Y}$*, both* $\in \mathcal{D}$ **do**

    for $i \leftarrow 1$ **to** L **do**

      Compute overlap $I(H_i(\boldsymbol{X}), H_i(\boldsymbol{Y}))$ by (4.40).;

      Update kernel $\mathcal{K}(\boldsymbol{X}, \boldsymbol{Y})$ by (4.42).;

    Output kernel $\mathcal{K}(\boldsymbol{X}, \boldsymbol{y})$.;

We explain this procedure by using toy examples. Again for simplicity we use $d = 1$, meaning that each element is a numerical scalar. Fig. 4.4 (a)–(d) correspond to the procedure of obtaining the pyramid match kernel:

(a): Two sets of real values are sorted, where one instance is shown by white circles and the other is by light gray circles.

(b): We can consider three bin sets with different bin sizes: From the bottom, bin sizes are 2, 4 and 8.

(c): We count the number of elements in each bin.

(d): We take the smaller number of elements in each bin between two instances, and sum this number up over all bins.

Fig. 4.4 shows the real results of performing the above procedure. The final result, i.e. (d), shows that when we use 8 bins, most bins show the minimum number is zero, and the sum of the minimum numbers is only 1, implying that there are no overlaps between two instances. Thus this bin-size setting might not be good. For 4 bins, the sum of the minimum numbers is three. On the other hand, using 2 bins, the minimum number was not zero, and the sum of the minimum numbers is 4.

From this result, we can see that bin sizes are important, implying that using a variety of bin sizes in pyramid match kernel will work well.

As the final step of pyramid match kernel is, as shown in (4.42), to compute the difference in the sum of minimum numbers between different bin sizes and sum up them with weights over all bins. That is, from the top to bottom of Fig. 4.4,

$1 = (1\text{-}0)$
$2 = (3\text{-}1)$
$1 = (4\text{-}3)$
are the differences and they are weighed by $\boldsymbol{w}$.

Algorithm 4.5 shows a pseudocode, which summarizes the algorithm of computing pyramid match kernel. Finally we prove the following proposition.

Proposition 4.2. *Pyramid match kernel satisfies the property of kernel functions.*

Proof. We need to prove that pyramid match kernel has three properties: symmetry, linearity and positive semidefiniteness (Section A.5). Pyramid match kernel is clearly symmetric, and so we focus on to prove that pyramid match kernel is positive semidefinite.

First since we focus on $d = 1$, instead of using matrices, $\boldsymbol{X}$ and $\boldsymbol{Y}$, we use vectors, such as $\boldsymbol{x}$ and $\boldsymbol{y}$ for instances. We then write the number of elements in each bin for instance $\boldsymbol{x}_i$, as a vector as follows:

$$H(\boldsymbol{x}_i) = (H^{(1)}(\boldsymbol{x}_i), \ldots, H^{(R)}(\boldsymbol{x}_i)). \tag{4.43}$$

We again show the procedure of computing $\boldsymbol{I}$, defined in (4.40), which is the sum of the overlap for each bin, over all bins. The overlap for each bin was computed by the smaller (minimum) number of elements in the bin between two instances.

Then when the overlap between $\boldsymbol{x}_i$ and $\boldsymbol{x}_j$ is computed, following (4.40), for each bin, we take the minimum one of the two instances, resulting in the following vector of overlaps:

$$(\min\{H^{(1)}(\boldsymbol{x}_i), H^{(1)}(\boldsymbol{x}_j)\}, \ldots, \min\{H^{(R)}(\boldsymbol{x}_i), H^{(R)}(\boldsymbol{x}_j)\})^\top. \tag{4.44}$$

Then all elements are summed up as the (i,j) element of matrix $\boldsymbol{I}$, as follows:

$$\boldsymbol{I}_{ij} = \sum_{r=1}^{R} \min\{H^{(r)}(\boldsymbol{x}_i), H^{(r)}(\boldsymbol{x}_j)\}. \tag{4.45}$$

This is consistent with (4.40).

Thus when we compute the overlap of one pair, say $\boldsymbol{x}_i$ and $\boldsymbol{x}_j$, as shown in (4.44), we compute the minimum number at each bin between two instances $\boldsymbol{x}_i$ and $\boldsymbol{x}_j$, i.e. a pair of instances.

Now let us replace this minimum number of pair $\boldsymbol{x}_i$ and $\boldsymbol{x}_j$ with the minimum number of all instances $\boldsymbol{x}_i (i = 1, \ldots, N)$.

Then, let $\bar{x}_r$ be the minimum number in bin r of all instances, as follows:

$$\bar{H}^{(r)} := \min_i \{H^{(r)}(\boldsymbol{x}_i)\}. \tag{4.46}$$

Thus when we use the minimum of all instances, the vector of overlaps, given by (4.44), can be changed as follows:

$$\bar{H} = (\bar{H}^{(1)}, \ldots, \bar{H}^{(R)})^\top. \tag{4.47}$$

The point is that this has nothing do with indices, such as i and j.

Then, as shown in (4.45), we can sum up the elements of the vector in (4.47), and the resultant value as the (i, j) element of matrix $\boldsymbol{J}$, as follows:

$$\boldsymbol{J}_{ij} = \sum_{r=1}^{R} \bar{H}^{(r)}. \tag{4.48}$$

The point is again the right-hand side has nothing to do with i and j, meaning that all elements of $\boldsymbol{J}$ has the same value given by (4.51).

Then first we can see, for arbitrary vector $\boldsymbol{a}$,

$$\boldsymbol{a}^\top \boldsymbol{I} \boldsymbol{a} \geq \boldsymbol{a}^\top \boldsymbol{J} \boldsymbol{a}. \tag{4.49}$$

This is obvious, because of (4.46): $\boldsymbol{I}_{ij} \geq \boldsymbol{J}_{ij}$ for all i and j.

Secondly, from (4.51) and (4.47), if we define the following:

$$\bar{\boldsymbol{v}} = (\sqrt{\bar{H}^{(1)}}, \ldots, \sqrt{\bar{H}^{(R)}})^\top. \tag{4.50}$$

Then the elements of $\boldsymbol{J}$ can be written as follows

$$\boldsymbol{J}_{ij} = \sum_{r=1}^{R} \bar{H}^{(r)} = \sum_{k=r}^{R} \sqrt{\bar{H}^{(r)}} \sqrt{\bar{H}^{(r)}} = \bar{\boldsymbol{v}}^\top \bar{\boldsymbol{v}}. \tag{4.51}$$

Again note that this indicates all elements of $\boldsymbol{J}$ are the same value.

Thus $\boldsymbol{J}$ itself can be written as follows:

$$\boldsymbol{J} = \begin{pmatrix} \bar{\boldsymbol{v}}^\top \\ \vdots \\ \bar{\boldsymbol{v}}^\top \end{pmatrix} \begin{pmatrix} \bar{\boldsymbol{v}} \cdots \bar{\boldsymbol{v}} \end{pmatrix} = \boldsymbol{V}^\top \boldsymbol{V}, \tag{4.52}$$

where

$$\boldsymbol{V} = \begin{pmatrix} \bar{\boldsymbol{v}} \cdots \bar{\boldsymbol{v}}. \end{pmatrix} \tag{4.53}$$

Thus using (4.52), again for arbitrary vector $\boldsymbol{a}$,

$$\boldsymbol{a}^\top \boldsymbol{J} \boldsymbol{a} = \boldsymbol{a}^\top \boldsymbol{V}^\top \boldsymbol{V} \boldsymbol{a} = (\boldsymbol{V}\boldsymbol{a})^\top (\boldsymbol{V}\boldsymbol{a}) = (\boldsymbol{V}\boldsymbol{a})^2 \geq 0. \tag{4.54}$$

Finally from (4.49) and (4.54), we can see that $\boldsymbol{I}$ is positive semidefinite. Then, as shown in (4.41), if we set $w_i - w_{i-1}$ always positive, the weighted sum of $\boldsymbol{I}$ also keeps positive semidefinite. Then we can say that pyramid match kernel is positive semidefinite and so satisfies the conditions of kernel functions. □

(a)

	A	B	C	D	E	F	G
1	3	-1	-2.5	1.3	2.4	1.6	-4
2	-1.2	1.8	2	-0.4	-2	1.2	3.9
3	2	1.5	-3	1.4	2	-0.5	1.8
4	-4	-2.1	2	-1.3	-2	3	1.5
5	2	2.1	2.8	2.9	1.5	-2.5	-1.2

⇒

(b)

1	A	D	E	F	
2	B	C	F	G	
3	A	B	D	E	G
4	C	F	G		
5	A	B	C	D	E

⇒ (c) FIM finds A D E ⇒

(d)

	B	C	A	D	E	F	G
2	B	C				F	G
1			A	D	E	F	
3	B		A	D	E		G
5	B	C	A	D	E		
4		C				F	G

⇒

(e)

	B	C	A	D	E	F	G
2	1.8	2	-1.2	-0.4	-2	1.2	3.9
1	-1	-2.5	3	1.6	2.4	1.6	-4
3	1.5	-3	2	1.4	2	-0.5	1.8
5	2.1	2.8	2	2.9	1.5	-2.5	-1.2
4	-2.1	2	-4	-1.3	-2	3	1.5

Figure 4.5: Procedure of applying frequent itemset mining (FIM) to gene expression matrix.

4.4 Applications to Life Sciences

4.4.1 Biclustering by Frequent Pattern Mining

For gene expression data measured for patients, i.e. (patients × genes)-matrix, we focus on genes with high expression values by setting some cut-off value against expression values. Then the data turns into sets: each instance has a unfixed number of gene expression values (or just gene names (items)), which is a set of continuous/discrete elements. This means that each patient is an itemset, and by applying frequent itemset mining to a set of patients, we can enumerate sets of genes with high expression values. From the application side, by this procedure, we can find a set of genes peculiar to the disorder shared by the patients of the data.

Fig. 4.5 shows a sample process of applying frequent itemset mining to (patients × genes)-matrix to find a bicluster with genes and also patients, using a toy example schematically. We will explain this process by using this figure.

(a) Input data is a matrix of five patients (1 to 5) and seven genes (A to G).

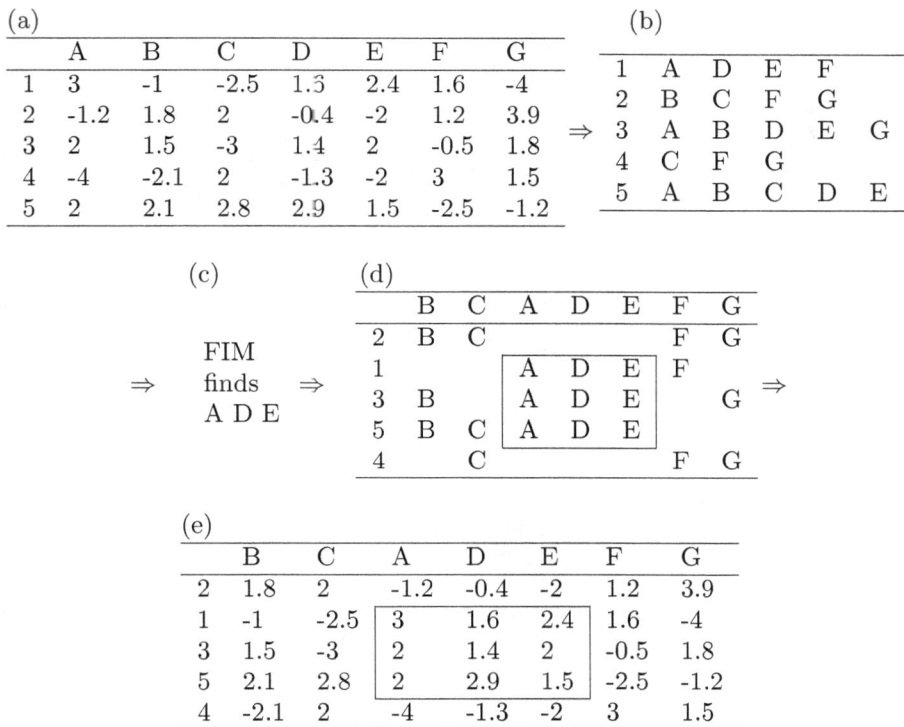

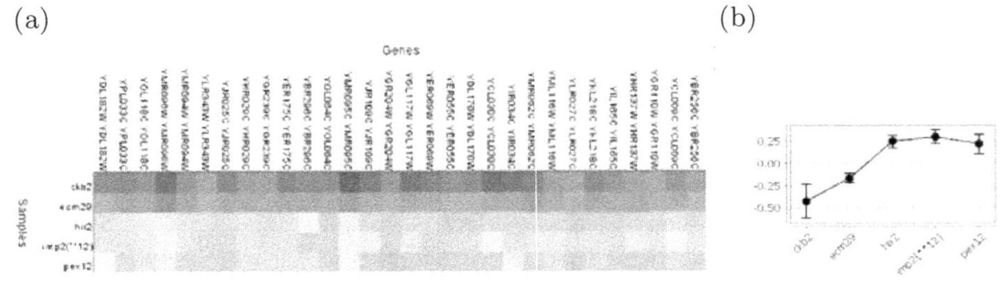

Figure 4.6: A sample bicluster obtained by frequent itemset mining over gene expression: (a) heat map and (b) line chart.

Elements are expression values.

(b) Cut-off value is set at zero, meaning that we focus on genes with positive expression values only. Then the matrix is transformed into sets of items, i.e. itemsets, where gene names are items, and the number of items for one instance is flexible.

(c) Applying frequent itemset mining (FIM) to (b), i.e. the sets of items, with the minimum support (min_sup) of three, finds a frequent itemset, ADE.

(d) Embed found frequent itemset to sets of items, i.e. (b) changing the order of both genes and patients, so that the found frequent itemsets can be seen as a cluster. More concretely, patients with the frequent itemsets are placed at neighboring each other, and also genes in this frequent itemset are also next to each other.

(e) Retrieve numerical values in (d) makes the original matrix (a) with different order in both patients and genes, showing a bicluster, which contains genes, A, D and E and also patients with higher expression values of all these three genes than the given cut-off value.

Thus this procedure shows that first by changing the original continuous values into discrete values, which are sets of itemsets. We can apply frequent item set mining to the processed data, resulting in finding biclusters [87]. In real applications, instead of taking larger values by a simple cut-off, we can use a more reasonable criterion, depending on applications like making genes with similar values into the same item. Fig. 4.6 shows a sample result obtained by applying the above frequent itemset mining to some gene expression dataset to enumerate biclusters of genes and experimental conditions. The bicluster obtained has around 20 genes under the five experimental conditions. In this figure, (a) is a heatmap showing expression values, and (b) shows the distribution of expression values under each condition. In this case, under each condition, genes with similar expression values are collected together in the cluster, while the similar value for each experimental condition can be different over five experimental conditions.

Chapter 5

Learning Sequences

If there is an order in elements of a set, this set becomes a *sequence*. In particular, if elements are discrete, each unique element is called a *symbol* or *letter*, a set of all unique letters is called *alphabet*, and a sequence of letters is called a *string*.

Again the difference between sets (itemsets) and sequences is that sequences have orders in itemsets. This means sequences are order-constrained sets, and a special case of sets, indicating that methods for sets can be applied to sequences if the order in elements in sequences is removed. However, we use data as sequences, meaning that the order is important or the key to the data.

Below we show examples of sequences in real-world.

Market basket: If items in a market basket have time stamp on when each item was put in the basket and the items are ordered according to the time stamps, the set of items becomes a sequence.

Accessed web sites: If we can add a time stamp on when a user accesses to each web site, we can order the web sites a user accessed, according to the time stamps. This then results in a sequence.

Words in natural language processing: In a sentence, a paragraph or any unit of documents, we keep the order of words in their appearances, which makes natural language sentences.

Gene sequences: Gene sequences with four letters (types of nucleic acids) are strings.

General time-series or temporary data can be also sequences by removing time stamps attached to the time-series data and keeping their order. In this chapter, we explain frequent pattern (subsequence) mining, probabilistic models and kernel learning for sequences,, as follows:

5.1 Frequent subsequence mining

5.2 Probabilistic model

5.3 Kernel learning

Notation in This Chapter

Suppose that a *substring* of string $\boldsymbol{x}$ is $\boldsymbol{z}$, we can write that as follows:

$$\boldsymbol{x} = \boldsymbol{u}\boldsymbol{z}\boldsymbol{v}. \tag{5.1}$$

For example, $\boldsymbol{x} = abcde$, $\boldsymbol{z} = bcd$, $\boldsymbol{u} = a$ and $\boldsymbol{v} = e$.

In (5.1), $\boldsymbol{u}$ is called a *prefix* and $\boldsymbol{v}$ is called a *suffix*.

A string with k letters is called k-*gram* in natural language processing, and called k-*mer* in bioinformatics. Part of a string is a substring, and similarly part of a sequence is called a *subsequence*. We write the substring from the i-th letter to the j-th letter of string $\boldsymbol{x}$, $\boldsymbol{x}[i : j]$.

5.1 Frequent Subsequence Mining

5.1.1 Preparation

One instance is one sequence. All terms and notation we defined in Section 4.1.1 are used in this section of frequent subsequence mining, particularly support, minimum support (min_sup) and frequent pattern. In this section, a frequent pattern is also called a *frequent subsequence* or a *frequent substring*, depending on the input. Also a sequence with the length of K is called K-sequence, similarly a subsequence with the length of K is called K-subsequence.

In frequent subsequence mining, non-decisive sequences are also allowed. In other words, multiple values are allowed at a position of a sequence. For example, $(ab)c$ means a string of two letters, where the first position can be a or b. For simplicity, however, we do not consider this situation explicitly.

Here is the problem setting:

Definition 5.1 (Frequent Subsequence Mining). *Given multiple sequences as input, frequent subsequence mining enumerates all possible frequent subsequences with supports larger than or equal to minimum support.*

Again a key technique to be used is the downward closure property. We will write this once again for frequent subsequence mining, as follows:

Proposition 5.1 (Downward Closure Property). *If the support of a subsequence is smaller than the minimum support, any sequence including this subsequence cannot be frequent.*

5.1.2 Two Types of Approaches

For frequent itemset mining, there are two approaches, candidate generation and pattern growth. Here we have two approaches also: candidate generation and pattern growth, each corresponding to the one with the same name for itemset mining. Here are the most representative method for each of the two approaches:

Candidate generation: A typical approach: generalized sequential patterns (GSP) [84] .

Pattern growth A typical approach: PrefixSpan [68].

Below we will explain these two approaches, i.e. GSP and PrefixSpan.

5.1.3 Generalized Sequential Patterns

Algorithm

Generalized sequential pattern (GSP), as well as Apriori, repeats the following two steps:

Expansion: adds one item (one letter) to an frequent K-subsequence to generate $K+1$-subsequences.

Examination: checks if each of the generated $K+1$-subsequences is frequent, and the sequence is left if frequent.

Notes of Generalized Sequential Pattern

In the second step, we need to check if the generated sequences are frequent or not. Simply speaking, in order to enumerate even a short frequent K-sequence, we need to scan the database K times to reach generating the frequent sequence with this length. This was already a problem in frequent itemset mining and is still a problem here. However a good news on handling sequences is that a sequence has an order in its elements. Thus expanding a frequent subsequence to a longer sequence has already a constraint from the data. That is, we should not generate candidates just by adding one arbitrary item. Instead the item to be added can be preselected from the data. This idea can be used for the pattern growth approach.

5.1.4 PrefixSpan

Background

We introduce a typical pattern growth-based method, called PrefixSpan. PrefixSpan focuses on prefix, and divides the search space by using the types and lengths of prefix to enumerate all frequent subsequences. This framework is not as time-efficient enough as FP-tree, which scans the entire database only twice, but PrefixSpan reduces the search space drastically and improves the search efficiency.

We strictly define the two key concepts of this algorithm.

prefix

Definition 5.2 (Prefix). *If given two sequences, $\boldsymbol{\alpha} = a_1a_2\dots a_n$ and $\boldsymbol{\beta} = b_1b_2\dots b_m (m \leq n)$, satisfy the following two:*

- $a_i = b_i \ (i \leq m - l)$.
- $b_m \subseteq a_m$.

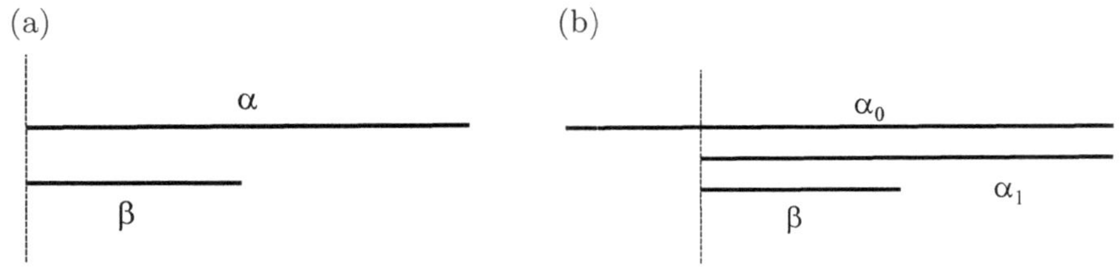

Figure 5.1: (a) prefix: $\boldsymbol{\beta}$ is a prefix of $\boldsymbol{\alpha}$, and (b) $\boldsymbol{\beta}$-projected: $\boldsymbol{\alpha}_1$ is a sequence that is $\boldsymbol{\beta}$-projected to $\boldsymbol{\alpha}_0$, when $\boldsymbol{\beta}$ is a prefix of $\boldsymbol{\alpha}_1$ and $\boldsymbol{\alpha}_1$ is a substring of $\boldsymbol{\alpha}_0$ from the beginning of $\boldsymbol{\beta}$ to the end of $\boldsymbol{\alpha}_0$.

$\boldsymbol{\beta}$ is called as prefix of $\boldsymbol{\alpha}$.

Fig. 5.1 (a) shows the relationships between $\boldsymbol{\alpha}$ and $\boldsymbol{\beta}$.

$\boldsymbol{\beta}$-projected

Definition 5.3 ($\boldsymbol{\beta}$-projected). *Subsequence $\boldsymbol{\alpha}_1$ of sequence $\boldsymbol{\alpha}_0$ is $\boldsymbol{\beta}$-projected to α_0, if this subsequence ($\boldsymbol{\alpha}_1$) satisfies the following:*

- *$\boldsymbol{\alpha}_1$ has prefix $\boldsymbol{\beta}$.*
- *In $\boldsymbol{\alpha}_0$, from the part the same as $\boldsymbol{\beta}$ to the last is the same as $\boldsymbol{\alpha}_0$.*

This concept would be understood by a figure more easily: the relationships among $\boldsymbol{\alpha}_0$, $\boldsymbol{\alpha}_1$ and $\boldsymbol{\beta}$ are shown in Fig. 5.1 (b).

Procedure by Example

We explain the procedure of PrefixSpan by using toy examples and then summarize the procedure as an algorithm later.

For simplicity, we use Table 4.1 and regard itemsets of this table as sequences, by keeping the order shown in this table. We show this table again in Table 5.1 (a) (Note that upper letters in Table 4.1 are changed to lower letter in Table 5.1 (a)). Thus we show to enumerate all frequent sequences out of the five sequences in Table 5.1 (a), with the minimum support of two.

The first step is to check the support of each of all 1-subsequences, and keep only frequent 1-subsequences. A 1-subsequence is the same as one item, and so this process is the same as enumerating frequent all itemsets with the size of one, i.e. the first step of Apriori, FP-growth, etc. Then as was done in FP-growth, these 1-subsequences are sorted by their supports (Table 5.1 (b)).

Then we first use this 1-subsequence as prefix $\boldsymbol{\beta}$ to find sequences which are $\boldsymbol{\beta}$-projected to each sequence in (a), one by one:

d**:** The first 1-subsequence is d, and Table 5.1 (c) (d is omitted) shows sequences, which are d-projected to each sequence in (a).

Table 5.1: Example for prefixspan (seq. and sup. mean sequence and support).

(a)	(b)		(c)	(d)	(e)	(f)
	seq.	sup.	d-projected	b-projected	bd-projected	bc-projected
$abcde$	d	5	e	cde	–	de
$abde$	b	4	e	de	e	–
$acde$	e	4	e	–	–	–
bcd	a	3	–	cd	–	d
bde	c	3	e	de	e	–

From (c), we can easily see that we can have only e with the support of four, meaning that we have frequent subsequence de. Also we are unable to explore another frequent subsequence more in this direction of d.

b: Table 5.1 (d) shows subsequences which are b-projected to each sequence in (a).

Then here we repeat the procedure we have conducted to reach (d) again. That is, out of four sequences in (d), we check the support of each of all 1-subsequences and keep frequent 1-subsequences. The result is d (support:4), e (support:3) and c (support:2).

We add each letter to b and in the order of the support, if the new string is frequent, we then regard the new string as $\boldsymbol{\beta}$ and find sequences which are $\boldsymbol{\beta}$-projected to each sequence in (a).

bd The first letter is d, and bd is frequent and (e) shows sequences that are bd-projected in each sequence of (a), showing that bde is frequent (the support is equal to min_sup).

be The string, be, is not frequent.

bc The bc is also frequent and (f) shows subsequences that are bc-projected to each sequence of (a), showing bcd is frequent.

The above procedure is conducted similarly for e, a and c in (b), since they are all frequent. The point is this manner of searching frequent subsequences is the depth-first-search.

Algorithm: Key Recursive Subroutine

Thus as we have seen the application of PrefixSpan to toy examples, the procedure of PrefixSpan has a particular pattern being repeated. We write that pattern here by sentences more than mathematical notations. The subroutine named Recursive Subroutine is recursively used, suppose that we have the following two already: 1) prefix $\boldsymbol{\beta}$ and 2) a set of subsequences that are $\boldsymbol{\beta}$-projected to all instances.

Algorithm 5.1: Recursive part of PrefixSpan: depth-first search for exhaustively enumerating frequent sequence patterns.

1 **Function** `PrefixSpan_recursive_part(`$\boldsymbol{\alpha}$, $\boldsymbol{S}/\boldsymbol{\alpha}$, Min_Sup`)`
  Data: subsequence: $\boldsymbol{\alpha}$, sequence set: $\boldsymbol{S}/\boldsymbol{\alpha}$, minimum support: Min_Sup
  Result: frequent subsequences
2   Generate set $\boldsymbol{A}$ of letters (sequences with the length of 1) which can follow $\boldsymbol{\alpha}$, by scanning over $\boldsymbol{S}/\boldsymbol{\alpha}$.;
3   **if** $\boldsymbol{A} == \varnothing$ **then**
4   stop ;
5   **else**
6   **foreach** $a \in \boldsymbol{A}$ **do**
7       $\boldsymbol{\alpha}_{new} \leftarrow \boldsymbol{\alpha}a$.;
8   **if** *Support of* $\boldsymbol{\alpha}_{new} \geq$ Min_Sup **then**
9   Output $\boldsymbol{\alpha}_{new}$ as frequent subsequences.;
10   `PrefixSpan_recursive_part(`$\boldsymbol{\alpha}_{new}$, $\boldsymbol{S}/\boldsymbol{\alpha}_{new}$, Min_Sup`)`

Recursive Subroutine:

1. **Generate 1-subsequences:** Given a set of subsequences, we check the support of all 1-subsequences and keep only frequent 1-subsequences, sorting them by their support.

2. **Check generated 1-subsequences:** Then we give each 1-subsequence, one by one, to 3, and if done, return to the place, which called this subroutine.

3. **Generate new $\boldsymbol{\beta}$:** We add one 1-subsequence to $\boldsymbol{\beta}$ to generate new $\boldsymbol{\beta}$ and check if $\boldsymbol{\beta}$ is frequent or not. If frequent keeps that and go to 4; otherwise go back to 2.

4. **Generate subsequences $\boldsymbol{\beta}$-projected:**
   We obtain sequences that are $\boldsymbol{\beta}$-projected to each of all given instances. Go back to 1.

Algorithm: Pseudocode

We can use the subroutine to generate the entire algorithm, which is based on depth-first search. **Algorithm 5.2** and **Algorithm 5.1** show pseudocodes of the algorithm of PrefixSpan. **Algorithm 5.2** is the main part, which starts by calling the subroutine **Algorithm 5.1**. **Algorithm 5.1** runs recursively to enumerate frequent subsequences.

Now we show the correspondence between the above subroutine Recursive Subroutine and the pseudocode of **Algorithm 5.1** (Note that lines 3–5 of the pseudocode are not considered in Recursive Subroutine):

1. **Generate 1-subsequences:** This corresponds to line 2. In the pseudocode, the set of all 1-subsequences is $\boldsymbol{A}$.

Algorithm 5.2: Main part of PrefixSpan: Initial mining for length-1 sequences and calling the recursive part.

Function `PrefixSpan_main_part(`$\boldsymbol{S}$, Min_Sup`)`
  Data: training sequences: $\boldsymbol{S}$, Minimum support: Min_Sup
  Result: all frequent subsequences
  $\boldsymbol{A} \leftarrow$ all unique letters appearing in $\boldsymbol{S}$.;
  foreach $a \in \boldsymbol{A}$ **do**
    if *Support of* $a \geq$ Min_Sup **then**
      Output a as a frequent subsequence.;
      `PrefixSpan_recursive_part(`a, $\boldsymbol{S}/a$, Min_Sup`)`

2. **Check generated 1-subsequences:** This just corresponds to the **foreach** loop of line 6.

3. **Generate new $\boldsymbol{\beta}$:** This corresponds to lines 7–9. In the pseudocode, each 1-subsequence is a, and the old and new $\boldsymbol{\beta}$ are shown as $\boldsymbol{\alpha}$ and $\boldsymbol{\alpha}_{new}$, respectively.

4. **Generate subsequences $\boldsymbol{\beta}$-projected:** This corresponds to 10. In fact in the pseudocode, sequences are obtained just by specifying the prefix in the original sequences.

Thus this means that each point of the above subroutine is properly implemented in the pseudocode **Algorithm 5.1**.

5.1.5 Notes on PrefixSpan

Number of scanning the database: In FP-growth, we need FP-tree only to enumerate all frequent itemsets, and to generate FP-tree, we had to scan the database only twice.

PrefixSpan scans the database more than two, and in fact much larger than two. However scanning the database by PrefixSpan is not necessarily search all sequences (or all itemsets). For example, in the above Recursive Subroutine, there are three points we might need to scan the database which are 1, 3 and 4 of the above procedure:

1: We check the support of all 1-subsequences.

3: We check the support of $\boldsymbol{\beta}$.

4: We enumerate subsequences $\boldsymbol{\beta}$-projected to all sequences.

There are two points to note:

1. The subsequences generated in the above 4 are used to compute the support of all 1-subsequences in the above 1. Thus, 1 can be computed in 4 at the same time. This means 1 and 4 needs only one database scan.

2. The computation in 3 is for the new $\boldsymbol{\beta}$, if we already know the place where the old $\boldsymbol{\beta}$ appear in given sequences, we can ignore the sequences which do not have the old $\boldsymbol{\beta}$. In other words, we can just check the place of the old $\boldsymbol{\beta}$ to count the support of the new $\boldsymbol{\beta}$. Thus this scan is computationally much lighter than the regular database scan.

In summary, PrefixSpan uses prefix efficiently to reduce the practical times we have to check the given database, by which practical computation can be more efficient than other straight-forward approaches, such as generalized sequential pattern.

5.2 Probabilistic Models for Sequences

In bag of words (BOW) of natural language processing, we ignore the order of words in sequences. However, if we keep the order of words, sentences, paragraphs or longer texts of natural language, all are sequences.

In general elements of sequences can be discrete- or continuous- values. We then call elements *outputs* hereafter. We suppose that there is some model behind sequences, and the model has several *states*, and *transition* over the states generates the outputs, turning into a sequence.

We call the property that the present state is depends on the previous K states *Kth-order Markov property*. For example, first-order Markov property shows that the current state depends on the last state. More concretely, the first-order Markov property focuses on neighboring two outputs, like that b is likely to come after a. However, in practice, the first-order Markov model is usually used, because even neighboring two outputs can be connected each other, eventually allowing to capture the effect by the old state on the current state. Again by contiguous dependencies between the two states, the first-order Markov property is in general thought to be good enough for real data. In this book, the Markov property or Markov model always mean "1st-order" Markov property.

Practically, neighboring outputs would be unable to be decided completely, and probabilistic models would be a reasonable choice for sequences.

We first show that we can make a wide variety of probabilistic models under the Markov property assumption. Below, we will start with the simplest model, then from a simpler to a more complex model. In particular, we will describe the last two models in detail.

1) Pair model (without prior knowledge) This is the simplest model, which has two states, one being dependent upon the other, to satisfy the Markov property. More concretely, two variables X_t and X_{t-1} have a dependency: $p(X_t|X_{t-1})$.

Model parameters can be estimated, for example, by using the frequency of two neighboring letters in given strings:

$$p(X_t = a|X_{t-1} = b) = \frac{f(ab)}{f(a)}, \tag{5.2}$$

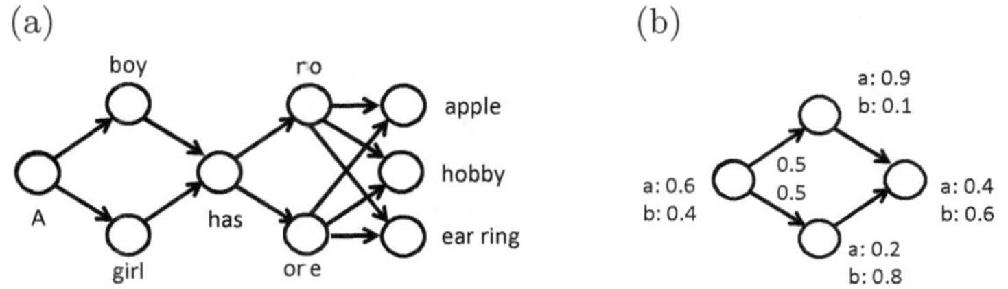

Figure 5.2: (a) A state transition diagram and (b) a hidden Markov model.

where $f(a)$ is the frequency of letter a.

We call this model a *pair model*, since its parameters can be estimated only by the frequency of neighboring letters.

An important extension is to introduce a latent variable behind the pair model, similar to a finite mixture model. We call this model a *mixture pair model*. This mixture pair model allows to do clustering, as well as a finite mixture model.

2) Markov model (with prior knowledge) The above two models do not assume any prior knowledge. However, there must be some prior knowledge in data, regarding the possibility of the sequence of letters. For example, in English, "of" and "at" cannot be neighbors. In Japanese, two neighboring sounds cannot be both consonants. Thus we can define possible state transition by a direct graph, which is called a *state transition diagram*, as an input. In the direct graph, nodes are states, and edges are possible state transition (prior knowledge), where outputs are emitted at states. Fig. 5.2 (a) shows a toy example.

Note that a series of state transition and a sequence have a one-to-one correspondence. For example, in (a), if the most top state transition is taken, the sequence emitted by states of this transition is "A boy has no apple", without any ambiguities.

When we regard this model as a probabilistic model, for example, we can assign conditional probabilities to edges: the two left-most edges in Fig. 5.2 (a) will have two probabilities $p(\text{boy}|\text{A})$ and $p(\text{girl}|\text{A})$,
where $p(\text{boy}|\text{A}) + p(\text{girl}|\text{A}) = 1$.

Hereafter if such a state transition diagram (like Fig. 5.2 (a)) is given as prior knowledge, we call this model a *Markov model*[1]

We can introduce a latent variable into the Markov model, which we call a *mixture Markov model* [17, 63]. This model allows to do clustering given sequences, within the capacity of the state transition diagram. That is, for example in (a), there are several possible paths (state transitions) from the

[1]A Markov model with no prior information can be regarded as a model with a special state transition diagram, where any state transition are possible.

Table 5.2: Different types of Markov models.

		Latent variable	
		No	Yes
Prior	No	Pair model	Mixture pair model
knowledge	Yes	Markov model	Mixture Markov model

begin node to the end node. From the training data, we can see what path is most used or what steps (even if they are distant) can be correlated each other, etc. We will describe this model more later, and also experimental results will be explained more in Section 5.4.2).

Table 5.2 summarizes the above four models (two models with the version of having latent variables).

3) Hidden Markov model (HMM) All four models in Table 5.2 emit one output at a node deterministically. In other words, there was a one-to-one correspondence between each node and its output.

We can change this idea of one output at one status/node to a probability distribution at each node/state. One example is Fig. 5.2 (b), in which any letter can be generated at each node with some probability. We then have multiple sequences at one state transition (path), meaning reversely that if a sequence is given, this sequence can be generated by all possible state transitions with the same length as the given sequence. Thus we cannot see by what state transition each sequence is generated. This would look like a sequence is hidden in the state transition diagram, and state transition itself becomes a latent variable. This model is called a *hidden Markov model (HMM)* in more detail.

Below, we will describe mixture Markov model and hidden Markov model in more detail.

5.2.1 Mixture Markov Model

Fig. 5.2 (a) shows a state transition model, and if we have probabilities over edges, the probability of generating $\boldsymbol{x}$ = A boy has no apple is given as follows:

$$p(\boldsymbol{x}) = p(\text{A})p(\text{boy}|\text{A})p(\text{has}|\text{boy})p(\text{no}|\text{has})p(\text{apple}|\text{no}). \tag{5.3}$$

We call this probability a *generation probability* of $\boldsymbol{x}$, and in general, the generation probability of sequence $\boldsymbol{x}$ is given as follows:

$$p(\boldsymbol{x}) = p(x_1)\prod_{i=2}^{|\boldsymbol{x}|} p(x_i|x_{i-1}). \tag{5.4}$$

We add a latent variable behind this model, resulting in a mixture Markov model. Thus each value of the latent variable has the same topology of the state transition diagram (for example, like that of Fig. 5.2 (a)), while their probability parameters are different, depending on values of the latent variable. This is equivalent to grouping (or clustering) given multiple sequences, where each group is called a *component*, which is similar to that of finite mixture model.

Then generation probability $p(\boldsymbol{x})$ of sequence $\boldsymbol{x}$ is, adding latent variable $\boldsymbol{z}$ to the probability computation, given as follows:

$$p(\boldsymbol{x}) = \sum_{\boldsymbol{z}} p(\boldsymbol{z})p(\boldsymbol{x}|\boldsymbol{z}) \tag{5.5}$$

$$= \sum_{\boldsymbol{z}} p(x_1|\boldsymbol{z}) \prod_{i=2}^{|\boldsymbol{x}|} p(x_i|x_{i-1}, \boldsymbol{z}). \tag{5.6}$$

Probability parameters can be estimated by, for example, the maximum likelihood criterion, for which a well-known approximation is the Expectation and Maximization (EM) algorithm.

EM algorithm for estimating parameters of this model, repeats the following two steps, until convergence of parameter values.

- E-step: Compute the posterior probability of latent variable.

$$p(\boldsymbol{z}|\boldsymbol{x}) = \frac{p(\boldsymbol{z})p(\boldsymbol{x}|\boldsymbol{z})}{\sum_{\boldsymbol{z}} p(\boldsymbol{z})p(\boldsymbol{x}|\boldsymbol{z})}. \tag{5.7}$$

- M-step: Update parameters by using posterior probabilities.

$$p(\boldsymbol{z}) \propto \sum_{\boldsymbol{x} \in \mathcal{D}} p(\boldsymbol{z}|\boldsymbol{x}), \tag{5.8}$$

$$p(x_i|\boldsymbol{z}) \propto \sum_{\boldsymbol{x} \in \mathcal{D}} N_{x_i}(\boldsymbol{x})p(\boldsymbol{z}|\boldsymbol{x}), \tag{5.9}$$

$$p(x_i|x_j, \boldsymbol{z}) \propto \sum_{\boldsymbol{x} \in \mathcal{D}} N_{x_i \leftarrow x_j}(\boldsymbol{x})p(\boldsymbol{z}|\boldsymbol{x}), \tag{5.10}$$

where $\mathcal{D}$ is given training sequences, and $N_{x_i}(\boldsymbol{x})$ is given as follows:

$$N_{x_i}(\boldsymbol{x}) \leftarrow 1 \quad \text{if} \quad x_i = x_1. \tag{5.11}$$

$$N_{x_i}(\boldsymbol{x}) \leftarrow 0 \quad \text{otherwise.} \tag{5.12}$$

And also $N_{x_i \leftarrow x_j}(\boldsymbol{x})$ is the number of times $\langle x_j, x_i \rangle$ (neighboring two letters) appears in $\boldsymbol{x}$. We skip the derivation of these steps, while using the analogy of finite mixture model (Section 3.1.4), the above EM algorithm would be reasonable.

Algorithm 5.3 shows a pseudocode of this learning algorithm.

Algorithm 5.3: EM algorithm for mixture Markov model.

Function `EM_algorithm_for_mixture_Markov_model`($\mathcal{D}$, K)
  Data: training data: $\mathcal{D}$, size of $\boldsymbol{z}$: K
  Result: trained parameters
  Initialize $p(\boldsymbol{z}|\boldsymbol{x})$ randomly.;
  repeat
    M-step: Compute (5.8)、(5.9)、(5.10).;
    E-step: Compute (5.7).;
  until *convergence*;
  Output the trained parameters.;

5.2.2 Hidden Markov Model (HMM)

Preparation

Hidden Markov model (HMM) has a long history in research of speech recognition, natural language processing and bioinformatics, since 1980s.

HMM has multiple states, and transition over them is constrained by a state transition diagram. A state transition diagram is a graph: (V, E), where V is the set of states and E is the set of directed edges.

As mentioned earlier, HMM is different from other models having the Markov property, in that each state emits an output with a probability.

Thus HMM has two types of parameters:

a_{ij}**:** state transition probability of states $i \to j$,

$b_i(O)$**:** output generation probability of output O at state i,

where $\sum_j a_{ij} = 1$ and $\sum_O b_i(O) = 1$.

Fig. 5.8 (a) shows a graphical model of HMM. In this figure, (a) has two types of edges: horizontal and vertical edges. Horizontal edges correspond to the above a, i.e. state transition probabilities: states being dependent upon the last states in a time-series manner. Vertical edges correspond to b, i.e. output generation probabilities, where outputs are emitted according to each state.

We can define a model by its probability parameters. Let Φ be a set of probability parameters to define a model. Given sequence $\boldsymbol{x} = x_1 \ldots x_{|\boldsymbol{x}|}$, let $\xi = (\xi_1, \ldots, \xi_{|\boldsymbol{x}|})$ be a state transition by $\boldsymbol{x}$, and let Ξ ($\xi \in \Xi$) be all possible state transitions. Each state transition is the latent variable of HMM which is often written as $\boldsymbol{z}$.

We will explain the following three points of HMM:

Computing likelihood of a given sequence.

Learning (Estimating) parameters of HMM, given multiple sequences.

Predicting the likelihood of unknown sequence after training parameters.

In general, the likelihood of the above first point is the sum of likelihoods over all possible state transitions, while that of the third point is the maximum likelihood over all state transitions.

HMM: Computing Likelihood

Given $\boldsymbol{x}$, over state transition ξ, we can compute likelihood $L_\Phi(\boldsymbol{x}, \xi)$ by multiplying all state transition probabilities and all output generation probabilities over ξ. For example, given $\boldsymbol{x} = abb$, over two possible state transitions (top:ξ_1 and bottom:ξ_2) in Fig. 5.2 (b), the likelihood over each of these two can be computed as follows:

1) ξ_1**:** $L_\Phi(abb, \xi_1) = 0.6 \times 0.5 \times 0.1 \times 1 \times 0.6 = 0.018.$

2) ξ_w**:** $L_\Phi(abb, \xi_2) = 0.6 \times 0.5 \times 0.8 \times 1 \times 0.6 = 0.144.$

We can further compute the likelihood of the model: $L_\Phi(\boldsymbol{x})$, by summing up the likelihood of each ξ of all state transitions Ξ:

$$L_\Phi(\boldsymbol{x}) = \sum_{\xi \in \Xi} L_\Phi(\boldsymbol{x}, \xi). \tag{5.13}$$

Then again, the likelihood of abb by the model of Fig. 5.2 (b) is

$$L_\Phi(abb) = 0.018 + 0.144 = 0.162. \tag{5.14}$$

The model of Fig. 5.2 (b) is very simple, and we have only two state transitions, but if the model is larger and more complex, the number of possible state transitions increases exponentially, according to the number of states. We then cannot compute the likelihood $L_\Phi(\boldsymbol{x}, \xi)$ simply by summing up over all ξ ($\in \Xi$). We need an efficient manner of computation, and this computation can be realized by the following *forward probability*:

Definition 5.4 (Forward Probability).
Forward probability $\alpha_i(x_t)$ $(= P(x_1, \ldots, x_t | z_t = s_i))$ *is the probability that from the first to the t-th outputs are generated and the current state is state i.*

We can compute the likelihood of sequence $\boldsymbol{x}$ by using the forward probability that from the first and the last outputs are already generated:

$$L_\Phi(\boldsymbol{x}) = \sum_i \alpha_i(x_{|\boldsymbol{x}|}), \tag{5.15}$$

where $|\boldsymbol{x}|$ is the last output of sequence $\boldsymbol{x}$.

Also a good news is the following:

Proposition 5.2. *Forward probability* $\alpha_i(x_t)$ *can be computed from forward probability* $\alpha_j(x_{t-1})(j = 1, \ldots, |V|)$.

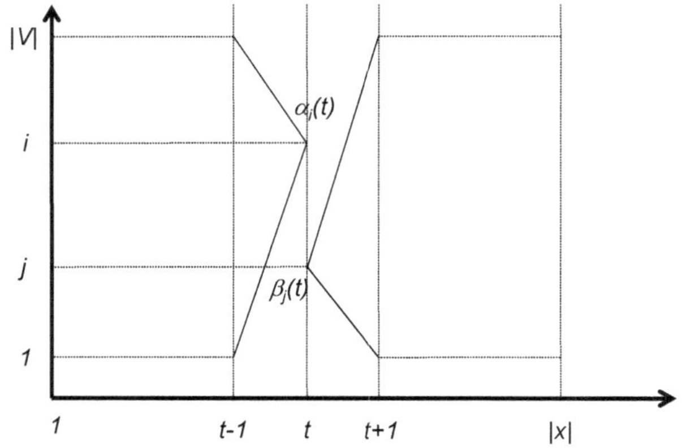

Figure 5.3: Forward and backward probabilities can be obtained recursively by dynamic programming.

Algorithm 5.4: Computing likelihood of a sequence $\boldsymbol{x}$, given a hidden Markov model.

Function `Likelihood_computation_by_hidden_Markov_model(`$\boldsymbol{x}$, D, Φ`)`

  Data: sequence: $\boldsymbol{x}$, state transition diagram: $D = (V, E)$, hidden Markov model: $\Phi = \{\boldsymbol{a}, \boldsymbol{b}\}$

  Result: $L_\Phi(\boldsymbol{x})$

  Initialize $\alpha_i(x_1)$ by checking if x_1 can be output at state i.;

  for $t \leftarrow 2$ **to** $|\boldsymbol{x}|$ **do**

    for $i \leftarrow 1$ **to** $|V|$ **do**

      $\alpha_i(x_t) \leftarrow \sum_{k|(k,i)\in E} \alpha_k(x_{t-1}) a_{ki} b_i(x_t)$.;

  $L_\Phi(\boldsymbol{x}) \leftarrow \sum_i \alpha_i(x_{|\boldsymbol{x}|})$.;

  Output the final $L_\Phi \boldsymbol{x}$.;

This can be seen from Fig. 5.3, which is called *trellis*: Forward probability $\alpha_i(x_t)$ is point (t, i) on the two-dimensional trellis, and due to the (first-order) Markov property, $\alpha_i(x_t)$ can be computed from $\alpha_j(x_{t-1})$ of all j, i.e. points $(t-1, j)$ of all j.

Thus, forward probabilities can be recursively computed as follows (increasing t):

$$\alpha_i(x_t) = \sum_{k|(k,i)\in E} \alpha_k(x_{t-1}) a_{ki} b_i(x_t). \tag{5.16}$$

In general this efficient computation is called *dynamic programming.*

Algorithm 5.4 shows a pseudocode of this computation of forward probability and the likelihood of sequence $\boldsymbol{x}$.

We now consider time complexity of computing the likelihood: computing $\alpha_i(x_t)$ of each i needs $\alpha_j(x_{t-1})$ of all $j(j = 1, \ldots, |V|)$, which needs $O(|V|^2)$, and this must be done through t, where $t = 1, \ldots, |\boldsymbol{x}|$. Thus entirely complexity is $O(|V|^2|\boldsymbol{x}|)$.

HMM: Learning Parameters

For simplicity, we focus on state transition probability a only here, and we will omit the explanation on output generation probability b, since this probability can be understood by using the analogy of a.

We use the maximum likelihood to estimate a, i.e. estimating Φ to maximize $L_\Phi(\boldsymbol{x})$, given $\boldsymbol{x}$. We can again use the EM algorithm (called *Baum–Welch* for HMM), and the key point is, for state transition probability a_{ij}, to compute the expectation value of states i to j, which we write $E_t(\#(i \to j)|\boldsymbol{x})$.

Then if the expectation value is computed, we can set up the update rule for state transition probabilities, as follows:

$$\hat{a}_{ij} \quad \leftarrow \quad \frac{\sum_t E_t(\#(i \to j)|\boldsymbol{x})}{\sum_j \sum_t E_t(\#(i \to j)|\boldsymbol{x})}, \tag{5.17}$$

where the denominator is the expectation value of using state i, and the sum the denominator over i is the likelihood that given $\boldsymbol{x}$ is generated:

$$L_\Phi(\boldsymbol{x}) \quad = \quad \sum_i \sum_j \sum_t E_t(\#(i \to j)|\boldsymbol{x}). \tag{5.18}$$

A good news is that the expectation value can be computed by using forward probability $\alpha_i(\boldsymbol{x}_t)$ and the following *backward probability* $\beta_i(x_t)$:

Definition 5.5 (Backward Probability).
Backward probability $\beta_i(x_t)$ $(= P(x_t, \ldots, x_{|\boldsymbol{x}|}|z_t = s_i))$ is the probability that from the t-th to the last outputs are generated and the current state is state i.

The backward probability keeps the same property as the forward probability, just covering the other part of the sequence of the forward probability and having the reverse direction:

Proposition 5.3. *Backward probability $\beta_i(x_t)$ can be computed from backward probability $\beta_j(x_{t+1})(j = 1, \ldots, |V|)$.*

Then similar to the forward probability, as indicated by Fig. 5.3, the backward probability can be recursively computed from the last to the first output of the sequence:

$$\beta_j(x_t) \quad = \quad \sum_{k|(j,k)\in E} \beta_k(x_{t+1}) a_{jk} b_j(x_t). \tag{5.19}$$

Then using these two probabilities, we can compute the expectation value from state i to j for the t-th to $t+1$-th positions of $\boldsymbol{x}$:

$$E_t(\#(i \to j)|\boldsymbol{x}) \quad = \quad \alpha_i(x_t) a_{ij} \beta_j(x_{t+1}). \tag{5.20}$$

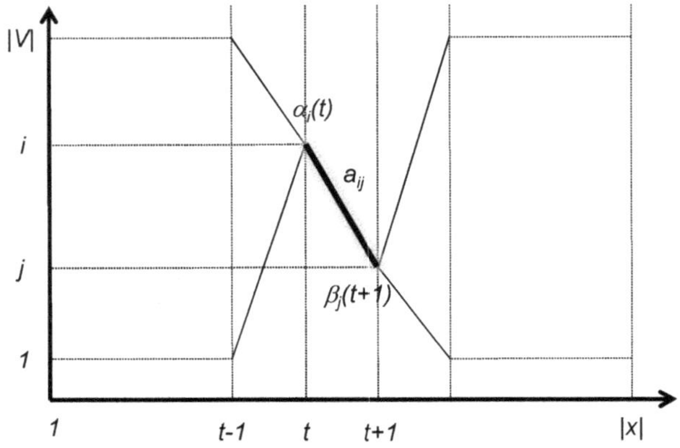

Figure 5.4: Computing expectation of transition from state i to state j.

Fig. 5.4 is a schematic picture of showing the computation of the expectation value by using α, β and a, particularly showing that α and β cover the first and last parts, respectively, of $\boldsymbol{x}$, a being connecting α with β. **Algorithm 5.5** shows a pseudocode of learning parameters of HMM.

Note that the backward probability can be computed efficiently by dynamic programming. Then, the time complexity for computing the backward probability is the same as computing the likelihood. Thus suppose that $|V|$ and $|\boldsymbol{x}|$ would be equivalent approximately, the time complexity of learning parameters is $O(|\boldsymbol{x}|^3)$, while due to the constraint of the state transition diagram (depending on applications), the time complexity would be around $O(|\boldsymbol{x}|^2)$. This complexity can be applied to all three types of computation of HMM.

Below we show the validity of the update rule by using expectation value.

Proposition 5.4. *The update rule of parameters, given by* (5.17), *guarantees to increase the likelihood of sequence* $\boldsymbol{x}$.

Proof. The objective of the EM algorithm is not to maximize the likelihood[2] but the following Q function, which has the two inputs: current parameter set Φ and future parameter set Φ'.

$$Q(\Phi; \Phi') = \sum_{\boldsymbol{z}} P_{\Phi}(\boldsymbol{x}, \boldsymbol{z}) \log P_{\Phi'}(\boldsymbol{x}, \boldsymbol{z}), \tag{5.21}$$

where P_{Φ} corresponds to the likelihood and $\boldsymbol{z}$ is the latent variable (state transition in HMM).

Then the key point of the EM algorithm is to estimate Φ' to maximize the Q function (see Section A.8.9 in more detail). First (5.21) can be written for HMM

[2]The likelihood cannot be maximized directly, because of the latent variable.

Algorithm 5.5: Baum–Welch: Algorithm for learning parameters of HMM.

Function `Baum--Welch_for_training_hidden_Markov_model`($\boldsymbol{x}$, D, Φ)
  Data: sequence: $\boldsymbol{x}$, state transition diagram: $D = (V, E)$, hidden Markov model: $\Phi = \{\boldsymbol{a}, \boldsymbol{b}\}$
  Result: trained hidden Markov model: $\Phi = \{\boldsymbol{a}, \boldsymbol{b}\}$
  repeat
    Initialize $\alpha_i(x_1)(i = 1, \ldots, |V|)$ by if x_1 is generated at state i.;
    for $t \leftarrow 2$ **to** $|\boldsymbol{x}|$ **do**
      for $i \leftarrow 1$ **to** $|V|$ **do**
        $\alpha_i(x_t) \leftarrow \sum_{j|(j,i)\in E} \alpha_j(x_{t-1}) a_{ji} b_i(x_t)$.;
    Initialize $\beta_j(x_{|\boldsymbol{x}|})$ by if $x_{|\boldsymbol{x}|}$ is generated at state j.;
    for $t \leftarrow |\boldsymbol{x}| - 1$ **to** *1* **do**
      for $j \leftarrow 1$ **to** $|V|$ **do**
        $\beta_j(x_t) \leftarrow \sum_{k|(j,k)\in E} \beta_k(x_{t+1}) a_{jk} b_j(x_t)$. ;
    for $i \leftarrow 1$ **to** $|V|$ **do**
      for $j \leftarrow 1$ **to** $|V|$ **do**
        $E_t(\#(i \to j)|\boldsymbol{x}) = \alpha_i(x_t) a_{ij} \beta_j(x_{t+1})$.;
        $\hat{a}_{ij} \leftarrow \frac{\sum_t E_t(\#(i\to j)|\boldsymbol{x})}{\sum_j \sum_t E_t(\#(i\to j)|\boldsymbol{x})}$.;
  until *convergence*;
  Output the trained parameters of HMM.;

as follows:

$$Q(\Phi; \Phi') = \sum_{\xi \in \Xi} L_\Phi(\boldsymbol{x}, \xi) \log L_{\Phi'}(\boldsymbol{x}, \xi), \tag{5.22}$$

where likelihood $L_\Phi(\boldsymbol{x}, \xi)$ of $\boldsymbol{x}$ under state transition ξ is obtained by multiplying all state transition probabilities and output generation probabilities along with ξ. For simplicity, we ignore the output generation probabilities, and consider state transition probabilities only. Then (5.22) can be written and transformed as follows:

$$\begin{aligned}
Q(\Phi; \Phi') &= \sum_{\xi \in \Xi} L_\Phi(\boldsymbol{x}, \xi) \log L_{\Phi'}(\boldsymbol{x}, \xi) && (5.23)\\
&= \sum_{\xi \in \Xi} L_\Phi(\boldsymbol{x}, \xi) \sum_{i=1}^{|\boldsymbol{x}|} \log a'_{\xi_i \xi_{i+1}} && (5.24)\\
&= \sum_{i,j} \log a'_{i,j} \sum_{\xi \in \Xi | (i \to j) \in \xi} L_\Phi(\boldsymbol{x}, \xi) && (5.25)\\
&= \sum_{i,j} \sum_{\xi \in \Xi | (i \to j) \in \xi} L_\Phi(\boldsymbol{x}, \xi) \log a'_{i,j}. && (5.26)
\end{aligned}$$

Then to maximize $Q(\Phi; \Phi')$, from (5.26), a'_{ij} can be updated to satisfy the following:

$$a'_{ij} \quad \leftarrow \quad \frac{\sum_{\xi \in \Xi | (i \to j) \in \xi} L_\Phi(\boldsymbol{x}, \xi)}{\sum_j \sum_{\xi \in \Xi | (i \to j) \in \xi} L_\Phi(\boldsymbol{x}, \xi)}. \tag{5.27}$$

The numerator of (5.27) is the expectation value of state transition from i to j, i.e. $\sum_t E_t(\#(i \to j))$. Note that (5.27) was derived from the following:

Proposition 5.5. *For non-negative real-valued vector* $\boldsymbol{x} = (x_1, \ldots, x_N)$, *if* $\sum_i p_i = 1$ *and* $p_i \leq 0$, $\sum_i^N x_i \log p_i$ *is maximized when* $p_i = \frac{x_i}{\sum_i x_i}$.

□

Similarly, estimating the output generation probability $b_i(O)$ needs the expectation value $E_t(\#(i, x_t = O)|\boldsymbol{x})$ for state i and output O at t-th position of sequence $\boldsymbol{x}$:

$$\text{if} \quad x_t = O \quad \text{then} \qquad E_t(\#(i, x_t = O)|\boldsymbol{x}) = \sum_j \alpha_i(x_t) a_{ij} \beta_j(x_{t+1}), \tag{5.28}$$

$$\text{otherwise} \qquad E_t(\#(i, x_t = O)|\boldsymbol{x}) = 0. \tag{5.29}$$

Using this expectation value, the update rule of output generation probability $b_i(O)$ can be computed as follows:

$$\hat{b}_i(O) \quad \leftarrow \quad \frac{\sum_t E_t(\#(i, x_t = O)|\boldsymbol{x})}{\sum_\Omega \sum_t E_t(\#(i, x_t = O)|\boldsymbol{x})}, \tag{5.30}$$

where Ω is all possible outputs or the entire range of outputs.

We have shown the learning from one sequence $\boldsymbol{x}$, while normally training data $\mathcal{D}$ has multiple sequences. Thus **Algorithm 5.5** needs the loop for the entire sequences in $\mathcal{D}$ within **while**, and update rules of a_{ij} and $b_i(O)$ can be given as follows:

$$\hat{a}_{ij} \quad \leftarrow \quad \frac{\sum_{\boldsymbol{x} \in \mathcal{D}} \sum_t E_t(\#(i \to j)|\boldsymbol{x})}{\sum_{\boldsymbol{x} \in \mathcal{D}} \sum_j \sum_t E_t(\#(i \to j)|\boldsymbol{x})}. \tag{5.31}$$

$$\hat{b}_i(O) \quad \leftarrow \quad \frac{\sum_{\boldsymbol{x} \in \mathcal{D}} \sum_t E_t(\#(i, x_t = O)|\boldsymbol{x})}{\sum_{\boldsymbol{x} \in \mathcal{D}} \sum_\Omega \sum_t E_t(\#(i, x_t = O)|\boldsymbol{x})}. \tag{5.32}$$

HMM: Predicting the Likelihood of Unknown Sequence

As shown in (5.13), we can compute the likelihood for any sequence, and this likelihood can be used as a prediction value for the unknown, test sequence. However, instead of this likelihood, among the likelihoods given by possible state transitions, the maximum (highest) likelihood (also the state transition which provides this likelihood is called *most likely path*) is usually used as a prediction value. In other words, given a new sequence, the likelihood of this sequence should be generated by one state transition not by all state transitions, and in this sense, considering the most likely path would be reasonable. The most likely path can be computed in a similar manner to the dynamic programming of computing likelihood. We can first define the *most likely path probability*:

Definition 5.6 (Most Likely Path Probability). *Most likely path probability* $\gamma(t,i)$ *is the probability of the most likely path, generating the first to the t-th letter (position) of the sequence and the current state is state i.*

We can compute the likelihood of the most likely path ξ by using the most likely path probability which generates the last output of $\boldsymbol{x}$:

$$L_\xi(\boldsymbol{x}) \quad = \quad \max_i \gamma(|\boldsymbol{x}|, i). \tag{5.33}$$

Also we can easily see that the most likely path can be recursively computed along the sequence, as we did for the forward probability:

$$\gamma(t,i) = \max_{k|(k,i)\in E} \gamma(t-1,k) a_{ki} b_i(x_t). \tag{5.34}$$

We can compute the likelihood by the most likely path probability, but in practice, the most likely path itself is often more important than the likelihood. For this purpose, we can record the following to retrieve the state transition of the most likely path:

Definition 5.7 ($M(t,i)$). *$M(t,i)$ is the state of the $t-1$ position, when from the first to t-th positions of the sequence are generated and the t-th position is state i*

In practice, $M(t,i)$ can be computed at the same time when the above $\gamma(t,i)$:

$$M(t,i) = \arg \max_{k|(k,i)\in E} \gamma(t-1,k) a_{ki} b_i(x_t). \tag{5.35}$$

Then once after we obtained the most likely path, we can go back from the last of the sequence to the first (we call this process *backtracking*), resulting in that we can have the state transition giving the most likely path. This algorithm is called *Viterbi*, being named after the developer. **Algorithm 5.6** shows a pseudocode of the Viterbi algorithm for HMM.

5.2.3 Various Extensions of HMM for Sequences

We introduce probabilistic models, which are extensions of HMM, having been developed in natural language processing, speech recognition and bioinformatics. The extension rises the time- and space-complexity, by which it might be not easy to apply to practical applications, particularly against big data, and so introduction on them here are rather brief. However, seeing the extension of HMM is useful to understand the property and limitation of HMM and also other probabilistic models.

Probabilistic Context-free Grammar

In linguistics, the hierarchy of formal languages are defined [22], and the complexity is increased as it goes to the higher class:

1. **formal grammar**

Algorithm 5.6: *Viterbi*: Algorithm for computing most likely state transition of a sequence $\boldsymbol{x}$, given a trained hidden Markov model.

Function `Viterbi_for_parsing_HMM(`$\boldsymbol{x}$, D, Φ`)`

    Data: sequence: $\boldsymbol{x}$, state transition diagram: $D = (V, E)$, trained HMM: $\Phi = \{\boldsymbol{a}, \boldsymbol{b}\}$

    Result: most likely state path: MostLikelyPath

    for $t \leftarrow 2$ **to** $|\boldsymbol{x}|$ **do**

        for $i \leftarrow 1$ **to** $|V|$ **do**

            $\gamma(i,t) \leftarrow \max_{k|(k,i)\in E} \gamma(k, t-1) a_{ki} b_i(x_t)$.;

            $\mathrm{M}(i,t) \leftarrow \arg\max_{k|(k,i)\in E} \gamma(k, t-1) a_{ki} b_i(x_t)$.;

    state $\leftarrow \arg\max_i \gamma_i(x_{|\boldsymbol{x}|})$.;

    MostLikelyPath $\leftarrow \langle$state$\rangle$.;

    for $t \leftarrow |\boldsymbol{x}|$ **to** 2 **do**

        state $\leftarrow$ M(state, t).;

        MostLikelyPath $\leftarrow \langle$state, MostLikelyPath$\rangle$.;

    Output MostLikelyPath.;

---

2. context free grammar

3. context sensitive grammar

HMM corresponds to the formal grammar, and from this hierarchy, one step more complex model is context-free grammar.

Markov models focused on the first-order Markov property, i.e. continuous or short-range interactions, while in sequences, *long-range interactions* can be found. For example, one natural language sentence can have one subject, one verb and one object, where these three are not necessarily continuous but sometimes distant in a sequence, while they should have interactions. This type of long-range interactions cannot be captured by HMM. Context free grammar (CFG) [44] is a model to capture a particular type of long-range interactions.

We explain this more by using toy examples, where upper letter S be a state, lower letter a,b,c be outputs. For example, a CFG can have the following set of *rewriting rules*:

$$S_0 \rightarrow \mathrm{a}S_1\mathrm{a} \tag{5.36}$$

$$S_1 \rightarrow \mathrm{b}S_2\mathrm{b} \tag{5.37}$$

$$S_2 \rightarrow \mathrm{c}, \tag{5.38}$$

Rewriting rules correspond to (one step) state transition in HMM, and a set of rewriting rules like the above corresponds to a state transition diagram. Starting with S_0, we can apply these rewriting rules, in the order of (5.36), (5.37) and (5.38), which generates a string: abcba. Fig. 5.5 (a) shows the derivation of this string from the rewriting rules. This tree is called *derivation tree* (or *parse tree*). In this string, two a and two b are generated from the same rewriting rule,

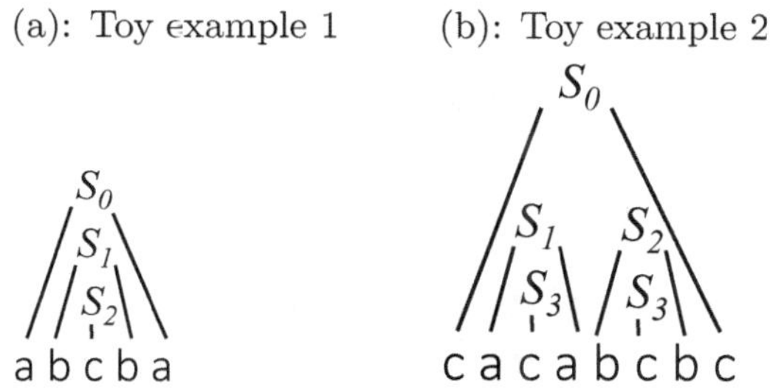

Figure 5.5: (a) and (b): derivation (parse) trees of two toy examples of CFG.

implying that each is a long-range interaction, which can be captured by CFG. A further example is, for example, rewriting rules are:

$$S \rightarrow \mathrm{c}S_0\mathrm{c} \tag{5.39}$$

$$S_0 \rightarrow S_1S_2 \tag{5.40}$$

$$S_1 \rightarrow \mathrm{a}S_3\mathrm{a} \tag{5.41}$$

$$S_2 \rightarrow \mathrm{b}S_3\mathrm{b} \tag{5.42}$$

$$S_3 \rightarrow \mathrm{c} \tag{5.43}$$

Starting with S, we can apply the above rewriting rules, in the order of (5.39), (5.40), (5.41), (5.42) and (5.43), which this time generates a string: cacabcbc. Again Fig. 5.5 (b) shows the derivation tree of this string from the rewriting rules. This indicates that various types of long-range interactions can be embedded in a sequence. Then, in short, as can be seen from Fig. 5.5, CFG can capture so-called *nested structure* in a derivation tree, which is long-range interactions and cannot be captured by HMM.

In summary, except probabilities, CFG has the following differences from HMM[3].

- In the right-hand side of rewriting rules, CFG generates multiple outputs, particularly both sides of the state.

Then the following two modifications make CFG a probabilistic model, a probabilistic CFG (PCFG):

Probabilistic rewriting rules: assign probabilities to rewriting rules, where the sum of probabilities must be 1 for the rewriting rules with the same state at the left-hand side. If the right-hand side has a state, this probability corresponds to the state transition probability of HMM.

[3]As shown in (5.40), multiple states an be generated from one state in CFG. Usually HMM cannot do this, but can sequentially emit the two states. Also the state without emitting any letter can be implemented by HMM practically. In fact Profile HMM (see Section 5.4.3) uses such states.

Sam and Tim ... vegetables and meats, respectively.

Figure 5.6: An example long range interaction to be captured by tree grammar.

Output generation probability: assign probabilities to rewriting rules which generate outputs, where again the sum of probabilities must be 1 for the rewriting rules with the same state at the left-hand side. This corresponds to output generation probabilities of HMM. This allows to generate multiple sequences from the same node sequence and also multiple state sequences for the same sequence, making estimating probability parameters a learning problem.

The above probability parameters can be estimated from data by using a straight-forward extension of Baum–Welch of HMM. One difference is, because of the next structure, the complexity of PCFG is higher than HMM. That is, PCFG needs $O(|\boldsymbol{x}|^4)$, while that was $O(|\boldsymbol{x}|^3)$ for HMM. Prediction is also the same manner as HMM, and so computing the most likely path is called *parsing* in natural language processing.

Stochastic Tree Grammar

In the hierarchy of formal grammars, context-sensitive grammar (CSG) is the next complex grammar to CFG, while there are grammars not CSG but more complex than CFG: tree grammar (TG) and graph grammar (GG). We briefly explain tree grammar[4].

We showed that CFG could capture the nested structure in two-dimensional (2D) space (Fig. 5.5), while below is an example, which is long-range interactions and cannot be captured by the nested structure:

Sam and Tim eat a lot of vegetables and meats, respectively.

In this example, there are two interactions: Sam $\leftrightarrow$ vegetables and Tim $\leftrightarrow$ meats, while as in Fig. 5.6, these relations are crossed, and cannot be the nested structure of CFG.

We will explain TG by a toy example. Fig. 5.7 shows a simple tree grammar (ranked node rewriting grammar with rank 1 (RNRG-1)) [1] example, where again upper letter S be a state and lower letter be outputs, while t is a special symbol showing the terminal and does not appear in the final string. By applying the rules in the order, Fig. 5.7 generates abbaabba, where the same letter by only one rule. In this string, abba is a nested structure, but interactions between two abba are crossed and cannot be captured by CFG but TG. TG has probability parameters in the same manner as HMM and CFG, turning into stochastic tree grammar (STG), and its learning algorithm can be implemented in the same manner as in

[4]Graph grammar is too complex for real data, and we skip the introduction of graph grammar

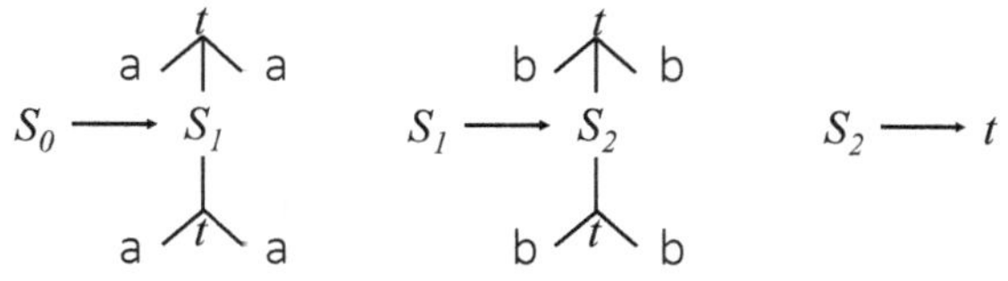

Figure 5.7: Tree grammar example.

Baum–Welch of HMM (we skip the explanation, while it will be easily understood by using the analogy of the extension from HMM to CFG).

The time complexity of tree grammar has to be higher than HMM ($O(|\boldsymbol{x}|^3)$) and CFG ($O(|\boldsymbol{x}|^4)$), reaching ($O(|\boldsymbol{x}|^6)$) even for RNRG-1, the simplest TG, because as in Fig. 5.7, there are four locations for outputs in TG, while only one location for HMM and two for CFG.

One interesting aspect of tree grammar (or in general highly complex grammars) is parallelization in computation. Again as in Fig. 5.7, a tree grammar generates outputs at four locations at the same time, which is controlled by the following quadruple loop (for computing forward probabilities etc., and this loop is just single for HMM and double for CFG.):

For $i := T$ to 1
    For $j := i$ to T
        For $k := T$ to $j + 1$
            For $l := k$ to T

You can see that each of these four loops depends on the upper loop, so looking hard to compute parallelly. However, we introduce $d = (j - i) + (l - k)$, where d shows the size of the region specified by the four locations, and the above quadruple loop can be changed into the following, keeping the inside of the loop the same and also the results totally the same:

For $d := T$ **to** 1
    For $i := 0$ to $T - d$
        For $j := i + 1$ to $i + d$
            For $l := i + d$ to T

From this change, j and l depend on d and i only, meaning that we can parallelize the computation, for each d, by dividing the computation by i and distribute each to an unique CPU. That is, suppose that we have Q CPUs, every time we reduce d, since the range i can take is $T - d + 1$ (the second **For** loop), we give $\lfloor \frac{N-d+1}{Q} \rfloor$ or $\lceil \frac{N-d+1}{Q} \rceil$ to each CPU. The computation of each CPU is, since d and i are specified, independent of each other. By keeping the computation of each CPU

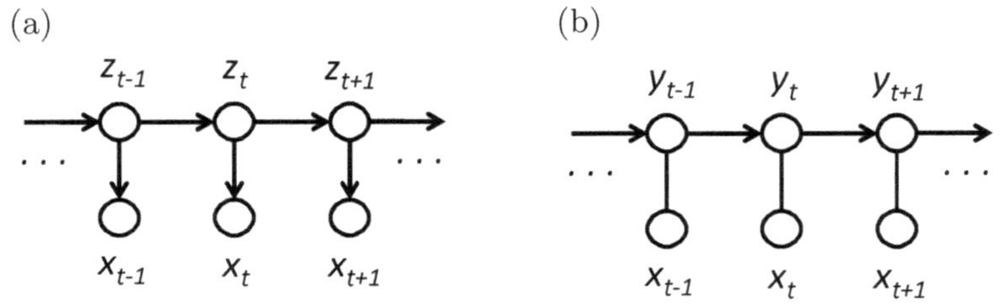

Figure 5.8: Graphical representation of (a) hidden Markov model and (b) conditional random field, where $\boldsymbol{z}$ are latent variables to specify states, $\boldsymbol{x}$ are observable variables like letters, and $\boldsymbol{y}$ are labels. In hidden Markov model, horizontal arrows can be labeled by state transition probabilities a and vertical arrows can be by output generation probabilities b.

at some shared memory space, the work at the current d can use the results of $d-1$. This parallelization has almost no overhead and works well just following the number of CPUs [1].

Prediction is also possible by an extension of the most likely path of HMM, where computing the most likely path is a rather simple modification of computing forward probabilities, and so the above parallelization is useful to reduce the high computational burden of the quadruple loop.

Conditional Random Field

Conditional random field (CRF) [56] is a probabilistic model, which is very similar to HMM, while the problem setting is different. For CRF, not only training sequences but also their true state transitions are given. That is, a real problem setting is that not only natural language sentences but also their labels, such as subjects, verbs and objects are given, and the objective is to predict the correct label annotation given an unknown sequence. Thus, state transition in CRF is not a latent variable but again the true label sequence.

Concretely, given training data $\mathcal{D}$ is a pair of sequence $\boldsymbol{x} = \langle x_1, \ldots, x_{|\boldsymbol{x}|}\rangle$ and true labels (state transition) $\boldsymbol{y} = \langle y_1, \ldots, y_{|\boldsymbol{x}|}\rangle$. Also Fig. 5.8 shows the difference between (a) HMM and (b) CRF by graphical models. HMM has state transition by latent variable $\boldsymbol{z}$, while CRF generates time-series labels $\boldsymbol{y}$. For both outputs $\boldsymbol{x}$ are observable values.

Learning of CRF maximizes the following conditional log-likelihood:

$$\sum_{(\boldsymbol{x},\boldsymbol{y})\in\mathcal{D}} \log p(\boldsymbol{y}|\boldsymbol{x}), \tag{5.44}$$

where

$$p(\boldsymbol{y}|\boldsymbol{x}) \quad = \quad \frac{p(\boldsymbol{y},\boldsymbol{x})}{\sum_{\boldsymbol{y}} p(\boldsymbol{y},\boldsymbol{x})}. \tag{5.45}$$

$p(\boldsymbol{y}, \boldsymbol{x})$ is the likelihood of $\boldsymbol{x}$ under the state transition $\boldsymbol{y}$, and so in a similar manner to $L_\Phi(\boldsymbol{x}, \xi)$ ($\xi \in \Xi$) of HMM, this can be written as follows:

$$p(\boldsymbol{y}, \boldsymbol{x}) = L_\Phi(\boldsymbol{x}, \boldsymbol{y}). \tag{5.46}$$

The denominator of (5.45) is a normalization term which can be written as $\prod_{\boldsymbol{x} \in \mathcal{D}} Z(\boldsymbol{x})$. Then the log-likelihood to be maximized can be rewritten as follows:

$$L(\Phi) = \sum_{(\boldsymbol{x},\boldsymbol{y}) \in \mathcal{D}} \log p(\boldsymbol{y}|\boldsymbol{x}) \tag{5.47}$$

$$= \sum_{(\boldsymbol{x},\boldsymbol{y}) \in \mathcal{D}} \log L_\Phi(\boldsymbol{x}, \boldsymbol{y}) - \sum_{\boldsymbol{x} \in \mathcal{D}} Z(\boldsymbol{x}). \tag{5.48}$$

Although the objective function of CRF is very similar to HMM, CRF has no latent variables, and so the algorithm of maximizing the log-likelihood cannot be like the EM algorithm but instead a more straight-forward (or regular) manner, in which a regularization term is added to (5.48) and a local maximum is obtained by gradient descent.

On the other hand, when labels are predicted, we can compute $p(\boldsymbol{y}|\boldsymbol{x})$. In (5.45), the denominator is a normalization term, and the numerator is the same as the likelihood computation of HMM. Thus, for prediction, CRF just obtains the most likely path as in HMM.

HMM and CRF are a generative model and a discriminative model, respectively, for the same problem setting, i.e. the likelihood under one state transition $\boldsymbol{y}$ (though usually it is $\boldsymbol{z}$ for HMM because of the latent variable) of sequence $\boldsymbol{x}$. Then the likelihood by HMM is

$$p(\boldsymbol{y}, \boldsymbol{x}). \tag{5.49}$$

On the other hand, CRF is

$$p(\boldsymbol{y}|\boldsymbol{x}). \tag{5.50}$$

This type of relation is already introduced for another machine learning models, e.g. naive Bayes classifier and logistic regression (see Section 3.2.5), where naive Bayes classifier and logistic regression correspond to HMM and CRF, respectively.

5.3 Kernel Learning

Again the advantage of kernel learning is that once after kernels are generated, regardless of the original inputs, we can use a variety of methods of kernel learning (see Section 3.4), such as SVM, kernel PCA, etc.

We explain kernels for sequences, particularly focusing on strings. The kernel function for strings is called a *string kernel*. A reasonable idea to generate a string kernel is if two given sequences share the same subsequences more, the

Table 5.3: (a) vectors by #appearances of 2-mer substrings and (b) kernel for "whe", "who", "woe" and "woo".

(a)

	he	ho	oe	oo	wh	wo
whe	1	0	0	0	1	0
who	0	1	0	0	1	0
woe	0	0	1	0	0	1
woo	0	0	0	1	0	1

(b)

K	whe	who	woe	woo
whe	2	1	0	0
who	1	2	0	0
woe	0	0	2	1
woo	0	0	1	2

similarity between two sequences should be higher. We introduce two kernels: *spectrum kernel* and *all subsequence kernel*, both being based on the above idea. Among the proposed string kernels so far, these two are basic kernels, which are rather easier to understand. In fact more complex kernels exist, for example, by considering the weights over subsequences shared by the two given sequences. However, these consideration depends upon practical applications. Thus we focus on the two basic string kernels, and explain a method for computing the string kernel efficiently by using suffix tree.

5.3.1 Spectrum Kernel

Given a sequence, we can generate a vector (called spectrum) by using all possible subsequences with the length of k, i.e. k-mers, which appear in the given sequence, and then compute the inner product of the two vectors for given two sequences. This is called spectrum kernel [59].

First we define the function of the number of appearances of k-mer $\boldsymbol{z}$ for given sequence $\boldsymbol{x}$ as follows:

$$\phi^k_{\boldsymbol{z}}(\boldsymbol{x}) = \#\boldsymbol{z}, \tag{5.51}$$

where $\boldsymbol{z}$ is a subsequence of $\boldsymbol{x}$ i.e. $\boldsymbol{x} = \boldsymbol{uzv}$, and $\boldsymbol{z} \in \Sigma^k$ where $\boldsymbol{u}$ and $\boldsymbol{v}$ are strings with arbitrary lengths.

Spectrum kernel can be defined, using the above function, as follows:

$$K(\boldsymbol{x}, \boldsymbol{x}') = \sum_{\boldsymbol{z} \in \Sigma^k} \phi^k_{\boldsymbol{z}}(\boldsymbol{x}) \phi^k_{\boldsymbol{z}}(\boldsymbol{x}'). \tag{5.52}$$

To understand this intuitively, we will raise an example: suppose that we have four string {whe, who, woe, woo} and $k = 2$, then Table 5.3 (a) shows the number of appearance of 2-mer: $\phi^2_{\boldsymbol{z}}(\boldsymbol{x})$. Then we can compute spectral kernel by taking the inner product of two row vectors, resulting in Table 5.3 (b).

5.3.2 All Subsequence Kernel

All subsequence kernel [60] is an extension of spectrum kernel. All subsequence kernel generates vectors by using all k-mers (not only one k but all possible k)

Table 5.4: vectors by # appearances of all possible substrings of "whe", "who", "woe" and "woo".

	e	h	o	w	he	ho	oe	oo	wh	wo	whe	who	woe	woo
whe	1	1	0	1	1	0	0	0	1	0	1	0	0	0
who	0	1	1	1	0	1	0	0	1	0	0	1	0	0
woe	1	0	1	1	0	0	1	0	0	1	0	0	1	0
woo	0	0	2	1	0	0	0	1	0	1	0	0	0	1

Table 5.5: All subsequence kernel for "whe", "who", "woe" and "woo".

K	whe	who	woe	woo
whe	6	3	2	1
who	3	6	2	3
woe	2	2	6	4
woo	1	3	4	8

and compute inner products of two of these vectors to generate a kernel.

$$K(\boldsymbol{x}, \boldsymbol{x}') = \sum_{k=1}^{\min\{|\boldsymbol{x}|,|\boldsymbol{x}'|\}} \sum_{\boldsymbol{z}\in\Sigma^k} \phi_{\boldsymbol{z}}^k(\boldsymbol{x})\phi_{\boldsymbol{z}}^k(\boldsymbol{x}'). \tag{5.53}$$

Again we raise a toy example. As well as spectrum kernel, suppose that we have four strings: whe, who, woe, woo. We compute the number of appearances for each of all possible subsequences in each of the four given sequences, as shown in Table 5.4. Then, using Table 5.4, we compute the inner product of row vectors, to generate all subsequence kernel, resulting in Table 5.5. From this example, we can see that all subsequence kernel representing by all k show the similarity in more detail than spectrum kernel with $k = 2$ only.

Note that both spectrum kernel and all subsequence kernel compute the kernel function from the inner product, and so clearly both satisfy the property of kernel.

Note also that depending on applications, we can consider many options of string kernel. We here raise two:

1. **Gaps** are allowed in subsequences (this case is also called all subsequence kernel sometimes). This case is also to take the inner product, by which this is also a kernel.

2. **Weights over subsequences** can be considered. This case is regarded as a weighted sum over kernels. Thus this also can satisfy the property of kernel.

However these consideration are rather application-specific. We then consider the problem of efficient computation of string kernel, more than variations of string kernel.

Before we explain time-efficient algorithms for string kernels, we first examine the time-complexity of the straight-forward computation of string kernel, when computing the kernel function between two sequences, $\boldsymbol{x}_i$ and $\boldsymbol{x}_j$: $\mathcal{K}(\boldsymbol{x}_i, \boldsymbol{x}_j)$. A naive method has the following three steps [59]:

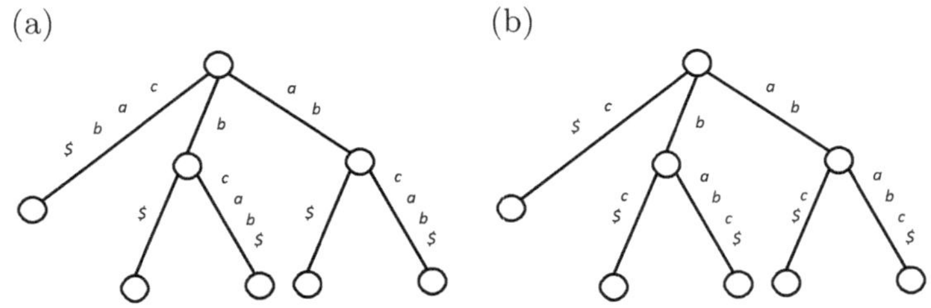

Figure 5.9: Two toy suffix tree examples: (a) *abcab* and (b) *ababc*.

1. **Extract all subsequences.** We cut out all possible k-mers out of a given sequence. This is possible just by moving from the first to the last of a given sequence. Also at the same time, when moving from the first to the last, we can cut out sequences with a different length (Totally k different lengths). This means we can extract all possible subsequences just by moving from the first to the last. This computation needs only $O(k|\boldsymbol{x}_i|)$. Kernel functions need computation over given two sequences, and so for a pair of sequences: $\boldsymbol{x}_i$ and $\boldsymbol{x}_j$: the time-complexity is $O(k|\boldsymbol{x}_i| + k|\boldsymbol{x}_j|)$.

2. **Sort all extracted subsequences.** Subsequences, extracted at the 1st step, are sorted to compare with each other, particularly to compute the inner product to generate a string kernel. From sequence $\boldsymbol{x}_i$, we cut out at most $k|\boldsymbol{x}_i|$ subsequences. The time complexity of sorting sequences cannot be linear, and for example, $O(k|\boldsymbol{x}_i|\log(k|\boldsymbol{x}_i|))$ for $k|\boldsymbol{x}_i|$ subsequences.

3. **Computing the inner product.** The vectors to be used to compute the inner product is already sorted, by which the complexity is just the length of the vectors. Since the number of subsequences is at most $k|\boldsymbol{x}_i|$ for $\boldsymbol{x}_i$. For two sequences $\boldsymbol{x}_i$ and $\boldsymbol{x}_j$, the complexity is $O(\max\{k|\boldsymbol{x}_i|, k|\boldsymbol{x}_j|\})$.

Thus anyway the step of sorting makes the entire time complexity more than linear time.

5.3.3 Fast Computation by Suffix Tree

We can compute the string kernel fast by using suffix tree, an efficient data structure for strings [94, 95].

Suffix Tree?

A suffix tree has labels over edges, and all suffixes of a given sequence can be represented by a tree. We first explain suffix trees by using toy examples. Fig. 5.9 (a) shows a suffix tree for all suffixes of sequence *abcab* (all suffixes are added by terminal symbol \$). We explain this tree. First Table 5.6 (a) shows the five suffixes of *abcab*. Each of these five suffixes corresponds to a unique path from the root

Table 5.6: All possible suffixes from (a) *abcab* and (b) *ababc*.

(a)	(b)
abcab$	*ababc*$
bcab$	*babc*$
cab$	*abc*$
ab$	*bc*$
b$	*c*$

to a leaf. For example, the suffixes from top to bottom in (a) correspond to the first, third, fifth, second and fourth path from the right in the tree (Fig. 5.9 (a)) with five leaves.

In a suffix tree, if a prefix is overlapped in multiple suffixes, the same edge is used for the prefix. Thus checking the above suffixes one by one, against the first suffix:*abcab*$, the prefix of the next suffix:*bcab*$ is not overlapped, and so this suffix starts with the root. Similarly the third suffix *cab*$ is also not overlapped with any of prefix of the previous suffixes, and this suffix also starts with the root. However, the fourth suffix:*ab*$ shares the same prefix:*ab* with the first suffix:*abcab*$, and then the edge corresponding to prefix:*ab* is shared with suffix:*abcab*$. In other words, the original edge for *abcab*$ is divided into two edges, and after *ab*, a node is added and from this node, two edges are extended to represent two different suffixes. By these operations, each suffix is a unique path from the root to a leave, and also (the number of) edges are not wasted. Similarly the last suffix:*b*$ shares prefix:*b* with the second suffix:*bcab*$, and this prefix is used as one edge and one node is inserted and then two edges are generated for two suffixes. Again these operations make suffixes have unique five paths, sharing prefix as common edges.

We raise another example to make understanding suffix tree further clear. This time, the string is *ababc*, and Fig. 5.9 (b) is its suffix tree. Also Table 5.6 (b) shows five suffixes (added by terminal symbol $) of *ababc*. As well as (a), each of the five suffixes in (b) corresponds to a unique path from the root to a leaf: From the top to bottom of Table 5.6 (b) correspond to the first, third, second, fourth and fifth path in Fig. 5.9 (b). Thus the length of string, the number of suffixes and the number of leaves are all the same, i.e. five (also five in (a)). In (b), suffix:*ababc*$ and suffix:*abc*$ share prefix:*ab*, and suffix:*babc*$ and suffix:*bc*$ share prefix:*b*. Thus each of these prefixes is shared as an edge by the two suffixes.

Suffix Tree can be Generated by Linear Time- and Space-Complexity

The biggest advantage of suffix trees is that a suffix tree can be generated with only linear time- and space-complexity ($O(|\boldsymbol{x}|)$ for sequence $\boldsymbol{x}$). We skip explaining the linear-time algorithm to generate a suffix tree, because this algorithm is complex, and note that the above procedure we described in the above is in fact an algorithm with $O(|\boldsymbol{x}|^2)$. Briefly the linear-time algorithm processes given sequence $\boldsymbol{x}$ from the first position to the last positive, generating a suffix tree.

In summary, a suffix tree can represent all suffixes of a given sequence, and holds the following important properties:

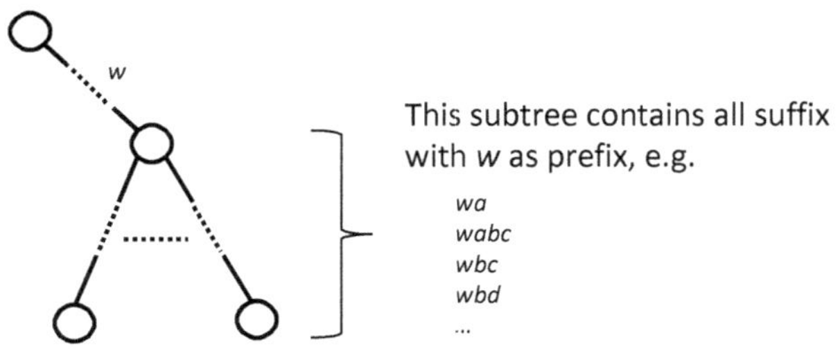
w
This subtree contains all suffix with w as prefix, e.g.
wa
wabc
wbc
wbd
...

Figure 5.10: Subtree with its root, which is connected from the suffix tree root by the edge(s) labeled by w, has all suffix with prefix w.

Linear time complexity for generating a suffix tree.

Linear space complexity for generating a suffix tree.

Each leave corresponds to a unique suffix in a suffix tree.

Compute #Appearances of Subsequence w in Sequence $\boldsymbol{x}$ by Suffix Tree

We focus on subtrees of a suffix tree. For example, in the suffix tree of Fig. 5.9 (a), we consider the subtree with the node (as the root) which is connected to the root, the edge being labeled by ab. This subtree is for suffix:$abcab\$$ and $ab\$$. That is, this subtree is for all suffixes with prefix:ab. More generally, as in Fig. 5.10, we can say the following:

Proposition 5.6. *A subtree correspond to all suffixes, which have subsequence w as their common prefix, which is attached to the edges from the root (of the suffix tree) to the root of the subtree.*

The key point of the above is "all" suffixes sharing the prefix attached to the edges between the root of the suffix tree and the root of the subtree. Also a suffix tree has all suffixes in a given sequence. Thus the above all suffixes mean "all" suffixes of a given sequence. Then another point is that the leaves of the subtree are all unique. Thus again the subtree contains all suffixes sharing the prefix, and these all suffixes uniquely appear at leaves. Thus, when we have a prefix, w, which is a label the edges coming to some node, say A, the number of times w appears among all suffixes can be counted by the number of leaves of the subtree with node A as its root. We can further summarize this as follows:

Proposition 5.7. *The number of appearances of subsequence w can be counted by the number of leaves of the subtree with the node (as its root), which is connected to the root of the suffix tree, being labeled by w.*

We can write the same more mathematically: Given sequence $\boldsymbol{x}$, we generate suffix tree:T, and let $S(t)$ be the number of leaves of the subtree with t as its root.

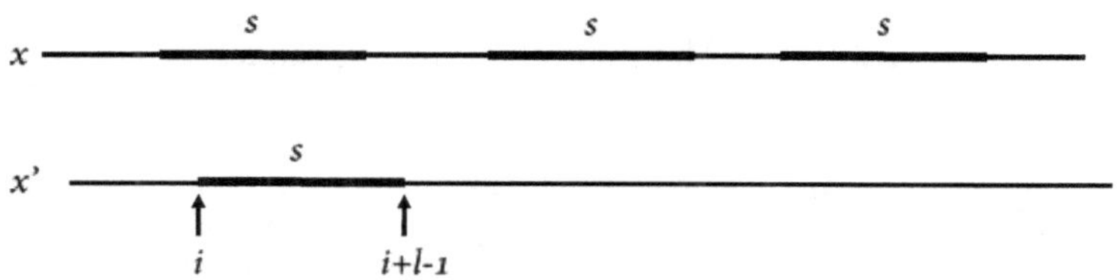

Figure 5.11: Commonly appearing substring $\boldsymbol{s}$ between $\boldsymbol{x}$ and $\boldsymbol{x}'$ appears at the i-th position of $\boldsymbol{x}'$. This substring can be found by enumerating all suffixes and then also all prefixes of each suffix.

Let lnode(w) be the node (of the leaf side) of the edge (labeled by w) from the root of T.

The number of appearances of w in sequence $\boldsymbol{x}$, count$_{\boldsymbol{x}}(w)$, can be given as follows:

$$\text{count}_{\boldsymbol{x}}(w) \quad = \quad S(\text{lnode}(w)). \tag{5.54}$$

Let us see our example, say Fig. 5.9 (a). For example, we regard ab as w, and then lnode(w) is the right-most node directly connected from the root. Then $S(\text{lnode}(w))$ is 2, and this means that ab appears in $abcab$ twice. Similarly, regarding b as w, lnode(w) is the center node directly connected from the root, and again $S(\text{lnode}(w))$ is 2, and b appears in $abcab$ twice. A further point is that, when we regard a as w, lnode(w) is the same as ab, and then a appears twice in $abcab$. Also, from Fig. 5.9 (b), we can see that ab appears twice in $ababc$, and a and b appear twice from the above computation manner.

The other important point is, if a suffix tree is once generated, we can compute the number of leaves of each subtree and save the number at the root of the subtree. We then do not have to compute this number again. Also this computation can be done from the leaves to the node, with at most the number of nodes of the suffix tree.

In summary, we can compute the number of leaves of each subtree, once after the suffix tree is generated. Also this number can be saved at the root of each subtree, and then no computation is needed any more about the number of appearances of any subsequence w.

Proposition 5.8. *The number of appearances of w in sequence $\boldsymbol{x}$ can be computed (presented) in $O(1)$ by using the suffix tree of $\boldsymbol{x}$.*

Compute the Number of Common Subsequences

We now consider to compute all subsequence kernel between sequence $\boldsymbol{x}$ and sequence $\boldsymbol{x}'$. The procedure is that we generate a suffix tree from $\boldsymbol{x}$ and count the number of appearances of all subsequences in $\boldsymbol{x}'$ by using the suffix tree of $\boldsymbol{x}$.

Then there is an important point:

Table 5.7: Maximum lengths of prefixes.

(a)

$\boldsymbol{x}'$	b	a	a	b	c
v_i	1	1	3	2	1

(b)

$\boldsymbol{x}'$	b	a	a	b	c
v_i	2	1	3	2	1

Proposition 5.9. *Any common subsequence between $\boldsymbol{x}$ and $\boldsymbol{x}'$ appears as a prefix in the suffixes of $\boldsymbol{x}'$.*

For example, as shown in Fig. 5.11, a common subsequence $\boldsymbol{s}$ with the length of l appears from the i-th position of $\boldsymbol{x}'$. Then, in the suffix starting with the i-th position to the last, its first l positions, i.e. the prefix with the length of l, corresponds to $\boldsymbol{s}$. Again in short, regardless that the common subsequence is located in any position in $\boldsymbol{x}'$, if you take all suffixes of $\boldsymbol{x}'$ and then all prefixes of each suffix, the common substring can be found.

Thus, to count the number of appearances of subsequences of $\boldsymbol{x}'$ in $\boldsymbol{x}$, we can do the following:

Examine all suffixes of $\boldsymbol{x}'$ in terms of prefixes.

Let $\boldsymbol{x}[i:j]$ be a subsequence starting with the i-th position and ending to the j-th position of sequence $\boldsymbol{x}$. Suppose that a prefix of the suffix starting with the i-th position of $\boldsymbol{x}'$ has, a common subsequence with $\boldsymbol{x}$, the maximum length of v_i (that is, $\boldsymbol{x}'[i:i+v_i]$ is not a common subsequence any more), we can (need) examine prefix:$\boldsymbol{x}'[i:i],\ldots,\boldsymbol{x}'[i:i+v_i-1]$. For example, $\boldsymbol{x}'=baabc$, for Fig. 5.9 (a) $abcab$ and (b) $ababc$, we can have v_i, i.e. the maximum length of prefix (common subsequence), in Table 5.7[5].

In fact, the number of appearances of the prefixes (in the suffix starting with the i-th location of $\boldsymbol{x}'$) in sequence $\boldsymbol{x}$ can be computed by using the suffix tree of $\boldsymbol{x}$ as follows:

$$\sum_{j=0}^{v_i-1} \mathrm{count}_{\boldsymbol{x}}(\boldsymbol{x}'[i:j]), \tag{5.55}$$

where the right-hand side can be computed by (5.54) and the suffix tree of $\boldsymbol{x}$.

Thus, the sum of the number of appearances of the common subsequences between $\boldsymbol{x}'$ and $\boldsymbol{x}$ can be computed the sum of (5.55) over all i (location of $\boldsymbol{x}'$), as follows:

$$\sum_{i}\sum_{j=0}^{v_i-1} \mathrm{count}_{\boldsymbol{x}}(\boldsymbol{x}'[i:j]). \tag{5.56}$$

[5]When we check v_i, again we can use the suffix tree of $\boldsymbol{x}$. We just show this by our toy example: for $\boldsymbol{x}'=baabc$, the first position is b, in Fig. 5.9 (a), the next is not a but only c or \$, meaning $v_1=1$. The second position is a and again in Fig. 5.9 (a), the next must be b, and so again $v_2=1$. The third position is a, which can be $abcab\$$ or $ab\$$ in Fig. 5.9 (a), so $v_3=3$.

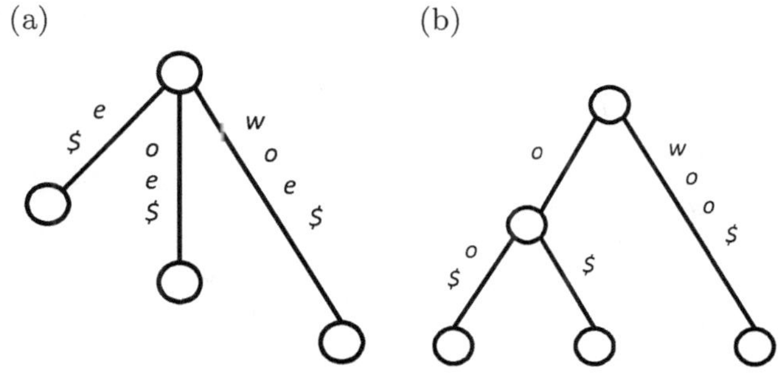

Figure 5.12: Two suffix trees: (a) *woe* and (b) *woo*.

Table 5.8: Computing kernel function between $\boldsymbol{x}$ and $\boldsymbol{x}'$:(a) $\boldsymbol{x} = woe$ and $\boldsymbol{x}' = woo$, and (b) $\boldsymbol{x} = woo$ and $\boldsymbol{x}' = woe$.

(a)

prefix	# apperances
woo	0
wo	1
w	1
oo	0
o	1
o	1
Total	4

(b)

prefix	# appearances
woe	0
wo	1
w	1
oe	0
o	2
e	0
Total	4

Applying Toy Examples

We now check if this works to compute all subsequence kernel by using our toy examples.

For example, $\boldsymbol{x} = woe$ and $\boldsymbol{x}' = woo$. First we generate the suffix tree of $\boldsymbol{x}$, and Fig. 5.12 (a) shows the suffix tree of $\boldsymbol{x}$. This suffix tree has no layers, and so the number of appearances of any subsequence found in the label of an edge is always 1.

Then from $\boldsymbol{x}' = woo$, we generate suffixes *woo*, *oo* and *o*, and then prefixes from each suffix. From the corresponding edge (in Fig. 5.12 (a)) of each prefix, we can compute the number of appearances. Table 5.8 (a) shows this result.

Then we switch between $\boldsymbol{x}$ and $\boldsymbol{x}'$, and generate the suffix tree from $\boldsymbol{x} = woo$. Fig. 5.12 (b) is this suffix tree. Then we generate suffixes from $\boldsymbol{x}' = woe$ and prefixes, and then compute the number of appearances of common subsequences, again the above suffix tree. Table 5.8 (b) shows the results. In particular, prefix:*wo* and *w* are assigned to the edge of *woo*\$ in the suffix tree of Fig. 5.12 (b), by which the number of leaves is both 1. prefix:*o* is assigned to the edge from the root to the

Algorithm 5.7: Compute string kernel by using suffix tree.

Function `Compute_string_kernel_with_suffix_tree(`$\mathcal{D}$`)`
  Data: string set: $\mathcal{D}$
  Result: string kernel: $\boldsymbol{K}$
  foreach *string* $\boldsymbol{x} \in \mathcal{D}$ **do**
    Generate suffix tree $T_{\boldsymbol{x}}$.;
    foreach *node* $v \in T_{\boldsymbol{x}}$*; from leaves to the root* **do**
      Count #leaves of the subtree (with v as the root) to record this as $\text{count}_{\boldsymbol{x}}(w)$ at v (w: string for the path from the root to v).;
  foreach *string pair, (*$\boldsymbol{x}$ *and* $\boldsymbol{x}'$*)* $\in \mathcal{D}$ **do**
    for $i \leftarrow 1$ **to** $|\boldsymbol{x}'|$ **do**
      Check $\text{count}_{\boldsymbol{x}}[i:j]$ for all possible j, to generate a table like 5.8.;
    Compute kernel $\boldsymbol{K}(\boldsymbol{x}, \boldsymbol{x}')$ from the table. ;
  Output kernel $\boldsymbol{K}$.;

internal node in the suffix tree (Fig. 5.12 (b)), resulting in the number of leaves is 2. Finally we can compute the sum of the column vector by using Table 5.8 (a) (or (b)), which we can see that is consistent with Table 5.5 of all subsequence kernel.

Algorithm and Time Complexity

Algorithm 5.7 shows a pseudocode of the above algorithm. In this pseudocode, we generate a table on the number of appearances of subsequences, like Table 5.8, and compute the kernel, but we can compute the kernel during obtaining the number of appearances of subsequences, which would be more efficient.

The suffix tree of sequence $\boldsymbol{x}$ can be generated in linear time complexity, and the suffixes of $\boldsymbol{x}'$ can be generated in also the linear time complexity. Thus the entire time complexity is only $O(|\boldsymbol{x}| + |\boldsymbol{x}'|)$.

Thus over all, we can see that we can enumerate all subsequences to be shared by two given sequences efficiently. Also all subsequent kernel treats all subsequences equally, while we can place weights over subsequences. For example, if we just want to focus on only k-mers, that is also possible and the result will be the spectrum kernel.

5.4 Applications to Life Sciences

5.4.1 Frequent Subsequence Mining for Sequence Motifs

Numerous gene and protein sequences are obtained in modern life sciences. Proteins with similar sequences and sharing the same function are called a *family*. Proteins in a family has a key *conserved sequence*, which is called a *motif*. It would be straight-forward but reasonable to apply frequent subsequence mining to protein sequences in the same protein family to find conserved motifs [71].

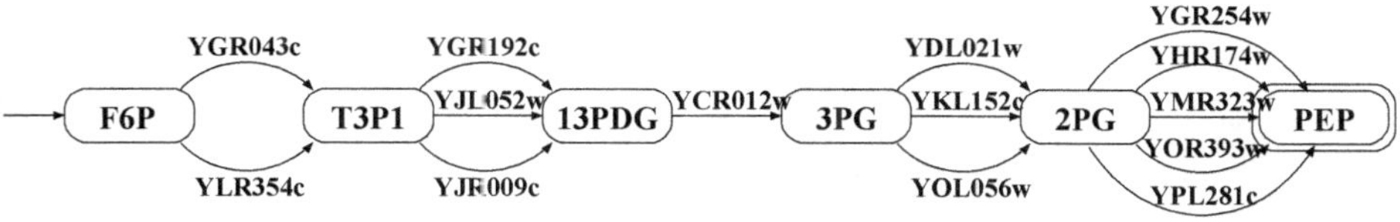

Figure 5.13: Part of glycolysis pathway: F6P → PEP.

5.4.2 Mixture Markov Model for Metabolic Network

Metabolic network is a graph with chemical compounds as unique nodes and edges (usually directed) are chemical reactions, labeled by genes working as enzymes. Fig. 5.13 shows an example network on glycolysis. For example, the first chemical reaction of this network produces T3P1 (3-Phosphate glyceraldehyde) from F6P (Fructose 6-phosphate), with two different types of enzymes, YGR043c and YLR354c, which are regarded as two different chemical reactions (edges). This network is a left-to-right type, simple network, but even so there are 90 ($= 2 \times 3 \times 1 \times 3 \times 5$) paths possible from F6P to PEP. We then regard the network of Fig. 5.13 as a state transition diagram. At the same time, the other input is a gene expression matrix of (genes × experimental conditions). We can set a cut-off value over the expression values, making them binary (+1 or -1), suppose that genes are expressed (enzymes are produced) when the expression value is +1. We can then, from each experimental condition, generate sequences, in which all genes of one of the 90 paths from F6P to PEP have +1, meaning that we think for this path all enzymes are active (genes are expressed because of +1). It would be easy to check this for 90 paths in Fig. 5.13, one by one, while in general we can do this efficiently by using dynamic programming from F6P to PEP. We omit the detail for generating these sequences, while the generated sequences are used as training for estimating a mixture Markov model we explained in Section 5.2.1.

We show experimental results on yeast gene expression [30] with the experimental conditions on Sporulation. We fixed the number of components at ten, and the results showed eight out of ten are similar, while the other two are different. These ten components were characterized by two steps: F6P → T3P1 and 3PG →2PG. Fig. 5.14 (a-c) show these results. Focusing on two enzymes, YGR043c and TDL021w of these two steps, the major pattern (a) has the anti-correlation between these two genes, indicating the correlation between YGR043c and YKL152c (or, YLR354c and YDL021w) (see the right-hand side of Fig. 5.14). In general these correlations are called *coexpression*). Whereas in minor patterns, (b) and (c), simply YGR043c and JDL021w both have larger probabilities than the counter part genes. Particularly in (c), both YGR043c and YDL021w have the probability of 1.0. In summary these two steps are mainly coexpressed but minor different patterns.

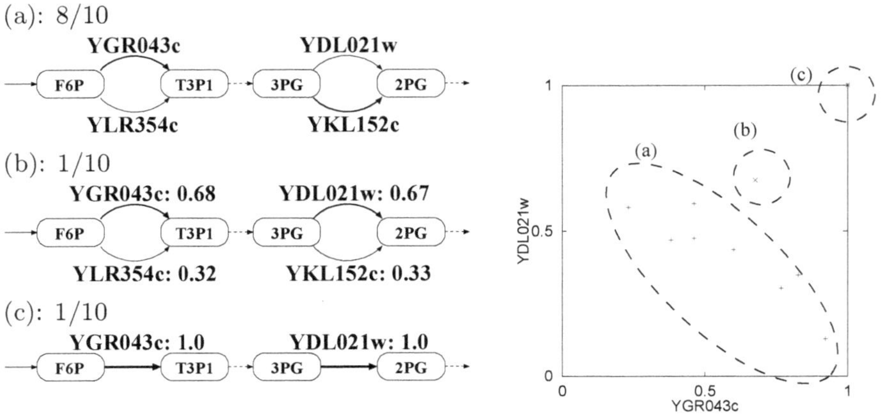

Figure 5.14: Three patterns obtained for Sporulation of [30].

5.4.3 Multiple Sequence Alignment by HMM

Gene (or amino acid) sequence alignment is the basic problem, which started in the early stage (in fact even in 1970s) of bioinformatics. There exist two problems: pairwise sequence alignment for two sequences and multiple sequence alignment for more than two sequences [29].

Pairwise Sequence Alignment: Introduction

Pairwise sequence alignment (PSA) is the problem that has two sequences as input and insert the gaps (–) or null characters so that the same characters (letters) correspond to each other. We show a toy example for intuitive understanding. Two sequences, ATCAG and ACAGT, are the input and can be aligned as follows.

$$\begin{array}{ccccc} \mathrm{A} & \mathrm{T} & \mathrm{C} & \mathrm{A} & \mathrm{G} \\ \mathrm{A} & \mathrm{C} & \mathrm{A} & \mathrm{G} & \mathrm{T} \end{array} \rightarrow \begin{array}{cccccc} \mathrm{A} & \mathrm{T} & \mathrm{C} & \mathrm{A} & \mathrm{G} & - \\ \mathrm{A} & - & \mathrm{C} & \mathrm{A} & \mathrm{G} & \mathrm{T} \end{array}$$

In sequence alignment, particularly PSA, we use the following terms for each position of the alignment:

match: The two characters are consistent. In the above example, four positions are *match*, including the first A.

mismatch or substitution: The two characters are inconsistent. No such a position in the above example.

insertion: Gap in the first sequence and some character in the second sequence. In the above example, the first T.

deletion: Some character in the first sequence and gap in the second sequence. In the above example, the second T.

indels: Insertion and deletion are equivalent by reversing the first and second sequences. Thus we can say both insertion and deletion as *indels*.

PSA is important, because soon after determining the gene sequence, similar sequences are searched over sequence databases, and from the found known sequences (genes), the function of the newly determined gene can be inferred. This type of sequence search has continued since the early days of life science.

Pairwise Sequence Alignment: Problem Setting

The above five types of a position has different scores, and PSA is an optimization problem to minimize the score. To explain this point intuitively, we treat all types of costs except match equal each other. That is, pairwise cost $C(a, b)$ for characters a, b (including gap) can be set as follows:

$$\text{if} \quad a = b \ (\neq -) \quad \text{then} \quad C(a, b) = 0, \tag{5.57}$$

$$\text{otherwise} \qquad C(a, b) = 1. \tag{5.58}$$

This setting says that if two characters are consistent, the cost is zero; otherwise one. We check the above example of PSA under this setting: Two positions are indels, where the cost is one, and the other four positions are match, where the cost is zero. Thus the cost sum is two. Clearly we are unable to increase the number of matches from the current pairwise sequence alignment. Thus this cost sum is smaller than any other alignment of these two sequences, and so is minimum. That is, the objective of PSA is to minimize the cost sum over all positions.

Pairwise Sequence Alignment: Dynamic Programming for Solving the Problem

Let $\boldsymbol{x} = \langle x_1, \ldots, x_N \rangle$ and $\boldsymbol{y} = \langle y_1, \ldots, y_M \rangle$ be the two input sequences. We now consider the method for obtaining the PSA with the minimum cost sum.

The simplest, bruteforce approach is to enumerate all possible pairwise sequence alignments, computing each of their costs and keeping the sequence alignment with the minimum cost. However the combination of the PSA increases exponentially, according to the length of the two sequences, and so enumerating all possible sequences is far from practical.

Then here suppose that for $\boldsymbol{x}$ and $\boldsymbol{y}$, we already know the optimum PSA between $\boldsymbol{x}' = \langle x_1, \ldots, x_{N-1} \rangle$ and $\boldsymbol{y}' = \langle y_1, \ldots, y_{M-1} \rangle$. Then to have the optimum PSA between $\boldsymbol{x}$ and $\boldsymbol{y}$, there are only two following choices on the last character pair, x_N and y_M, and so we check both cases and can have the optimum PSA between $\boldsymbol{x}$ and $\boldsymbol{y}$:

match. If x_N and y_M are the same character, the cost does not increase, and we use these two as match and have the optimal PSA.

indels. if x_N and y_M are not the same, we can make them mismatch or indels, and we add their cost to the current cost, and have the minimum cost and the optimal PSA.

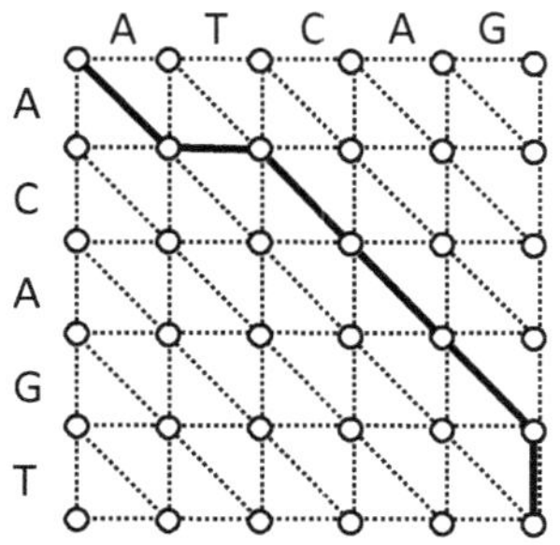

Figure 5.15: Pairwise alignment is equivalent to the problem of finding the least cost path on the two-dimensional grid.

The above observation indicates that $A(i,j)$, which is the optimum cost when we align from the first to the i-th character of $\boldsymbol{x}$ and from the first to the j-th character of $\boldsymbol{y}$, can be obtained from the three: $A(i,j-1), A(i-1,j)$ and $A(i-1,j-1)$. Also along with this line, starting with $A(0,0)$, if we compute $A(N,M)$, we can have the optimum cost of PSA. Below we can show the algorithm which can be obtained by summarizing the ideas so far.

Initialization: First at initialization, $A(0,0) = 0$ and compute $A(i,0)$ from $A(i-1,0)$ $(i = 1 \rightarrow N)$. Similarly compute $A(0,j)$ from $A(0,j-1)$ $(j = 1 \rightarrow M)$.

Repetition: Then $A(i,j)$ can be computed using the previous $A(*,*))$ as follows:

$$A(i,j) = \min \begin{cases} A(i-1,j-1) + C(x_i, y_j), & \text{(match)} \\ A(i-1,j) + C(x_i, -), & \text{(deletion)} \\ A(i,j-1) + C(-, y_j). & \text{(insertion)} \end{cases} \tag{5.59}$$

We repeat this by $i = 1 \rightarrow N$ and $j = 1 \rightarrow M$ and obtain the minimum cost sum by the last $A(N,M)$.

Traceback: Also we need traceback the computation process of obtaining the minimum cost, which gives the optimal PSA (called *traceback*).

Algorithm 5.8 shows a pseudocode of this algorithm, which is the basic PSA algorithm, called *Needleman-Wunsch*. This is a dynamic programming approach, almost the same as computing the forward probabilities of HMM: In this algorithm, $C(\cdot,\cdot)$ can be not only binary but also any numerical value depending on the input two characters. Note that as shown in Fig. 5.15 (which shows the alignment between ATCAG and ACAGT), the traceback is the thick line in this figure of the two dimensional (2D) grids by input two sequences.

Pairwise Sequence Alignment: Time Complexity

The complexity of computing $A(i,j)$ by dynamic programming reaches $O(NM)$, and the complexity of traceback is $O(N+M)$. Thus totally $O(N^2)$, assuming

Algorithm 5.8: Needleman-Wunsch algorithm for pairwise sequence alignment.

Function `Needleman_Wunsch_for_pairwise_alignment`($\boldsymbol{x}$, $\boldsymbol{y}$, C)
- **Data**: two sequences: $\boldsymbol{x}$, $\boldsymbol{y}$, cost function:$C(\cdot,\cdot)$
- **Result**: optimal alignment cost sum:$A(m,n)$, Optimal pairwise alignment:OPA

$A(0,0) = 0.$;
for $i \leftarrow 1$ **to** N **do**
  $A(i,0) \leftarrow A(i-1,0) + C(x_i,-)$; $\mathrm{BT}(i,0) \leftarrow (i-1,0)$.;
for $j \leftarrow 1$ **to** M **do**
  $A(0,j) \leftarrow A(0,j-1) + C(-,y_j)$; $\mathrm{BT}(0,j) \leftarrow (0,j-1)$.;
for $i \leftarrow 1$ **to** N **do**
  for $j \leftarrow 1$ **to** M **do**
    $A(i,j) \leftarrow \min\{A(i-1,j-1) + C(x_i,y_j), A(i-1,j) + C(x_i,-), A(i,j-1) + C(-,y_j)\}$.;
    $\mathrm{BT}(i,j) \leftarrow \arg\min\{A(i-1,j-1) + C(x_i,y_j), A(i-1,j) + C(x_i,-), A(i,j-1) + C(-,y_j)\}$.;
$i \leftarrow N; j \leftarrow M$; $\mathrm{OPA}[k=0] = (N,M)$.;
while $i \neq 0$ **and** $j \neq 0$ **do**
  $(i,j) \leftarrow \mathrm{BT}(i,j)$.
  $\mathrm{OPA}[k++] = (i, j)$.;
Output $A(\cdot,\cdot)$ and OPA.;

$N \sim M$. There are a variety of modifications of this algorithm, particularly on the cost function $C(*,*)$, while the time complexity of them is at least $O(N^2)$. In practice, sequence database is huge, and to implement a faster search, not only dynamic programming but also many heuristics are introduced into the currently well-used algorithm and tools, such as BLAST [3] and PSIBLAST [4], while these algorithms use various heuristics, which we do not touch in this book.

Multiple Sequence alignment: Background

Multiple sequence alignment (MSA) has more than two sequences as input and insert gaps into the more than two sequences so that characters are consistent over all sequences. Below we show one example to understand this problem intuitively.

A	T	C	A	G		A	T	C	A	–	G	–
A	C	A	G	T	→	A	–	C	A	–	G	T
A	C	C	G	T		A	–	C	–	C	G	T

In this alignment, three characters at each position are consistent completely. However, depending on the types of characters, mismatches can be allowed. For example, below can be allowed.

A	T	C	A	G		A	T	C	A	G	–
A	C	A	G	T	$\rightarrow$	A	–	C	A	G	T
A	C	C	G	T		A	–	C	C	G	T

In this alignment, the fourth position allow two different characters (A and C) to be at the same position. It depends on biological knowledge which alignment is better than the other. Thus using prior knowledge in life sciences, we can adjust the cost function $C(\cdot,\cdot)$, deciding which alignment is better than the other[6].

PSA is useful for searching a sequence similar to a sequence in hand. If we have however multiple sequences having their similarity, it would be more precise to use them to identify what characters (amino acids or nucleic acids) are at each position, because of using multiple sequences rather than only two sequences.

Thus if we have a set of sequences, e.g. amino acid sequences in the same protein family, we align these sequences to find motifs, i.e. conserved regions, which keep the same physico-chemical properties. Also found motifs are used to identify the family for unknown sequences, which is an important problem in bioinformatics, and also life sciences. For this purpose, HMM would be more useful than simple MSA. In fact, protein family database, Pfam [31] uses HMM for MSA of each family.

If we extend the dynamic programming of PSA to MSA, PSA was over 2D grids, and so MSA would need three-dimensional grids for three sequences and n-dimensional grids for n sequences. So the complexity of PSA was $O(N^2)$, while $O(N^n)$ for n sequences for MSA, which is practically intractable. To solve this practically, current two heuristics are used:

progressive alignment: The algorithm is similar to bottom-up (agglomerative) hierarchical clustering (see Section 3.1.5). That is, this algorithm repeats merging the closest cluster (sequence) and finally merges all sequences into one cluster. When the cluster (or sequence) distance is computed, pairwise alignment is used. A representative method is called ClusterW [91].

hidden Markov model (HMM): A state transition diagram is designed to allow MSA, and using the model trained from sequences in the same family, the optimal MSA can be obtained.

Below we explain MSA by HMM. The state transition diagram by HMM, called *Profile HMM* has a special state transition diagram for multiple sequence alignment. Fig. 5.16 shows an example of Profile HMM, where this model has three different types of states:

Match state (M): A regular state, from which a character, say an amino acid, is generated with some probability. This state corresponds to a conserved position in MSA. Thus the number of M should be set at the number of conserved positions in MSA. However, in reality, we do not know the number of M, before doing MSA, and then the number of M is finally fixed by a trial and error.

[6] This point was not explained in PSA, but the problem of choosing mismatch or indels can be controlled by the cost function $C(\cdot,\cdot)$ of pairwise sequence alignment also.

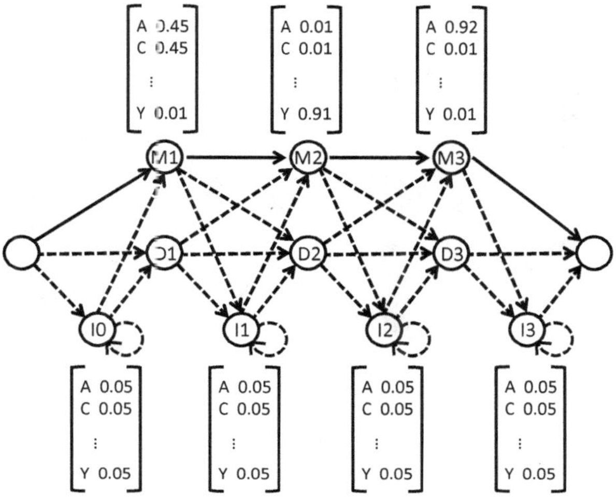

Figure 5.16: An example of Profile HMM.

Deletion state (D): No characters are generated, corresponding to a gap in MSA.

Insertion state (I): A character is generated, with a fixed uniform probability (0.05 for amino acids, because of 20 types of amino acids ($0.05 = 1/20$)). This position corresponds to a non conserved position.

As shown in 5.16, profile HMM has the main state transition from left to right, with the above three types of states parallelly. From M and D, it is possible to transit to any of M, D and I, while the transition from I to I is prohibited, because the role of I is to insert a character and it needs avoid to generate characters only by I.

Suppose that profile HMM in Fig. 5.16 is trained from the following sequences:

A	Y	C	A
C	W	Y	A
C	Y	T	A
K	A	Y	A
C	Y	A	

As mentioned earlier, to obtain MSA from the trained profile HMM, we compute the most likely path by dynamic programming for each sequence which corresponds to traceback.

Then we can see that for example, AYCA takes the following most likely path:

$$\text{(begin)} \rightarrow \text{M1} \rightarrow \text{M2} \rightarrow \text{I2} \rightarrow \text{M3} \rightarrow \text{(end)}$$

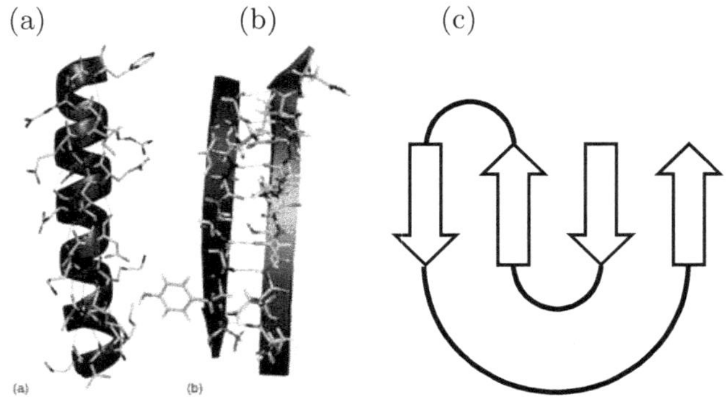

Figure 5.17: Typical examples of (a) α-helix, (b) β-sheet and (c) Greek-key motif of β-sheet.

Similarly CWYA takes the following most likely path:

$$(\text{begin}) \rightarrow \text{M1} \rightarrow \text{I1} \rightarrow \text{M2} \rightarrow \text{M3} \rightarrow (\text{end})$$

That is, from the most likely paths, we can see if each position should be M, I or D. Then from the most likely paths of all sequences, we can have the MSA:

–	A	–	Y	C	A
–	C	W	Y	–	A
–	C	–	Y	T	A
K	A	–	Y	–	A
–	C	–	Y	–	A

In summary HMM has the following advantages:

MSA: Given multiple sequences, estimating HMM and prediction by HMM (machine learning by HMM) achieve MSA.

Probabilities at conserved positions: We can see the probabilities of characters (amino acids or nucleic acids) at a conserved position from the probabilities at the corresponding match state (M).

Cost function estimation: As mentioned earlier, cost functions can be adjusted to control the priority between mismatch and indels, etc. by using prior knowledge. However, this would be not easy without any prior knowledge, by which priority among mismatch, indel, combination of characters would be hard to decide. On the other hand, HMM has no input on such costs. Instead, under the constraint of the state transition diagram of profile HMM, such costs are automatically trained from given sequences. This is a benefit. For example, the cost functions should be changed, depending on each position or some contiguous positions in MSA, while this is hard for progressive alignment. HMM already allows to do this careful adjustment

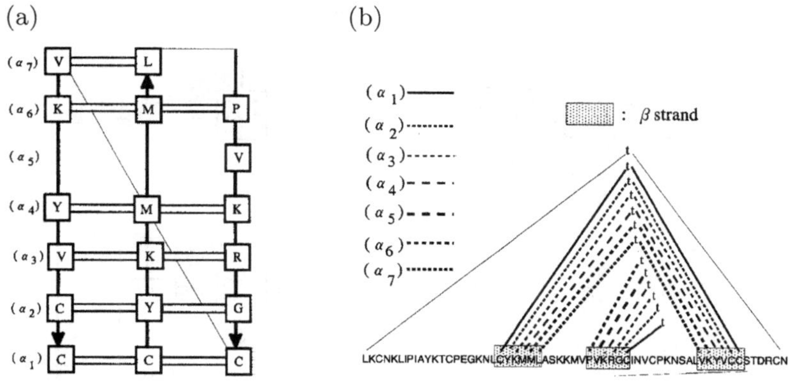

Figure 5.18: (a) Obtained rewriting rules, and (b) a parsing result over a small protein.

automatically through training profile HMM by sequences. In reality, probabilities at M and state transition probabilities are estimated by training sequences, according to these probabilities, each position of a sequence is decided to be at Insertion, Deletion or Match state. This is equivalent to the cost function is adjusted by local information of MSA.

Difference from progressive alignment Similar to hierarchical clustering, progressive alignment builds MSA starting with the most similar sequences, which has the possibility of the optimization being easily affected by some local optimality[7]. The MSA by HMM has no such "progressive" property, by which there are no possibility of being affected by local optimization, though EM algorithm itself is a local optimization algorithm.

5.4.4 Predicting β-sheet by Stochastic Tree Grammar

Typical patterns of protein three-dimensional (3D) structure are called *secondary structures*, where *α-helix* and *β-sheet* are the most major, representative. Fig. 5.17 shows examples of (a) α-helix and (b) β-sheet. In particular, β-sheet has multiple straightly-connected several amino acids, called *β-strands*, in parallel. Fig. 5.17 (c) shows a typical β-sheet, called *Greek-key*, which has also several β-strands in parallel. Note that this structure is more complex than the nested structure, meaning that Greek-key cannot be captured by CFG. However, β-sheet, more complex than the nested structure, can be captured by tree grammar with changing its complexity. In fact we gathered sequences in a protein family having a part of Greek-Key (Fig. 5.17 (c)) and trained parameters of stochastic tree grammar. Fig. 5.18 (a) shows examples of rewriting rules obtained and (b) is a prediction (parsing) result on an amino acid sequence in the same family but not used in training. This

[7]Latest progressive alignment-based method is carefully tuned to avoid this type of defects [52].

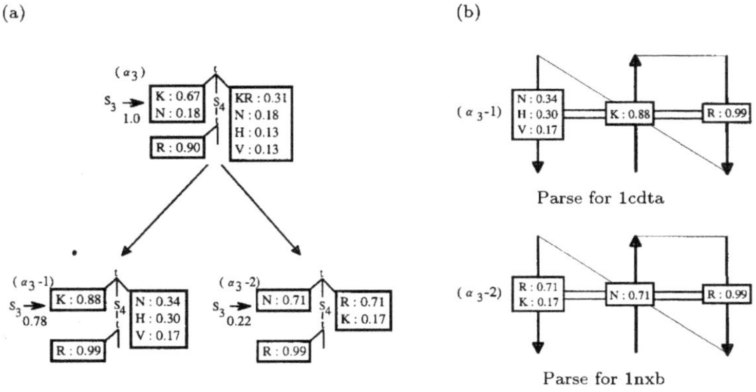

Figure 5.19: An example of capturing long-range interactions by rewriting rules of stochastic tree grammars.

clearly shows that β-strand regions and their directions are predicted correctly, and also tree grammar allows to capture the long-range interaction in a sequence.

Furthermore, we can see further long-range interactions in probabilities of characters (amino acids), by increasing the number of the same rewriting rules with different probabilities. The top of Fig. 5.19 (a) is the result we used only one rewriting rule, showing that the upper-left end node of S_4 has K with 0.67 and N with 0.18, and the upper-right end node has many amino acids, such as R, K, N, H and V. On the other hand, the bottom of Fig. 5.19 (a) is the result we used two rewriting rules with different initial probabilities: the bottom-left rule shows the upper-left end node of S_4 has only K with 0.88 without N. Also bottom-right rule has the upper-left end node of S_4 has only N with 0.71 without K. Furthermore, the upper-right of S_4 have N, H and V for left-hand and R and K for right-hand, though they are mixed at right-hand side of the top rule.

$$\{\{\text{K,N}\} \Longleftrightarrow \{\text{K,R,N,H,V}\}\} \longrightarrow \left\{ \begin{array}{ccc} \{\text{K}\} & \Longleftrightarrow & \{\text{N,H,V}\} \\ \{\text{N}\} & \Longleftrightarrow & \{\text{K,R}\} \end{array} \right\}. \tag{5.60}$$

This means the first rule was split into 1) K for left and N, H and V for right and 2) N for left and R and K for right. In fact, when we apply the set of rules with split ones to real sequences for parsing, as shown in Fig. 5.19 (b), each split rule is used, depending on proteins (one for 1cdta and the other for 1nxb). This indicates that stochastic tree grammar could capture different types of long-range interactions even for the same positions, in parallel.

5.4.5 Identifying Timing Difference of Gene Expression by HMM

There are numerous HMM which are developed for some specific application, in speech recognition, natural language processing and bioinformatics, We explain

(a) (b)

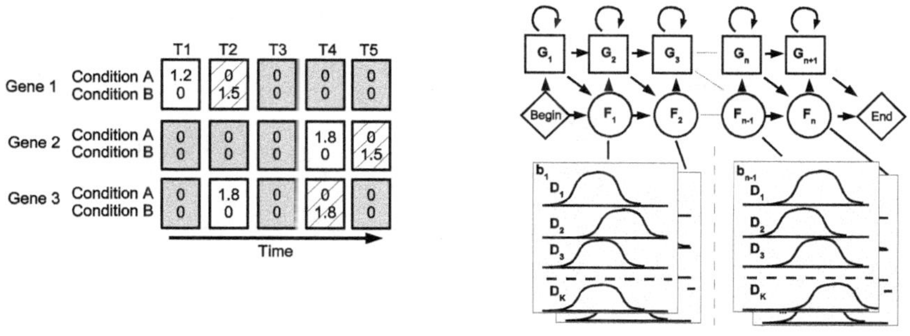

Figure 5.20: (a) Sample data and (b) state transition diagram.

(a) (b)

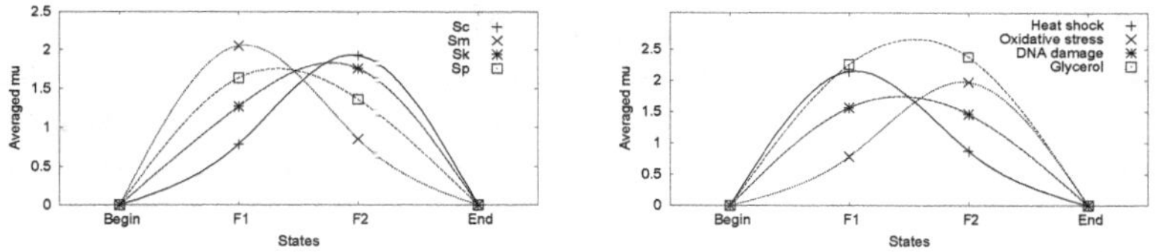

Figure 5.21: Result obtained from real yeast expression data.

one example, where HMM has multiple sequences as one example.

We consider time-series gene expression, where the data is a matrix with (time × time-series experiments). Fig. 5.20 (a) shows an example of the input transformed from such data. For simplicity, in this input dataset, there are two experimental conditions (Condition A and B), three genes (Gene 1, 2 and 3) and five time stamps (T1, T2, T3, T4 and T5). The objective is to detect the timing difference among conditions. More in detail, genes are instances, and for each instance, we have time-series (time-dependent) multiple conditions, which have difference in expression. For example, in Fig. 5.20 (a), expression in Condition A is always earlier than Condition B. The state transition diagram (Fig. 5.20 (b)) is designed to capture this difference. Here we consider only two conditions, while we can increase this number to an arbitrary number.

Fig. 5.21 shows experimental results, checking the difference of four conditions. From this figure, we can see that (a) for example, Sm expressing quicker than others and Sc slower and entirely expression appears in the order of Sm → Sp → Sk → Sc and (b) For example, Heat shock is quicker but Oxidative stress is slower, while DNA damage and Glycerol have no difference in speed of expression reaction. This is an example HMM can extract patterns (corresponding to some biological knowledge) from time-series data.

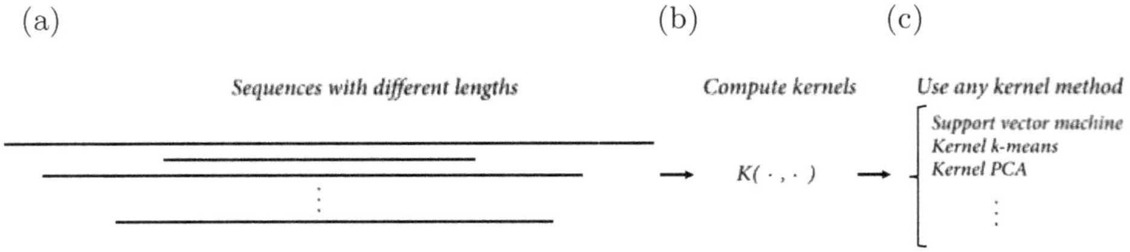

Figure 5.22: From (a) given sequences, (b) if once kernels can be computed over all pairs of sequences, (c) any kernel method can be used for the given sequences.

5.4.6 Sequence Analysis by String Kernel

In life sciences, gene or protein sequences are abundant data, and string kernel is widely used in a lot of problems. As mentioned earlier, for sequence analysis, once kernel functions are computed among sequences, a variety of methods in kernel learning can be used (Fig. 5.22). In particular, one of the major bioinformatics paradigm is a sequence classification problem (see Section 3.5.2). In this case, a classical procedure was to extract subsequences from the original input to generate a dataset of vectors. However, by using kernel function, particularly string kernel, now we do not have to cut out subsequences from the input data. Instead, we can generate kernel functions from input sequences, and compare different instances over the newly generated kernel space. In particular, this approach is useful for using the "I. entire sequence with one label" more than "II. each position with one label" (Section 3.5.2). That is, protein localization prediction and MHC binding peptide prediction should have more merits than "protein secondary structure prediction" and "calpain substrate cleavage site prediction", because the entire sequences can be compared more directly. For example, spectrum kernel is used to predict a protein family for a given sequence [59]. This is also a typical problem of the entire sequence with one label. Fig. 5.22 shows a schematic picture of dealing with sequences with different lengths by string kernel.

Chapter 6

Learning Trees

A tree is a branched sequence and at the same time a graph without any cycles. Thus we have a possibility of developing machine learning methods of sequences for trees, and machine learning method for graphs can be applied to trees. We describe methods for graphs in the next two chapters. Thus, in this chapter, we rather focus on the development on the methods of sequences for learning trees.

In this chapter, mainly we consider *labeled rooted trees*, which have the following two relations on the order.

parent-child: The order is from the root to leaves, i.e. ancestors to descendants and locally from the parent to a child.

Also if labels have a *lexicographical order*, children with the same parent can be ordered:

siblings: Among the nodes with the same parent, there is an order from elder siblings to younger ones.

A labeled ordered tree with lexicographical ordered labels has both relations on parent-child and sibling. However the order of the entire tree cannot be decided by these two only. That is, there is still a problem on which of the two orders should be prioritized.

To solve this problem, there are two solutions: breadth-first and depth-first:, Fig. 6.1 (a) and (b) show an example of each of these two orders: *breadth-first* and *depth-first*, respectively. The breadth-first (a) decides order at each layer and then goes to the next layer Thus in this case, children with the same parent are ordered consecutively, and all children are neighboring each other in their order. On the other hand, the depth-first (b) repeats going (from the root) to a leaf, and so even children with the same parent, the order can be totally different from other children with the same parent. As such a tree with children ordered over the entire tree is called an *ordered tree*, and the above depth-first and breadth-first are the methods to order nodes in the ordered tree.

In this chapter, as well as sets and sequences, we describe three machine learning approaches, frequent pattern mining, probabilistic models and kernel learning.

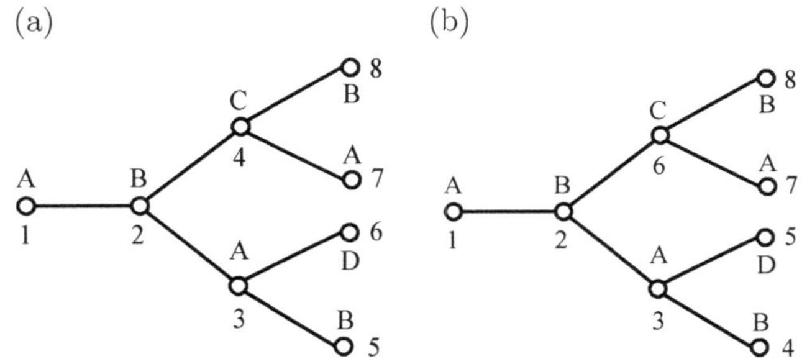

Figure 6.1: Examples of (a) breadth-first and (b) depth-first ordered trees.

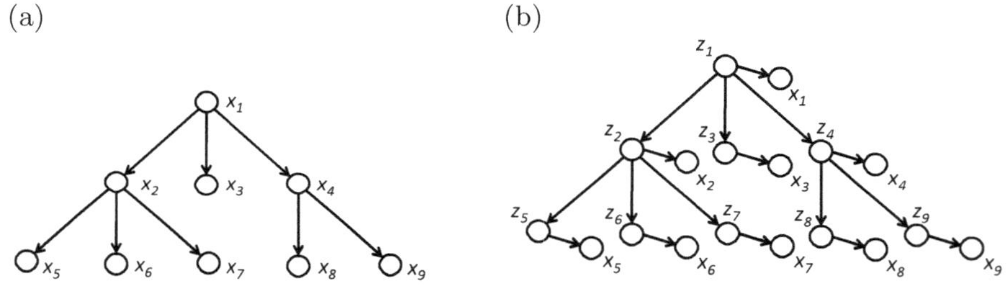

Figure 6.2: (a) Tree example and (b) graphical models of hidden tree Markov model for the tree example.

Frequent pattern (subtree) mining has two ideas: using the idea of frequent subsequence mining and part of frequent subgraph mining. Thus we explain these three approaches in the order of: 1) probabilistic models, 2) kernel learning and 3) frequent subtree mining.

Note that the input of probabilistic models is *labeled rooted ordered trees*, while in kernel learning and frequent subtree mining, we treat, more general, labeled ordered trees.

6.1 Probabilistic Models

We have studied HMM for learning sequences, where a state transition diagram (Fig. 5.2 (b)) is given and HMM generates a sequence by state transition as shown in Fig. 5.8 (a).

Similarly, given ordered trees like Fig. 6.2 (a) (Note that in this figure, $x_1, \ldots, x_9$ on nodes show the order of nodes in the tree), we have a probabilistic model, following the node order of the tree, with state transitions, generating tree labels which finally become a tree. Below we explain the probabilistic models for trees and methods for estimating probability parameters of the models.

6.1.1 Hidden Tree Markov Model

HTMM: Preliminaries

The first intuitive model is called *hidden tree Markov model* (HTMM) [26, 21], which incorporates, out of the two orders, only the first, parent-child (and not the second, siblings). Fig. 6.2 (b) shows a graphical model of HTMM. From this figure, we can see that HTMM captures vertical relations (parent-child) only and horizontal relations (siblings) are ignored. For example, the root has three children with three latent variable z_1, z_2 and z_3, while each of these three latent variables depends on z_1 only. Thus, HTMM considers dependencies just along with the edges of the given tree. Thus trees allow branches, which make trees look complex, while the dependency has nothing to do with branches, and the left-to-right dependency of HMM is just transformed into the up-to-down (parent-to-child) dependency by HTMM.

Also we can give a state transition diagram for HTMM, as in HMM. For simplicity, however, we just think the case that any state transition is possible, meaning that the current state transition diagram allows all possible state transitions.

HTMM has the following two probability parameters, which are totally the same as HMM:

$a(i,j)$: state transition probability from state i to state j.

$b(i,O)$: output generation probability of output O at state i.

Now we note that these two probability parameters are short forms that:

$a(i,j)$ is $p(z_c = s_j|z_p = s_i)$, i.e. the probability that two latent variables with the parent-child relation, z_p (parent) and z_c (child) take states s_i and s_j, respectively (Note that p and c are any parent and its child in the tree).

$b_i(O)$ is $p(o_c = O|z_c = s_i)$, i.e. the probability that the output is O at node c, given that latent variable z_c takes state s_i.

We can consider the following three, as we have done in HMM:

Computing likelihood of a given tree.

Learning (Estimating) parameters of HTMM, given multiple trees.

Predicting the likelihood of unknown tree after parameters are trained.

We explain, after HTMM, ordered tree Markov model (OTMM). These two models are overlapped in many points regarding the above three procedure, and so we skip showing the pseudocodes of HTMM on the above three procedures.

HTMM: Computing Likelihood

HMM used forward (or backward) probabilities for computing the likelihood. In trees, dependencies are up-to-down, and so we define upward probabilities, as follows:

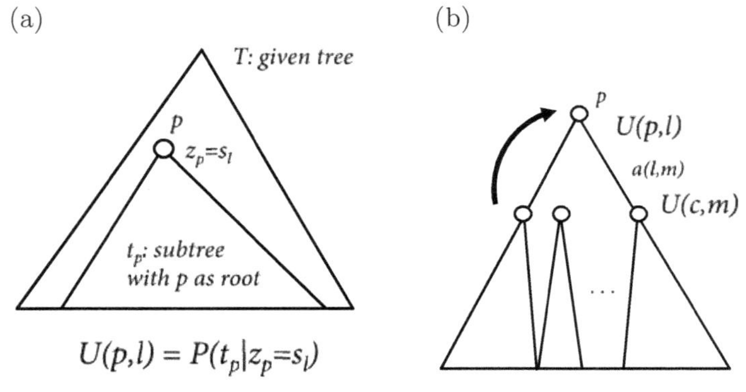

Figure 6.3: Schematic diagram of (a) upward probability for the subtree with p as its root and (b) recursively computing upward probabilities at the nodes of p and c as the parent and child.

Definition 6.1 (Upward Probability of HTMM). *Upward probability* $U(p,l)$ *(*$= P(t_p|z_p = s_l)$*) is the probability that subtree* t_p *with node* p *as its root is generated (and the current node is* p*) and the state of node* p *is* s_l.

Fig. 6.3 (a) shows a schematic picture of upward probabilities. The upward probability holds the following:

Proposition 6.1 (Upward Probability of HTMM). *Upward probability* $U(p,*)$ *in which the subtree with node* p *as its root is already generated can be computed from upward probabilities* $U(c,*)$ *that the subtree has child* c *(of node* p*) as its root is generated.*

Using this property, the upward probabilities can be computed from leaves of the tree in a bottom-up manner. Then as well as HMM, dynamic programming can be applied to this computation, as follows (let $C(p)$ be the set of children of node p and $\boldsymbol{S}$ be the set of all states):

$$U(p,l) = \begin{cases} b(l,o_p) & \text{if } p \text{ is a leave,} \\ b(l,o_p)\prod_{c\in C(p)}\sum_{m\in \boldsymbol{S}} a(l,m)U(c,m) & \text{otherwise,} \end{cases} \tag{6.1}$$

where the right-hand side shows the computation using upward probabilities $U(c,m)$ of the subtree with node c as its root, where c is a child of node p. Fig. 6.3 (b) shows a schematic picture of this dynamic programming computation.

We then recursively compute the upward probabilities following the order of the ordered tree T up to the root, and then the likelihood over T can be computed as follows:

$$L(T) = \sum_{l=1}^{|\boldsymbol{S}|} \pi(l)U(r,l), \tag{6.2}$$

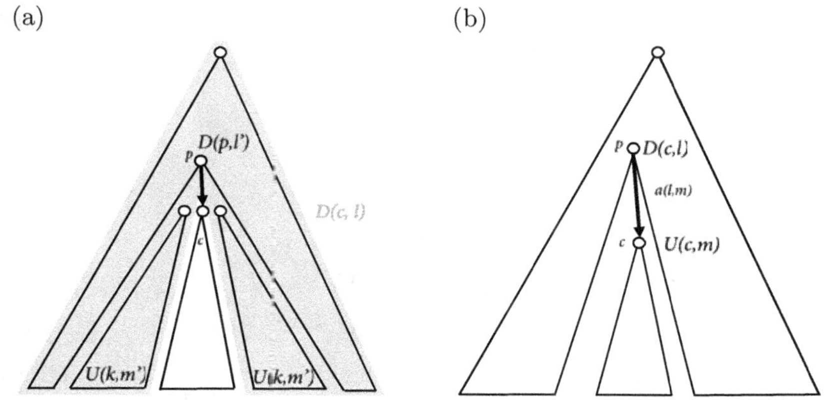

Figure 6.4: Schematic diagram of (a) computing downward probability $D(c, m)$ from upward probabilities $U(k, m)$ (where k are siblings of c) and downward probability $D(p, l)$, and (b) computing expectation of state transition: $m \to l$ at the nodes of p and c as the parent and child.

where r is the root, and $\pi(\cdot)$ is the initialization probability of the root.

Furthermore, we have a set of trees $\boldsymbol{T}$, and the likelihood over the set of trees is given as follows:

$$L(\boldsymbol{T}) = \prod_{T \in \boldsymbol{T}} L(\boldsymbol{T}). \tag{6.3}$$

HTMM: Learning (Estimating) Parameters

As well as HMM, for example, for a state transition probability, we need to compute the expectation value of the corresponding state transition, i.e. all cases of using the corresponding state transition.

Then, as we defined backward probabilities against forward probabilities in HMM, we define downward probabilities against upward probabilities:

Definition 6.2 (Downward Probability of HTMM). *Downward probability $D(c, l)$ $(= P(\neg t_c | z_p = s_l))$ is the probability that all nodes of the tree except subtree t_c with node c as its root are generated and the state of parent p of node c is s_l.*

Downward probabilities are in the opposite direction to upward probabilities, and so the computation is from the root to leaves. However, for example, when node p is the parent of node c, the parent of c is uniquely determined as p, while the child of p cannot be determined uniquely. Thus, to compute $D(c, l)$ from $D(p, l')$, we need to use upward probabilities $U(k, m')$, where k is the children of p, except c. This can be summarized as follows:

Proposition 6.2 (Downward Probability of HTMM). *The downward probability of node c can be computed from the downward probability of node p, which is the parent of node c, and also the upward probabilities of siblings of node c.*

This (complex) computation is needed exactly because of trees (or branches), and this is mathematically written as follows:

$$D(c,l) = \begin{cases} b(l,o_p) & \text{if } p \text{ is the root,} \\ LL, & \text{otherwise.} \end{cases} \quad (6.4)$$

where

$$LL = b(l,o_p)\sum_{l'=1}^{|\boldsymbol{S}|} a(l',l)D(p,l') \prod_{k\in C(p)|k\neq c} \sum_{m'=1}^{|\boldsymbol{S}|} a(l,m')U(k,m') \quad (6.5)$$

Fig. 6.4 shows a schematic picture of computing the downward probabilities.

Hereafter for simplicity we skip output generation probability. First given T, the expectation value that for parent node p and child node c, the states of p and c are l and m, respectively, can be computed as follows:

$$E_{(p,c)}(\#(l \to m)|T) = D(c,l)a(l,m)U(c,m), \quad (6.6)$$

where note that $D(c,l)$ must be used since parent p cannot uniquely determine its child. In other words, the definition of $D(c,l)$ says that the parent of node c is at state s_l ($z_p = x_l$ if the parent of c is p). This is not a problem, because the parent of node c can be uniquely determined, but not vice versa. Thus we need specify c instead of p.

We can estimate state transition probability $\hat{a}(i,j)$ from a given set of trees, $\boldsymbol{T}$:

$$\hat{a}(i,j) \leftarrow \frac{\sum_{T\in\boldsymbol{T}}\sum_{(p,c)\in\boldsymbol{P}_T} E_{(p,c)}(\#(i\to j)|T)}{\sum_j\sum_{T\in\boldsymbol{T}}\sum_{(p,c)\in\boldsymbol{P}_T} E_{(p,c)}(\#(i\to j)|T)}, \quad (6.7)$$

where $\boldsymbol{P}_T$ is a set of all parent-child pairs of a given tree T.

HTMM: Predicting the Likelihood of Unknown Tree

As well as HMM, in prediction the likelihood is given by the most likely path which can be computed by modifying the algorithm of computing the likelihood.

We first define the *most likely path probability* by modifying the upward probability:

Definition 6.3 ($V(t,i)$ of HTMM). *$V(t,i)$ is the probability of the most likely path, which already generates the subtree with node t as its root, and the current state is i.*

We can then compute the likelihood of a given tree T by using this most likely path probability $V(t,i)$, as follows:

$$L(T) = \max_{i\in\boldsymbol{S}} \pi(i)V(r,i), \quad (6.8)$$

where r is the root of T. Also the most likely path probability can be recursively computed from leaves to the root which is a dynamic programming algorithm:

$$V(p,l) = \begin{cases} b(l,o_p) & \text{if } p \text{ is a leave,} \\ b(l,o_p)\prod_{c\in C(p)} \max_{m\in \boldsymbol{S}} a(l,m)V(c,m) & \text{otherwise.} \end{cases} \quad (6.9)$$

In particular, the above second equation can be written as follows (m can be the most outside):

$$V(p,l) = \max_{m\in \boldsymbol{S}} b(l,o_p) \prod_{c\in C(p)} a(l,m)V(c,m). \quad (6.10)$$

Then, as well as HMM, most likely path itself is more important than the likelihood sometimes, and so we need save not only the likelihood but also the states which give the most likely path, as $M(p,l)$:

$$M(p,l) = \arg\max_{m\in \boldsymbol{S}} b(l,o_p) \prod_{c\in C(p)} a(l,m)V(c,m). \quad (6.11)$$

Then, again as well as HMM, we retrieve the most likely path by using M (so-called backtracking).

HTMM: Summary

HTMM considers the parent-child order only in given trees. In particular, since the parent of a child can be uniquely determined, as shown in Fig. 6.3, the forward (or backward) probability of HMM can be extended directly to the upward probability. On the other hand, a child of a parent cannot be determined uniquely, since a parent can have not only one child. Then the downward probability need to be computed not only from the downward probability of its parent but also the upward probabilities of its siblings. This is a feature of trees, different from sequences.

Trees have branches which makes tress look complex, while the complexity of HTMM is totally the same as HMM (assuming that the number of nodes in a tree and the number of outputs in a sequence are the same).

Note that HTMM considers only the parent-child order, and it would be worth to examine the probabilistic model which considers both parent-child and siblings orders.

6.1.2 Ordered Tree Markov Model

Background and Model Description

We here consider a probabilistic model, which deals with the both orders (relations) of parent-child and siblings. However another point to consider is the time complexity, which should be kept the same as HMM and HTMM. For this purpose, we design the probabilistic model, in which the current latent variable depends up on only one past latent variable.

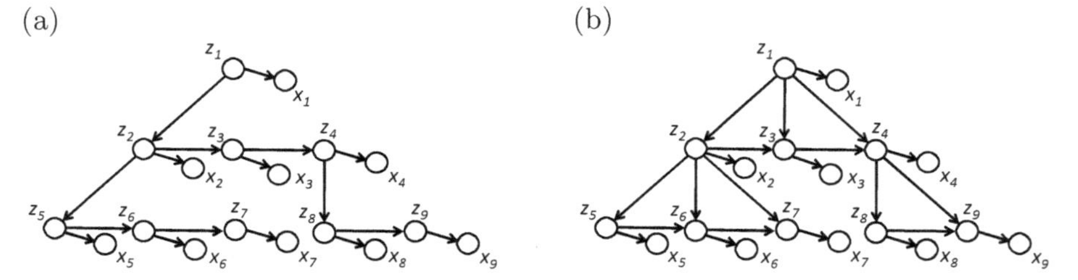

Figure 6.5: Graphical models of (a) ordered tree Markov model (OTMM) and (b) sibling tree Markov model (which has a higher complexity than HTMM and OTMM, and not mentioned in this book) for the tree example in Fig. 6.2.

In fact, in HMM, the current outputs depended on only the left last output. Also in HTMM, the current node depended on one left node only. Then when we consider the sibling order, for example, if the current node depends on both the parent and elder nodes, the complexity becomes more than HMM and HTMM.

Then we consider a model, in which only the eldest sibling depends on the parent, and each of other younger sibling is dependent upon its left sibling. We call this model *ordered tree Markov model* (OTMM) [40, 39].

Thus, for the tree of Fig. 6.2 (a), OTMM has a graphical model of Fig. 6.5 (a). HTMM just traced the given tree of Fig. 6.2 (a), as shown in Fig. 6.2 (b), while OTMM does not consider parent-child relations except between a parent and its eldest sibling, and instead all sibling relations, which are not in the edges of the original input tree, are considered. As such, the model structure of OTMM is totally different from HTMM, but in this model, since each variable depends on only one variable, the complexity is still the same as HMM and HTMM.

As well as HMM and HTMM, OTMM also can use a state transition diagram to control state transitions, while for simplicity we here allow all state transitions.

First this model also has two types of probability parameters, as well as HMM and HTMM:

$a(i, j)$: state transition probability from state i to state j.

$b(i, O)$: output generation probability of output O at state i.

Here again the above are short forms, and they can be written more formally:

$a(i, j)$ can be two types:

> $p(z_c = s_j | z_p = s_i)$: the probability that two latent variables with the parent-child relation, z_p (parent) and z_c (child), take states s_i and s_j, respectively. (For example, $z = 1$ and $z = 2$ in Fig. 6.5 (a))
>
> $p(z_y = s_j | z_o = s_i)$: the probability that two latent variables with the neighboring sibling relation, z_o (elder sibling) and z_y (younger sibling) take states s_i and s_j, respectively. (For example, $z = 2$ and $z = 3$ in Fig. 6.5 (a))

$b(i, O)$: is $p(o_c = O|z_c = s_i)$, i.e. the probability that the output is O at node c, given that latent variable z_c takes state s_i.

As well as HTMM, we can consider the following three:

Computing likelihood of a given tree.

Learning (Estimating) parameters of OTMM, given multiple trees.

Predicting the likelihood of unknown tree after parameters are trained.

OTMM: Computing Likelihood

HTMM considers only the parent-child relation, and the likelihood was obtained by computing the upward probabilities from leaves to the root by dynamic programming. On the other hand, OTMM considers not only parent-child but also sibling dependencies. Thus we introduce backward probabilities along with the direction of siblings. The definition of the backward probabilities is basically the same as that of HMM. We first write the definition upward probability once again.

Definition 6.4 (Upward Probability of OTMM). *Upward probability $U(p, l)$ ($= P(t_p|z_p = s_l)$) is the probability that subtree t_p with node p as its root is generated (and the current node is p) and the state of node p is s_l.*

We then define backward probability:

Definition 6.5 (Backward Probability of OTMM). *Backward probability $B(c, m)$ ($= P(t_c, \ldots, t_{c\rightarrow}|z_c = s_m)$) is the probability that subtree t_c with node c as its root and the younger siblings of node c and the subtrees with those siblings as their roots all are generated (and the current node is c) and the state of node c is s_m,*

where $c \rightarrow$ is the youngest sibling of c. OTMM considers the parent-child relation for the eldest child only, and so the upward probability holds the following:

Proposition 6.3 (Upward Probability of OTMM). *Upward probability $U(p, *)$ in which the subtree with node p as its root is already generated can be computed from backward probabilities $B(c, *)$ of node c, where c is the eldest child of p.*

Using this property, upward probability can be computed, from leaves to the root, following the order of the ordered tree by dynamic programming:

$$U(p, l) = \begin{cases} b(l, o_p) & \text{if } p \text{ is a leave,} \\ b(l, o_p) \sum_{m \in \boldsymbol{S}} a(l, m) B(c, m) & \text{otherwise,} \end{cases} \tag{6.12}$$

where $\boldsymbol{S}$ is the set of states and c is the eldest child of p. Again from the definition of OTMM, we use the eldest child node c of node p for computing the upward probability. Fig. 6.6 (a) shows a schematic picture of computing the upward probability.

On the other hand, backward probability at node p can be computed from its child and younger sibling. Thus the following holds:

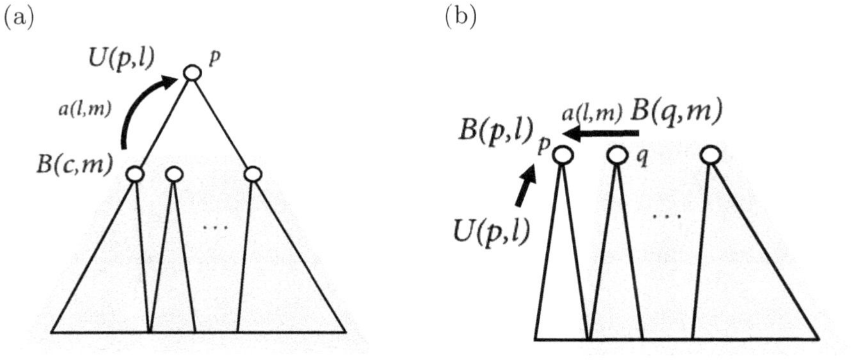

Figure 6.6: Schematic diagram of recursively computing (a) upward probability $U(p,l)$ and (b) backward probability $B(p,l)$.

Proposition 6.4 (Backward Probability of OTMM). *Backward probability $B(p,*)$ that the subtree with node p as its root and one younger sibling q (of node p) and its subtree are already generated can be computed from upward probability $U(p,*)$ and backward probability $B(q,*)$ of the neighboring younger sibling.*

Using this property, backward probability can be computed, following the order of the ordered tree, by using dynamic programming:

$$B(p,l) \quad = \quad \begin{cases} U(p,l) & \text{if } p \text{ is the youngest sibling,} \\ U(p,l)\sum_{m\in\boldsymbol{S}} a(l,m)B(q,m) & \text{otherwise,} \end{cases} \tag{6.13}$$

where q is one younger sibling of p. Fig. 6.6 (b) shows a schematic picture of this computation.

As well as HTMM, we can recursively compute the upward and backward probabilities, from leaves to the root, following the order of the given ordered tree, and finally using the upward probability at the root, we can compute the likelihood of given labeled ordered tree T:

$$L(T) = \sum_{m\in\boldsymbol{S}} \pi(m)U(r,m), \tag{6.14}$$

where r is the root.

Furthermore, if a set of trees, $\boldsymbol{T}$, is given, the likelihood of the given set of trees can be computed as follows:

$$L(\boldsymbol{T}) = \prod_{T\in\boldsymbol{T}} L(\boldsymbol{T}). \tag{6.15}$$

Algorithm 6.1 shows a pseudocode of computing the likelihood of OTMM.

Algorithm 6.1: Likelihood computation for ordered tree Markov model.

Function `Likelihood_computation_of_ordered_tree_Markov_model(`x, y, C`)`
- **Data**: tree set: $\mathcal{T}$, state set: $\mathcal{S}$, parameter: $\Phi = \{\boldsymbol{a}, \boldsymbol{b}\}$
- **Result**: likelihood $L(\mathcal{T})$ for $\mathcal{T}$
- .
- Initialize parameters.;
- **for** $T \in \mathcal{T}$ **do**
  - **for** $p \leftarrow |V_T|$ **to** 1 **do**
    - **foreach** $l \in \mathcal{S}$ **do**
      - Compute $U_T(p, l)$ by (6.13).;
    - **foreach** $l \in \mathcal{S}$ **do**
      - Compute $B_T(p, l)$ by (6.12).;
  - Compute $L(T)$ by (6.14).;
  - Update $L(\mathcal{T})$ by (6.15).;
- Output $L(\mathcal{T})$.;

As shown in the update rule (6.13) of backward probabilities, output generation probabilities do not appear in backward probabilities. So although we used both backward and upward probabilities to make understanding the procedure easier, we can see that these two probabilities can be represented by only one, say the backward probability, by changing its definition. However in this case, we need to check node situations in more detail, by which the procedure is, on the contrary, more complex and not easy to understand [40].

OTMM: Learning (Estimating) Parameters

Again for estimating parameters, we need to compute the expectation value corresponding to the parameter. We define downward and forward probabilities, which are the counter part of upward and backward probabilities, respectively. However, the definition of downward probability is different from HTMM, because in HTMM, nodes depend upon their parent and the parent has multiple children, and the downward probability cannot be defined without specifying the child for the parent node. However in OTMM, only the eldest child depends on the parent, and the parent can always specify the eldest child. That is, the downward probability is defined only for between the parent and the eldest child.

Definition 6.6 (Downward Probability of OTMM). *Downward probability $D(p, l)$ $(= P(\neg t_c | z_p = s_l))$ is the probability that all tree except subtrees with children of node p as their roots (node p is also generated) is generated and the state of node p is s_l.*

Also forward probability can be defined as follows:

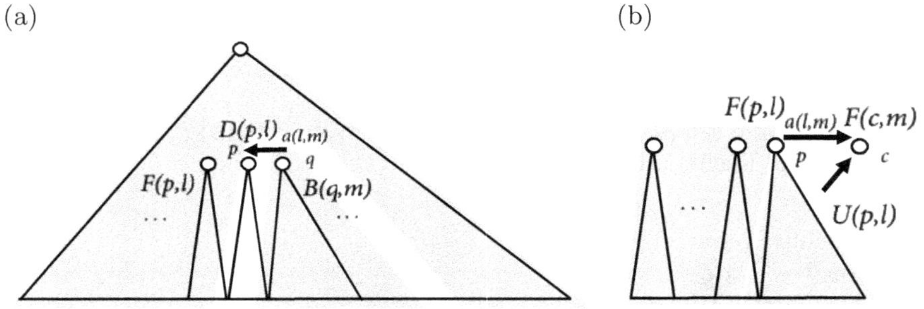

Figure 6.7: Schematic picture of recursively computing (a) downward probability $D(p, l)$ and (b) forward probability $F(c, m)$.

Definition 6.7 (Forward Probability of OTMM). *Forward probability $F(p, l)$ $(= P(\neg(t_q, \ldots, t_{q\rightarrow}|z_p = s_l))$ is the probability that all nodes except all younger siblings, $q, \ldots, q \rightarrow$ of node p, and their subtrees $t_q, \ldots, t_{q\rightarrow}$ are generated and the status of node p is s_l.*

The downward probability $D(p, *)$ does not have to consider the child of node p but needs consider the siblings of node p. Thus the downward probability can be computed from the elder side, forward probability $F(p, *)$ (note that $F(p, *)$ does not contain subtree with node p as its root) and from the younger side, backward probability $B(q, *)$ (node q is one younger than node p).

Proposition 6.5 (Computing downward probability of OTMM). *Downward probability $D(p, *)$ of node p can be computed from forward probability $F(p, *)$ and backward probability $B(q, *)$, where q is one younger sibling of node p.*

From this property, downward probabilities can be considered from the root to leaves by dynamic programming:

$$D(p, l) = \begin{cases} \pi[l] & \text{if } p \text{ is the root,} \\ F(p, l) & \text{if } p \text{ is the youngest sibling,} \\ F(p, l) \sum_{m \in |\boldsymbol{S}|} a(l, m) B(q, m) & \text{otherwise,} \end{cases} \tag{6.16}$$

where q is one younger sibling of node p. Fig. 6.7 (a) shows a schematic picture of this computation.

Forward probability $F(c, *)$ is simply the reverse of backward probability and rather easy to derive. $F(c, *)$ can be computed from forward probability $F(p, *)$ (where p is one elder sibling of node c) and upward probability $U(p, *)$ of node p. However if c is the eldest sibling, p becomes the parent of node c, and $F(c, *)$ can be computed from downward probability $D(p, *)$ of parent p.

Proposition 6.6 (Forward probability of OTMM). *Forward probability $F(c, *)$ of node c can be computed from forward probability $F(q, *)$ of node q (is one elder*

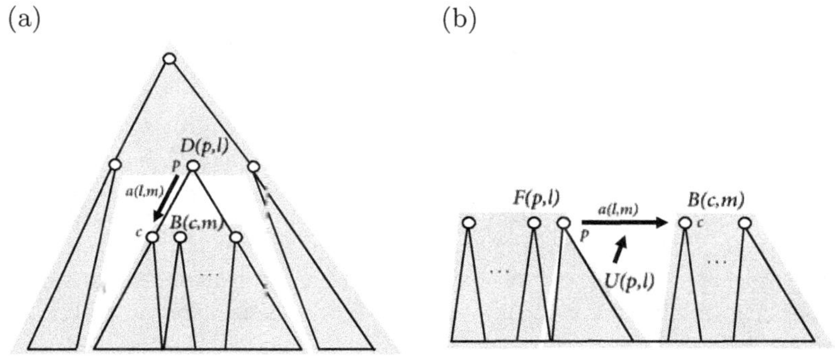

Figure 6.8: Schematic diagram of expectation of transition probabilities of (a) parent-child relationships and (b) siblings relationships.

sibling of node c) and upward probability $U(q,*)$ of node q. However, if node c is the eldest sibling, forward probability $F(c,*)$ of node c can be computed from downward probability $D(p,*)$ of parent p of node c.

From this property, forward probability can be computed as follows:

$$F(c,m) = \begin{cases} \sum_{l \in \boldsymbol{S}} b(l,o_p)a(l,m)D(p,l) & \text{if } c \text{ is the eldest sibling,} \\ \sum_{l \in \boldsymbol{S}} b(l,o_p)a(l,m)F(p,l)U(p,l) & \text{otherwise,} \end{cases} \tag{6.17}$$

where p is one elder sibling of node c, while if c is the eldest sibling, p is the parent of node c. Fig. 6.7 (b) shows a schematic picture of this computation.

Thus now we obtain the expectation value from the above four probabilities. As well as HTMM, we omit the output generation probability for simplicity. We focus on state transition probability, and this probability has two types: I) parent-child and II) siblings.

I) parent-child: In OTMM, state transition for a parent-child relation is only between parent and the eldest sibling. Thus the expectation value that the states of parent p and the eldest sibling c among the children of p are s_l and s_m, respectively can be computed as follows:

$$E^{(P)}_{(p,c)}(\#(l \to m)|T) = D(p,l)a(l,m)B(c,m). \tag{6.18}$$

Fig. 6.8 (a) shows a schematic picture of this computation of expectation values. For state transition of a pair of a parent and its eldest child, there are the remaining two parts: 1) the parent side is covered by the downward probability covering the entire tree except the subtree with node p as the root, and 2) the child side is covered by the backward probability. Thus the expectation value can be computed by these two probabilities and the corresponding state transition probability.

II) siblings: For state transition over siblings, the expectation value that the states of elder and younger siblings p and c are s_l and s_m, respectively is given as follows:

$$E^{(B)}_{(p,c)}(\#(l \to m)|T) \quad = \quad F(p,l)U(p,l)a(l,m)B(c,m). \tag{6.19}$$

Fig. 6.8 (b) shows a schematic picture of this computation. In this computation, we consider neighboring two nodes, and except the subtree of the elder node side by the upward probability, this computation is the same as the expectation value computation of HMM, which uses forward and backward probabilities.

By using the expectation values, we derive the update rule of state transition probabilities. First given a tree T and its set $\boldsymbol{T}$, let $\boldsymbol{P}_T$ be all pairs of the parent and eldest child in tree T, and let $\boldsymbol{B}_T$ be all neighboring sibling pairs in T. State transition probability $a(i,j)$ can be updated by using above expectation values (6.18) and (6.19), as follows:

$$\hat{a}(i,j) \leftarrow$$
$$\frac{\sum_{T\in\boldsymbol{T}}(\sum_{(p,c)\in\boldsymbol{P}_T} E^{(P)}_{(p,c)}(\#(i\to j)|T) + \sum_{(p,c)\in\boldsymbol{B}_T} E^{(B)}_{(p,c)}(\#(i\to j)|T))}{\sum_j \sum_{T\in\boldsymbol{T}}(\sum_{(p,c)\in\boldsymbol{P}_T} E^{(P)}_{(p,c)}(\#(i\to j)|T) + \sum_{(p,c)\in\boldsymbol{B}_T} E^{(B)}_{(p,c)}(\#(i\to j)|T))}. \tag{6.20}$$

Algorithm 6.2 shows a pseudocode of learning parameters of OTMM, where for tree T, forward, backward, downward and upward probabilities are written in the manner of $F_T(a,b)$, $B_T(a,b)$, $D_T(a,b)$ and $U_T(a,b)$.

OTMM: Predicting the Most Likely Path and Likelihood of Test Tree

As well as HMM and HTMM, prediction is given by the most likely path, which provides the highest likelihood against a given test tree.

The most likely path and the highest likelihood can be obtained by using the following probabilities:

Definition 6.8 (Most likely backward and upward probabilities). *$H(t,i)$ is the most likely upward probability that the subtree with node t as its root is generated and the current state is i.*

$V(t,i)$ is the most likely backward probability that the subtree with node t as its root and younger siblings of t and their subtrees are all generated and the state of the current node t is i.

By using $H(t,i)$, the likelihood of test tree T can be computed as follows:

$$L_\xi(T) = \max_{m\in\mathbf{S}} \pi(m)H(r,m), \tag{6.21}$$

where r is the root. We here note that both the most likely backward and upward probabilities can be recursively computed (hereafter node c is always the eldest

Algorithm 6.2: Learning algorithm for ordered tree Markov model.

Function `Learning_ordered_tree_Markov_model`($\mathcal{T}$, $\mathcal{S}$, Φ)
  Data: tree set: $\mathcal{T}$, state set: $\mathcal{S}$, parameter: $\Phi = \{\boldsymbol{a}, \boldsymbol{b}\}$
  Result: estimated parameter: $\hat{\Phi} = \{\hat{\boldsymbol{a}}, \hat{\boldsymbol{b}}\}$
  Initialize parameters.;
  repeat
    for $T \in \mathcal{T}$ **do**
      for $p \leftarrow |V_T|$ **to** 1 **do**
        foreach $l \in \mathcal{S}$ **do**
          Compute $U_T(p,l)$ by (6.13).;
        foreach $l \in \mathcal{S}$ **do**
          Compute $B_T(p,l)$ by (6.12).;
      for $p \leftarrow 1$ **to** $|V_T|$ **do**
        foreach $l \in \mathcal{S}$ **do**
          Compute $D_T(p,l)$ by (6.16).;
        foreach $m \in \mathcal{S}$ **do**
          compute $F_T(p,m)$ by (6.17).;
    for $l \in \mathcal{S}$ **do**
      for $m \in \mathcal{S}$ **do**
        $E^{(P)}_{(p,c)}(\#(l \to m)|\mathcal{T}) = \sum_{T \in \mathcal{T}} D_T(p,l) a(l,m) B_T(c,m)$.;
        $E^{(B)}_{(p,c)}(\#(l \to m)|\mathcal{T}) = \sum_{T \in \mathcal{T}} F_T(p,l) U_T(p,l) a(l,m) B_T(c,m)$.;
    for $i \in \mathcal{S}$ **do**
      for $j \in \mathcal{S}$ **do**
        $\hat{a}(i,j) = \frac{\sum_{T\in\mathcal{T}}(\sum_{(p,c)\in \boldsymbol{P}_T} E^{(P)}_{(p,c)}(\#(i\to j)|T) + \sum_{(p,c)\in \boldsymbol{B}_T} E^{(B)}_{(p,c)}(\#(i\to j)|T))}{\sum_j \sum_{T\in\mathcal{T}}(\sum_{(p,c)\in \boldsymbol{P}_T} E^{(P)}_{(p,c)}(\#(i\to j)|T) + \sum_{(p,c)\in \boldsymbol{B}_T} E^{(B)}_{(p,c)}(\#(i\to j)|T))}$.;
  until *convergence*;
  Output the estimated parameters.;

child of node p, and q is also always one younger sibling).

$$H(p,l) = \begin{cases} b(l,o_p) & \text{if } p \text{ is a leave,} \\ b(l,o_p)\max_{m\in \boldsymbol{S}} a(l,m)V(c,m) & \text{otherwise.} \end{cases} \tag{6.22}$$

$$V(p,l) = \begin{cases} H(p,l) & \text{if } c \text{ is the youngest sibling,} \\ H(p,l)\max_{m\in \boldsymbol{S}} a(l,m)V(q,m) & \text{otherwise.} \end{cases} \tag{6.23}$$

Also the states giving the most likely path for each of the above two backward

Algorithm 6.3: Recursive backtracking algorithm.

```
Function Backtrack(p, s)
    Data: node: p, state: s
    Result: backtracking state transition
    if p has a younger sibling then
        state ← M_V(p, s).;
        Add (q, state) to MostLikelyPath.;
        p ← q.;
        if q is not the youngest sibling then
            Backtrack(q, state).;
        return.;
    if p has a child then
        state ← M_H(p, s).;
        Add (c, state) to MostLikelyPath.;
        p ← c.;
        if c is not a leaf then
            Backtrack(c, state).;
        return.;
```

and upward probabilities, are recursively saved in $\boldsymbol{M}_H$ and $\boldsymbol{M}_V$, respectively:

$$\boldsymbol{M}_H(p,l) = \begin{cases} l & \text{if } p \text{ is a leave,} \\ b(l,o_p) \arg\max_{m \in \boldsymbol{S}} a(l,m)V(c,m) & \text{otherwise.} \end{cases} \tag{6.24}$$

$$\boldsymbol{M}_V(p,l) = \begin{cases} l & \text{if } c \text{ is the youngest sibling,} \\ H(p,l) \arg\max_{m \in \boldsymbol{S}} a(l,m)V(q,m) & \text{otherwise.} \end{cases} \tag{6.25}$$

We can capture the sequence of states, which generate the most likely path, by backtracking from the final state using these $\boldsymbol{M}_H$ and $\boldsymbol{M}_V$. **Algorithm 6.4** shows a pseudocode of the algorithm for predicting the most likely path and likelihood of a given test tree. Note that this pseudocode uses recursive subroutine **Function** `Backtrack`, which is in **Algorithm 6.3**. Here suppose that $\boldsymbol{M}_H$, $\boldsymbol{M}_V$ and MostLikelyPath are already computed and shared by the subroutine of backtracking.

6.1.3 Summary of Probabilistic Models for Trees

Parent-to-child dependency: Hidden Tree Markov Model: The first probabilistic model, Hidden Tree Markov Model (HTMM), an extension of HMM, captures the parent-child dependencies, by regarding the dependency of neighboring two outputs in a sequence as the dependency between a parent and children in a tree. This model keeps the same complexity as HMM, but dependency among siblings are not considered.

Algorithm 6.4: Computing most likely state transition of tree T, using recursive function `Backtrack`(see **Algorithm 6.3**).

Function `Computing_most_likely_path_for_OTMM(`T, $\mathcal{S}$, Φ`)`
    Data: tree: T, state set: $\mathcal{S}$, trained parameter: $\Phi = \{\boldsymbol{a}, \boldsymbol{b}\}$
    Result: most likely state path: MostLikelyPath
    for $p \leftarrow |T|$ **to** 1 **do**
        for $l \in \mathcal{S}$ **do**
            Compute $H(p, l)$ according to (6.22).;
            Compute $\boldsymbol{M}_H(p, l)$ according to (6.24).;
        for $l \in \mathcal{S}$ **do**
            Compute $V(p, l)$ according to (6.23).;
            Compute $\boldsymbol{M}_V(p, l)$ according to (6.25).;
    state $\leftarrow \arg\max_m \pi(m) H(r, m)$.;
    MostLikelyPath $\leftarrow (r, \text{state})$.;
    $p \leftarrow r$.;
    `Backtrack(`p, *state*`)`.;
        Output MostLikelyPath.;

Adding sibling dependency: Ordered Tree Markov Model:
Keeping the complexity of HMM and HTMM, Ordered Tree Markov Model (OTMM) considers dependency among siblings. In OTMM, dependency among siblings are the same as sequences, while dependency among parent-child is focused on between parents and eldest siblings, due to the time complexity. We can see OTMM performs well in real experimental results shown in the later sections.

6.2 Kernel Learning

6.2.1 All Subtree Kernel

If reasonable kernel functions are defined between two trees, the kernel functions can be used in a lot of methods in kernel learning. A reasonable idea to generate kernel functions for trees is to check how many subtrees are shared between given two trees: if the number of the shared subtrees is larger, the similarity should be higher.

If two subtrees are the same we say these subtrees are *isomorphic* or *equivalent* (more precise definition will be done later). We consider labeled ordered trees. Given two labeled ordered trees, T and T', we can define a kernel function, which considers the number of possible subtrees shared between T and T', can be defined as follows [94]:

$$K(T, T') := \sum_{t \models T, t' \models T} w_t \delta_{t,t'}, \tag{6.26}$$

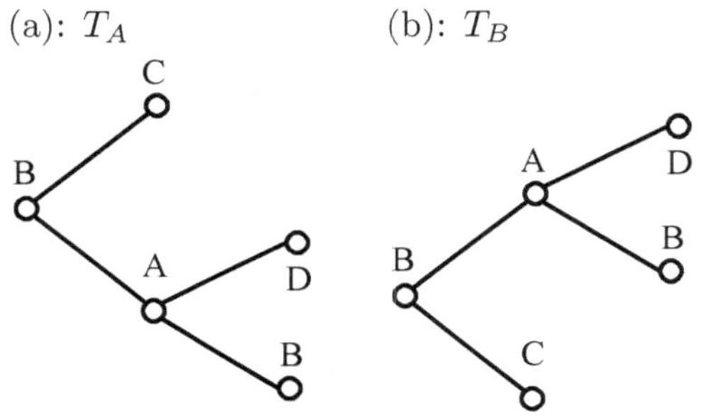

Figure 6.9: Example of equivalent trees.

where $t \models T$ means that t is a subtree of T and w_t is a weight over subtree t (all one if there are no prior knowledge), $\delta_{t,t'}$ is 1 if t and t' is isomorphic; otherwise zero. This kernel is called *all subtree kernel*.

6.2.2 Searching Isomorphic Subtrees: Transforming a Tree Following a Lexicographical Order

A first problem in this computation is to find isomorphic subtrees. For example, Fig. 6.9 (a) (T_A) and (b) (T_B) are isomorphic. From the observation over T_A and T_B in Fig. 6.9, we can see an important property to find the isomorphic subtrees:

Proposition 6.7 (Isomorphic Trees). *Two isomorphic trees (subtrees) can be the same trees (subtrees) by changing the order of the subtrees for one tree, while fixing the order of the subtrees for the other tree.*

For example, in Fig. 6.9, the root B has two children and their subtrees. We can switch the subtrees with the two children as roots for one tree, say T_A, and then T_A and T_B become the same tree. Thus, given two trees, we will search the isomorphic subtrees by changing the order of subtrees. However changing the subtree order randomly is just a waste of time. Instead, we change the subtrees, according to some order, which can be commonly used for the given two trees, and then by finishing this change beforehand for both two trees, comparison of the two trees would be much easier.

We explain this procedure by using toy examples. We use a lexicographical order for labels, i.e. simply A→B→C→D, where A is the most prioritized, and the tree topology can be changed so that node labels follow the lexicographical order, for example, in Fig. 6.9, from bottom (more prioritized) to top (less prioritized).

In Fig. 6.9, the nodes of T_A already satisfy the lexicographical order, while in T_B, two children of the root B, i.e. A and C, do not follow the lexicographical order. Then we change the order of the children with their subtrees. In fact, if we change this order, T_B and T_A are same. As such, if we order all nodes in a tree, according to the same criterion, particularly the lexicographical order, all trees can be compared with each other very easily.

The tree, which is already consistent with some criterion, say again the lexicographical order, is called a *canonical form.* Thus simply we can say that we first transform each input tree by using a lexicographical tree into a canonical form so that comparison of two trees become much easier.

6.2.3 All Subtree Kernel Computation Reduced to All Subsequence Kernel Computation

We first transform each given input tree into a canonical form, and then we can now enumerate subtrees shared by two trees, to examine the similarity between the two nodes. The issue here is how efficiently we enumerate subtrees shared by the input trees. For simplicity, we consider binary trees. For a binary tree, let n_p be a parent, and t_{c1} and t_{c2} be subtrees with ordered children of n_p, i.e. $c1$ and $c2$. Then we write the subtree with node n_p as its root (including and two children $c1$ and $c2$ and their subtrees t_{c1} and t_{c2}) as $(n_p\ (t_{c1})\ (t_{c2}))$. That is, we write a subtree by the order of its root (parent) and then its elder children to younger children. Then a subtree is written as a string in which each subtree is surrounded by the two parentheses: "(" and ")", meaning that the tree can be written by a string, keeping the order of its subtrees recursively. More simply, we can write the nested structure of a tree as a string. For example, Fig. 6.9 (a) can be written as (B (A (B D)) C). Most importantly, the resultant string keeps each subtree in the input tree as a consecutive subsequence. For example, in Fig. 6.9 (a), the subtree with node A as its root, i.e. (A (B D)), is a consecutive subsequence in the resultant string. Also if node C has some subtree, like (C (B E)), this would appear in a consecutive subsequence in the transformed substring. Thus again in the transformed string, any subtree in the input tree of a canonical form is a consecutive subsequence. From this we can see the following:

Proposition 6.8 (Equivalence to All Subsequence Kernel). *All subsequent kernel, which enumerates all subtrees shared between two input trees, is equivalent to all subsequence kernel, which enumerates all subsequences share between two sequences transformed from the canonical forms of the two trees.*

We omit the proof but this will be understood intuitively.

6.2.4 Algorithm

This means, when given two input trees, we can take the following three steps:

1. **Transform into a canonical form:** Each of the two trees can be transformed into a canonical form.

2. **Transform into a string:** The canonical form can be further transformed into a string, keeping each subtree as a subsequence within parentheses.

3. **Run computing all subsequence kernel:** Finally we can run computing the all subsequence kernel over the obtained two strings.

This can be summarized into **Algorithm 6.5**, which is a pseudocode of the above algorithm, by which we can compute all subtree kernel, which is defined by (6.26).

Algorithm 6.5: Computing a tree kernel, obtained by enumerating all possible subtrees shared between given two labeled rooted trees.

```
Function Computing_tree_kernel(T_1, T_2)
    Data: two trees: T_1, T_2
    Result: kernel function: K(T_1, T_2)
    Transform T_1 and T_2 into canonical forms T_1^* and T_2^*, respectively.;
    Transform T_1^* and T_2^* into string S_1^* and string S_2^*, respectively.;
    Compute string kernel K(S_1^*, S_2^*) by running computing all subsequence
    kernel Compute_string_kernel_with_suffix_tree(D), where
    D = {S_1^*, S_2^*}.;
    Output K(S_1^*, S_2^*).;
```

6.2.5 Time Complexity

The most important point of this computation is time complexity, which is linear time regarding the size of the input tree:

Transformation into a canonical form: In this transformation, the procedure is, for each node, to check the order of the children in terms of the lexicographical order and to change the order if it is inconsistent with the lexicographical order. So the entire procedure is done from the root to leaves, examining at most the size of nodes in a tree, meaning that the time-complexity is linear.

Transformation into a string: This also results in the procedure of examining all nodes of a given tree from the root to leaves sequentially, and so the time complexity is linear.

Running all subsequence kernel: We have already shown that all subsequence kernel can be computed by linear time-complexity in Section 5.3.

So again the time complexity of computing all subtree kernel is only the linear computation time: $O(|T_1| + |T_2|)$ where $|T_i|(i = 1, 2)$ is the number of nodes in tree T.

6.3 Frequent Subtree Mining

6.3.1 Problem Setting

We can set up the problem of enumerating all frequent subtrees from a given set of trees, as well as other problem settings of enumerating frequent patterns, such as items and subsequences.

Definition 6.9 (Frequent Subtree Mining). *Frequent subtree mining, i.e. frequent pattern mining from given trees, enumerates all frequent subtrees, where each of the frequent subtrees has a support, which is equal to or larger than the minimum support.*

Algorithm 6.6: Mining frequent subtrees, where a subtree for some node is defined as all descendants of the node.

Function `Mining_frequent_subtrees(`$\mathcal{T}$`, Min_Sup)`
  Data: tree set: $\mathcal{T}$, minimum support: Min_Sup
  Result: all frequent subtrees
  foreach *tree* $T \in \mathcal{T}$ **do**
    Transform T into canonical form T^*.;
    Transform T^* into string S^*.;
    Save S^* into $\mathcal{S}$.;
  Run frequent subsequence mining (such as `PrefixSpan_main_part(`$\mathcal{S}$`, Min_Sup)`), counting only substrings in parentheses, i.e. $(\ldots)$.;
  Output the results.;

We here explain the approaches to solve this problem.

6.3.2 Solutions

Subtrees are in general parts of a given tree. Depending on the exact definition of subtrees, we now have two different solutions.

Regular Definition of Subtrees

The regular definition of subtrees are derived from the definition of subsequences for sequences, and this regular definition was already mentioned in Chapter 2:

Definition 6.10 (Regular Definition of Subtree)**.** *A subtree has a node as the root, and the subtree is all nodes on the paths from the root to all possible leaves.*

We explained all subtree kernel in the last section on kernel learning, where subtrees in kernel learning follow the above definition. Then for the regular definition, we can use the idea of kernel learning here as well. More concretely, in kernel learning we can transform given two trees into two canonical forms, which are further transformed into two strings with subtrees are explicitly shown by parentheses. Then in kernel learning, by computing all subsequence kernel over these two strings, we can have a kernel function of two strings, i.e. given two trees. Thus we can follow these ideas, but we do use multiple trees, instead of only two trees, and so in the final step we use frequent subsequence mining, which are already explained in Section 5.1. In summary in this approach, we have the following three steps:

1. **Transformation to canonical forms:** Given multiple trees, we transform each of these trees into a canonical form, by using a lexicographical order.

2. **Transformation to strings:** We further transform each tree in a canonical form into a string.

3. Running frequent subsequence mining: We run frequent subsequence mining over a set of generated string, to enumerate all frequent subtrees. Note that we follow regular definition of subtrees, and so frequent subgraph must be subgraphs in terms of the regular definition. Thus frequent subgraphs are always specified by two parentheses "(" and ")".

Algorithm 6.6 shows a pseudocode of the above procedure. Obviously this approach is as fast as frequent subsequence mining.

Regarding Subtrees as Arbitrary Tree Pattern

The above regular definition of subgraphs was from the definition of subsequence for sequences. On the other hand, another more flexible definition of subtrees is from the definition of subgraphs:

Definition 6.11 (Flexible Definition of Subtree). *A subtree is a part of a tree. If the part is connected, it is called a connected subtree; otherwise an unconnected subtree.*

Even a connected subtree under this definition cannot be a subtree under the regular definition of a subtree. For example, in T_A of Fig. 6.9, if the edge from node A to node B and node B are removed, the remaining part (we call T_C) has the root B and its children A and C, and also node D. This part, T_C, cannot be a subtree of T_B under the regular definition of a subtree, while T_C can be a subtree under the flexible definition.

Then the key point is if we think this flexible definition of subtrees, we cannot use **Algorithm 6.6**, because this algorithm cannot enumerate T_C. Below we show the reason.

This can be obvious if we change the tree to the string. That is, T_A is already a canonical form, and transformed into string B (A (B D) C). On the other, T_C is B (A (D) C). Clearly the problem is, in strings, the part corresponding to T_C in that of T_A is "unconnected". This is not a subsequence, and we cannot use any algorithm for frequent subsequence mining. Then we cannot use **Algorithm 6.6**.

More generally we can say that under this definition of subtrees, trees (subtrees) cannot be dealt with by the algorithm for sequences, since this flexible definition of subtrees are from the definition of subgraphs. Thus we have to apply to an algorithm for frequent subgraph mining, thinking trees as a special case of graphs. We will explain ideas and methods for frequent subgraph mining in the next chapter.

6.4 Applications to Life Sciences

6.4.1 Glycans: Typical Biological Molecules as Trees

In life sciences, trees are glycans or carbohydrate sugar chains. Building blocks of glycans are monosaccharides, and they are connected each other to generate a tree. The number of types of monosaccharides is unclear (not like four of nucleotides

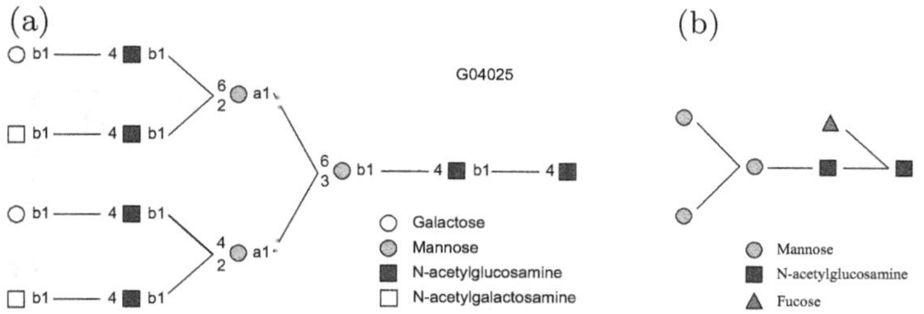

Figure 6.10: (a) Example carbohydrate sugar chain, and (b) N-glycan core-structure.

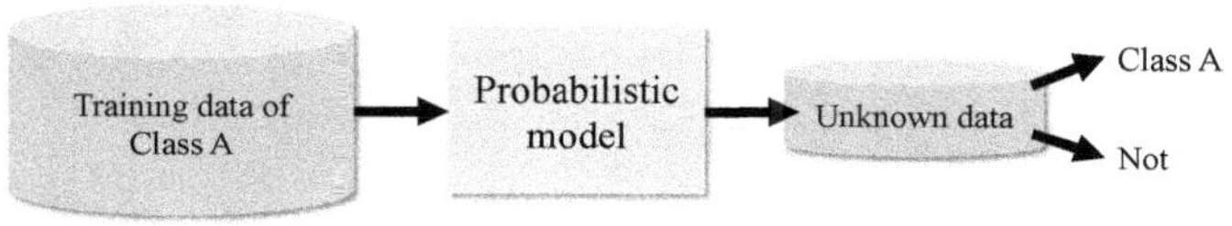

Figure 6.11: Probabilistic model for supervised learning.

and twenty of amino acids), and around 15 monosaccharides are well known and they can be labels. Usually glycans are connected to an amino acid, and so the connected monosaccharide can be regarded as the root of the glycan tree. That is, glycans can be regarded as labeled rooted trees (ordered trees). Fig. 6.10 shows two examples of glycans, where the most left-hand node of each tree is the root.

Below, we explain the applications of probabilistic models and frequent pattern mining to glycans.

6.4.2 Application of Probabilistic Models to Glycans

Predefined classes are usually used by supervised learning, while as shown in Fig. 6.11, after learning a probabilistic model from glycans in one class, we can use the trained model to predict if a given unknown glycan is in the class or not. That is, we apply a probabilistic model to supervised learning, using the class called "N-glycan". It is already known that N-glycan often has a subtree, called *core-structure* (Fig. 6.10 (b)). N-glycans are rather easy to discriminate from other glycans, and so for negatives in test data, we use synthetically generated trees, which all have the core-structure (Fig. 6.10 (b)) of N-glycans.

We used a state transition diagram with all state transitions possible, and changed the number of states in the experiments. Fig. 6.12 (a) shows AUC (see Section 10.1) of HTMM and OTMM, changing the number of states from 4 to 16. From this figure, OTMM gave better performances than HTMM, at any number of states. Also Fig. 6.12 (b) shows p-values obtained by checking paired t-test over the performance difference between HTMM and OTMM, showing the significance of the difference.

(a)

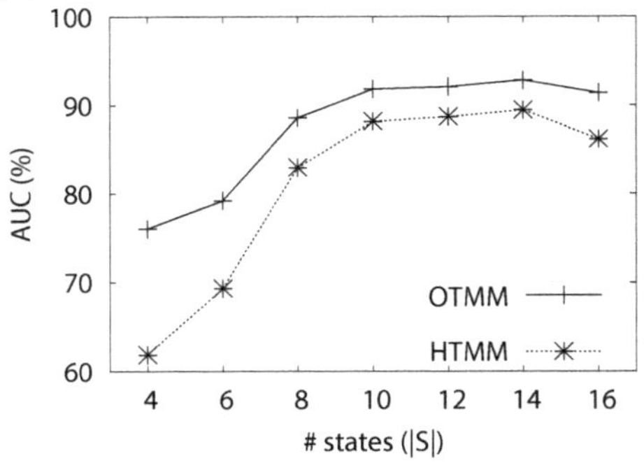

(b)

# states	OTMM vs. HTMM
4	0.00028
6	0.00306
8	0.000601
10	0.00120
12	0.00143
14	0.00125
16	4.38E-05

Figure 6.12: (a) Predictive performance (AUC) of OTMM and HTMM, and (b) the performance significance test by paired t-test.

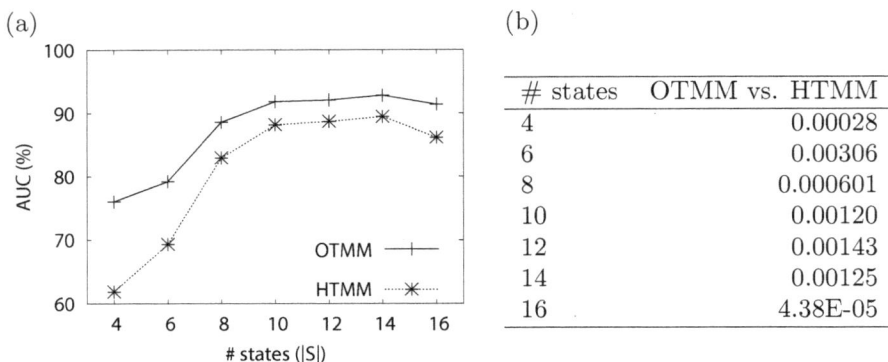

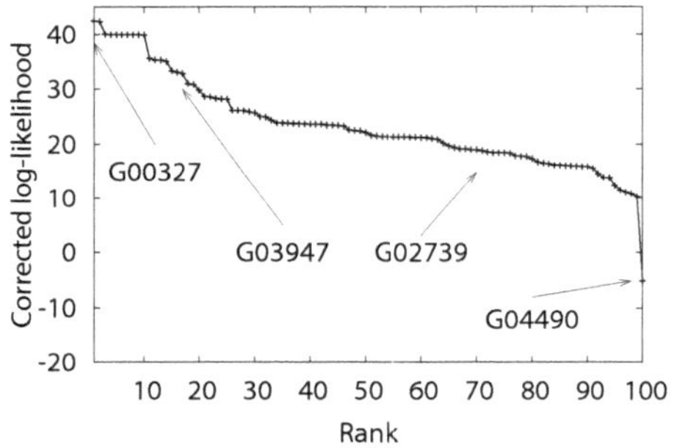

Figure 6.13: Log-likelihood distributions of 100 random N-glycans.

We then computed the log-likelihood and most likely path over randomly selected 100 N-glycans by using OTMM. Then the 100 N-glycans are sorted (ranked) according to the computed log-likelihood (Fig. 6.13). We expect that in this ranking, more highly ranked glycans are more truly N-glycans and lower ranked glycans are a bit more different from N-glycans. Fig. 6.14 shows the tree (right) and resultant most likely path (left) of glycan (ID: G00327), ranked 1st, achieving the highest log-likelihood, implying that this glycan, G00327 has the most typical N-glycan. In fact this tree has the "core-structure", shown in Fig. 6.10 (b), indicating this "core-structure" is a typical pattern of N-glycan. Also we can see four paths: N-acetylglucosamine (GlcNAc) $\rightarrow$ Galactose $\rightarrow$ sialic acid (Neu5Ac) (which we call "pattern paths" hereafter), implying that this might be another feature of N-glycan.

Fig. 6.15 shows the tree and resultant most likely path of G03947, which was ranked at 15th of 100 glycans. This glycan also has the "core-structure" of Fig. 6.10 (b) and three "pattern paths", which were found in G00327. However, Mannose has three children, and the same state (7) is assigned to two (different labels) of the three children in the most likely path (right of 6.15), which we think made the log-likelihood of G03947 smaller than G00327.

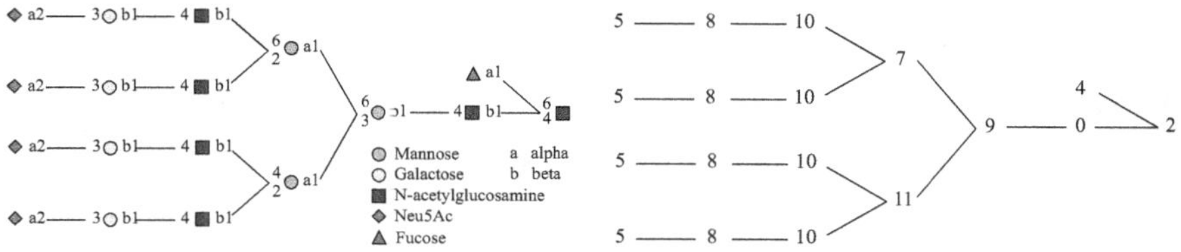

Figure 6.14: An actual glycan G00327 (left) and its most likely state path (right).

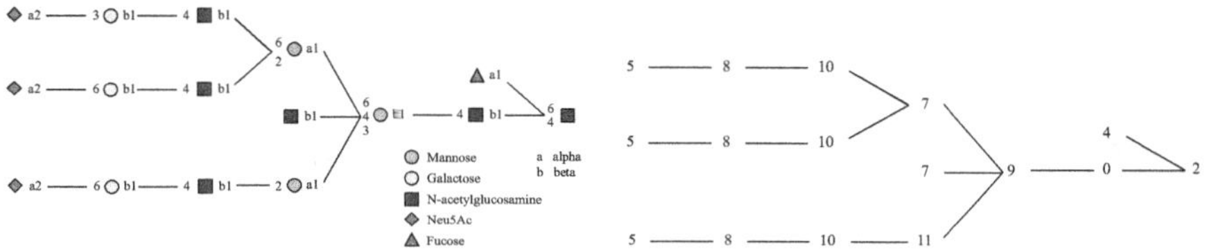

Figure 6.15: (left) An actual glycan G03947 and (right) its most likely state path.

Fig. 6.16 shows the tree and most likely path of G02739, ranked at 70th. This tree also has the "core-structure", while this tree has no "pattern paths", which were found in G00327 and G03947. We think this point makes the log-likelihood of G03947 much smaller than G00327 and G03947, although from the root to the fourth layer are totally the same as Fig. 6.14.

Fig. 6.17 shows the tree and the most likely path of G04490, which is ranked at 100th, i.e. the last of 100 N-glycans. This tree does not have the "core-structure" and also not have any "pattern paths" found in G00327 and G03947. These would be the reason why this glycan was ranked the last. Also from these, this glycan may not be in the class of N-glycan. In fact this is not examined visually by human, but this was done automatically by OTMM, indicating the effectiveness of learning and prediction of OTMM.

Also Figs. 6.14 to 6.16 suggest that two pairs of parent and child both labeled by Mannose would be another feature of N-glycan. In fact these two pairs cannot be separated by the tree (left-hand side of each figure), but the most likely path (right-hand side of each figure) shows that the state of parent Mannose is 9 and the states of children are 7 and 11, revealing that three Mannose are distinguished in the most likely state transition. This result implies that N-glycans are recognized by these three states, (9, 7. 11) in OTMM. Then tree of G04490 ranked the last also has the pair of parent and child by Mannose, but the state assigned them was both 9, being different from the typical pattern of N-glycan: (9, 7, 11). Thus, in this sense, G04490 might not be in the class of N-glycan.

Furthermore, regarding Mannose, three subclasses are known: 1) High-mannose: glycans with many Mannose, 2) Complex: glycans with various monosaccharides, and 3) Hybrid: intermediate of High-mannose and Complex. Fig. 6.18 shows the tree of a typical glycan of each class and the the most likely path of the corre-

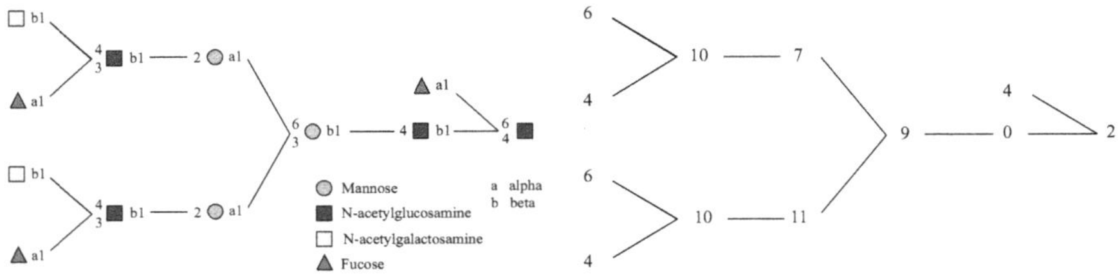

Figure 6.16: (left) An actual glycan G02739 and (right) its most likely state path.

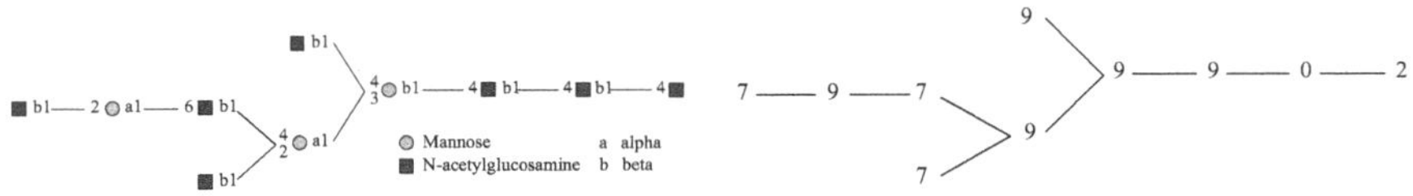

Figure 6.17: (left) An actual glycan G04490 and (right) its most likely state path.

sponding tree, obtained by OTMM trained by different training dataset from that for Fig. 6.17 and earlier. From the most likely state transition, for the glycan in High-mannose, the state of Mannose is always 1, particularly 1 and 1 for the parent and two children of Mannose. On the other hand, the glycan in Complex has 1 and 7 for parent and two children of Mannose. Interestingly, in the glycan of Hybrid, these two are mixed, and the parent state is 1 and two children states are split into 1 and 7. From this we can see that the most likely state transition diagram shows the property of glycan well by the state assignment to labels.

6.4.3 Application of Frequent Subtree Mining to Glycan

We have seen several interesting results by OTMM which are relevant to biological knowledge, while probabilistic models are unable to provide the acquired patterns so explicitly. On the other hand, frequent pattern mining is useful to show the frequent patterns in data directly.

As mentioned already, subtrees can be defined in two different ways. Here we use the flexible definition, meaning that we applied frequent subgraph mining (see the next chapter))to a set of trees.

If we use the support for frequent patterns, many small patterns appear. Then to remove insignificantly appearing small patterns, we generated negatives (parent-child pairs in positives randomly switched). Then we generate the (2×2) contingency table for each pattern and apply Fisher's exact test over the table to check the bias in the table. Then finally frequent patterns were ranked by p-values of Fisher's exact test (see Section 10.1.1 for the Fisher's exact test). This test checks the patterns peculiar to positives, and more peculiar, lower the p-value is. Thus smaller patterns are likely to have higher p-values and removed.

We used around 7,500 glycans as given instances. Table 6.1 shows δ-tolerance

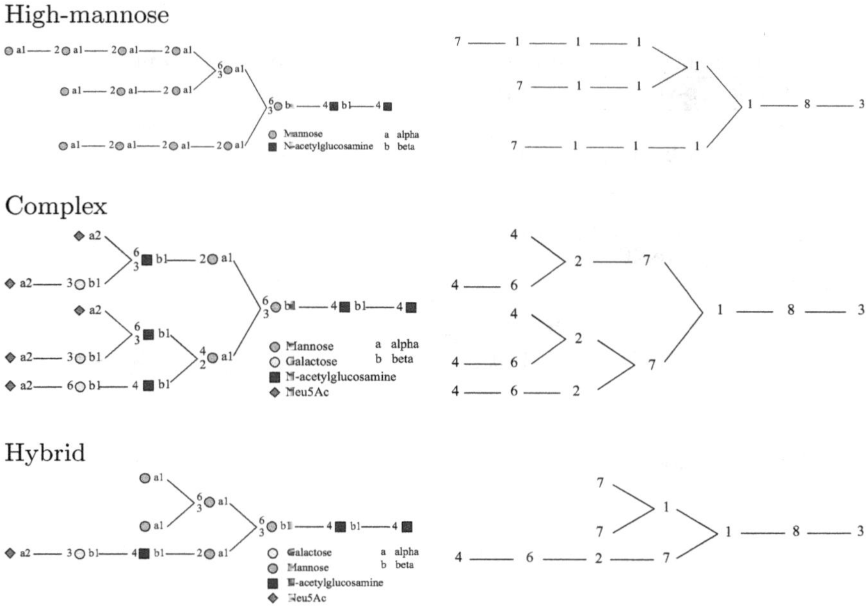

Figure 6.18: (left) glycans representing each of the three N-glycan sub-classes (High-mannose, Complex and Hybrid), and (right) their most likely state paths.

closed frequent subtrees (see Section 7.1.4 for δ-tolerance closed frequent pattern mining) ($\delta = 0.6$) ranked by p-values of Fisher's exact test under the minimum support of 0.5%(= around 38). For example, the sixth pattern is a typical subtree found in High-mannose glycans (ex. Fig. 6.18), and the seventh to ninth are all frequent patterns of N-glycans, implying that they are major glycans.

Table 6.1: Top ten frequent subtrees, sorted by p-values.

Rank	p-value	Support	Subtree
1	1.6e-46	381	○β1 / ▲α1 ⟩ ⁴₃ ■
2	1.1e-40	164	○β1—4■β1 ⟩ ⁶₃ □ ; ○β1
3	5.0e-26	109	○β1—3■β1—3○β1—4●
4	5.6e-26	233	○β1—3■β1—3○
5	8.2e-26	83	▲α1 / ○β1 ⟩ ⁴₃ ■β1—3○
6	1.3e-24	79	○α1 ⟩⁶₃○α1 ; ○α1 ⟩⁶₃ ; ○α1—2○α1—2○α1 ⟩⁶₃○β1—4■
7	2.7e-24	78	○β1—4■β1 ⟩⁴₂○α1 ; ○β1—4■β1 ; ○β1—4■β1—2○α1 ⟩⁶₃○β1—4■β1—4■
8	2.9e-21	68	○β1—4■β1—2○α1 ; ○β1—4■β1 ⟩⁴₂○α1 ; ◆α2—6○β1—4■β1 ⟩⁶₃○β1—4■β1—4■
9	2.9e-21	68	○β1—4■β1 ⟩⁶₂○α1 ; ○β1—4■β1 ; ○β1—4■β1 ⟩⁴₂○α1 ; ○β1—4■β1 ⟩⁶₃○β1—4■β1 ⟩ ▲α1 ⁶₄■
10	3.2e-20	74	▲α1—2○β1—3■β1—3○

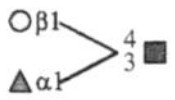

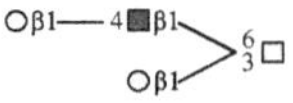

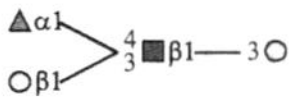

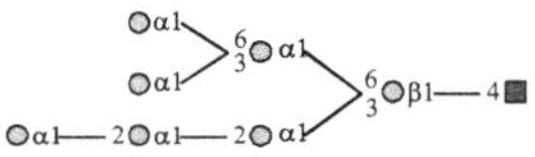

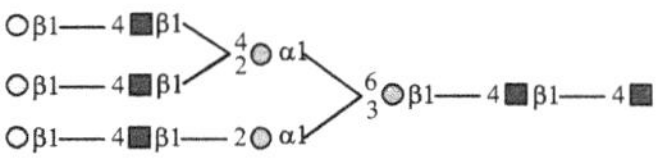

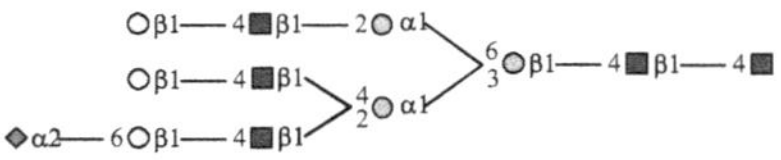

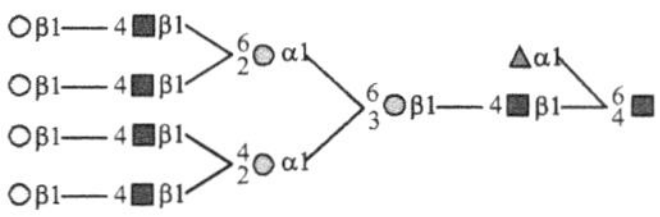

▲ Fucose (Fuc)
◔ Mannose (Man)
○ Galactose (Gal)
□ N-acetylgalactosamine (GalNAc)
● Glucose (Glc)
■ N-acetylglucosamine (GlcNAc)
◆ Neu5Ac

Chapter 7

Learning Graphs

A graph consists of nodes and edges, usually written as $G = (V, E)$, where V is a set of nodes and E is a set of edges. In machine learning graphs are used as data in two different ways: 1) one graph is one instance, where input data is multiple graphs and 2) one node in each graph is an instance, where input data is one graph (or multiple graphs sharing a node set). In this chapter, we focus on the first way, and the second one is considered in the next chapter.

A typical example of graphs in life science is chemical compounds, where chemical structure of each compound is a graph, called a *molecular graph*. Nodes of a molecular graph is labeled by elements, such as hydrogen, carbon and oxygen. The number of chemical compounds is huge, while recently public databases of chemical compounds are made, such as PubChem of NIH (National Institute of Health) [53] and ChEMBL of EMBL (European Molecular Biology Laboratory) [10]. Throughout these databases, we can have the chemical structures of a huge number of chemical compounds (molecular graphs). In fact the number of entries of PubChem reaches around 90 millions, although each of them might not be totally unique. In general the complexity of molecular graphs might have limitations, say the number of edges per graph [99], while we consider general labeled graphs without any constraints.

Given multiple graphs as input data, we can think about three types of learning: probabilistic models, frequent pattern mining and kernel learning. However, probabilistic models cannot be more complex than directed acyclic graphs (because cycles cannot be represented by probabilistic models), as explained in Section 3.1.7. Also complexity is very high even for directed acyclic graphs. Thus we focus on frequent pattern (subgraph) mining and kernel learning.

We first define graphs.

Definition 7.1 (Graph). *A graph is a set of nodes and a set of edges, where one edge connects two nodes, and both nodes and edges can be labeled.*

As explained in the preceding chapter, subtrees can be defined in two ways, and regular definition is derived from subsequences, while flexible definition is from subgraphs. Here we define subgraph.

Definition 7.2 (Subgraph). *A subgraph is a part of a graph, usually specified by a set of nodes. If nodes in a subgraph are all connected (any node can be reached from all other nodes by edges), this subgraph is called a connected subgraph; otherwise the subgraph is unconnected.*

We focus on connected subgraphs. We start with frequent subgraph mining and then move to kernel learning.

7.1 Frequent Subgraph Mining

7.1.1 Problem Setting

First we define *support*.

Definition 7.3 (Support). *Given multiple graphs, the number of graphs, where a subgraph appears in each of the multiple graphs, is called the support of the subgraph. Let $\mathcal{G}$ be a set of given graphs, and G be a graph in the set. Let g be a subgraph, we can write*

$$support(g) = \sum_{G \in \mathcal{G} | g \text{ is a subgraph of } G} 1. \tag{7.1}$$

Multiple graphs and minimum support are the input of the problem, which will be defined, as follows:

Definition 7.4 (Frequent Subgraph Mining). *Frequent pattern mining for graphs, i.e. frequent subgraph mining, is to enumerate all frequent subgraphs, where their supports are equal to or larger than the minimum support (min_sup).*

Let $\mathcal{F}$ be a set of all frequent subgraphs, and using subgraph g, we can write $\mathcal{F}$, according to the above definition, as follows:

$$\mathcal{F} = \{g | support(g) \geq min\_sup\}. \tag{7.2}$$

Also the most important property of frequent pattern mining can be written for graphs, as well:

Proposition 7.1 (Downward Closure for Frequent Subgraph Mining). *If subgraph g is not frequent any more, any subgraph having g is not frequent any more.*

The basic idea of frequent itemset mining was to start with a small itemset, increasing the size of itemsets, and if the itemset is not frequent, stop increasing the itemset in that direction, according to the downward closure property. As well as frequent itemset mining, frequent subgraph mining also can have a similar algorithm: we can start with small graphs, increasing the size of graphs and checking if the increased graphs are frequent, and stop increasing the graph if the current graph is not frequent any more.

Thus the definition and the procedure are both very similar between frequent itemset mining and frequent subgraph mining, and it might look that we can solve the problem of frequent subgraph mining rather easily, while we have to point out that we have to deal with much more larger data space in subgraphs than itemsets, from at least the following two points:

Not only node combinations but edge combinations: Even for the same set of nodes, usually many subgraphs are possible. For example, for itemsets, {A, A, B, C} is just only one itemset, while if we regard this as a graph with four nodes (A, A, B, C), There are various possibilities of generating graphs out of these four nodes, depending on which nodes are connected by edges.

One-edge extension with the possibility of the number of nodes: When increasing the size of an itemset, we can just add one item to an itemset, resulting in another itemset, which has both the items in the original itemset and the new item. That is, adding one item to an itemset generates only one itemset. On the other hand, when we add a node to some node of subgraph g with a new edge, we have $|g|$ possibilities (nodes) to add the new node, where $|g|$ is the number of nodes of g.

Thus entirely the number of possible combinations is always much larger in frequent subgraph mining than frequent itemset mining. In other words, exploring frequent subgraphs has to deal with a huge data space, which makes frequent subgraph mining harder, and we use more ideas to reduce the data space than frequent item set mining. Below we will describe how this high combination issue is tackled.

7.1.2 gSpan Algorithm

We introduce an algorithm for frequent subgraph mining, called *gSpan*[1] [100, 88], which enumerates all possible frequent subgraphs.

Minimum DFS Code – Uniquely Assign One Code to One Graph

The first attempt of gSpan is to assign an unique code to each graph. In fact as mentioned above, the data space of subgraphs is huge, and so we need a code which can define each graph uniquely; otherwise when enumerating subgraphs, we may be unable to count a subgraph properly.

gSpan scans a given graph based on the depth-first search (DFS), and the priority in search is decided by the lexicographical order (pregiven) over labels of nodes and also those of edges. According to DFS, one edge (u, v) is represented by the following four-tuple:

$$((\text{index}(u),\text{index}(v)),\ \text{label}(u),\ \text{label}(u,v),\ \text{label}(v)),$$

where index(u) is the number, attached to each node in the order of DFS. label(u) and label(u,v) are labels of node u and edge (u, v), respectively. We use this four-tuple to define the following code, which can be uniquely assigned to a subgraph:

Definition 7.5 (DFS Code). *The DFS code of a graph is a sequence of the four-tuple of each edge of the edges retrieved in the order of DFS over the graph.*

[1] We note that the algorithm to be introduced here is slightly modified from the original algorithm for explanation.

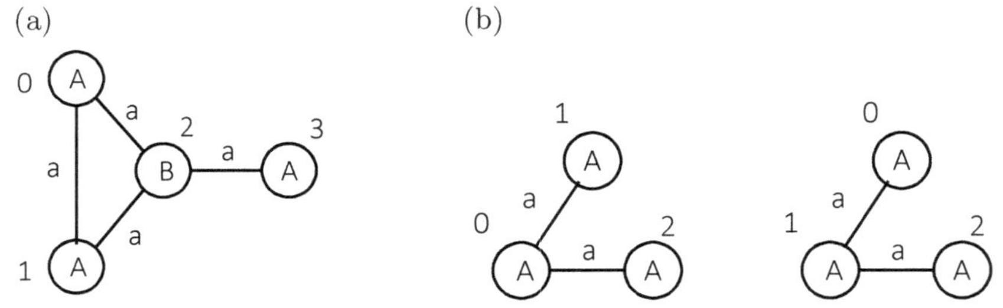

Figure 7.1: DFS code examples.

Fig. 7.1 shows simple examples. If we show DFS codes randomly, the number of DFS codes would be very large, and so we use that regarding node labels, A is more prioritized than B, according to the lexicographical order of labels.

Then in (a), for example, gSpan scans the graph, starting with the top left-sided node A, and then the next node is the bottom-left node A (because A is more prioritized than B), meaning that the edge connected by two As is more prioritized among edges, and starting with this way is better than starting from the node A of the right-hand side, which has to go to node B (A-A is more prioritized than A-B). Then from the bottom-left node A, the next node is node B. Thus from node B, there are two choices: going to the right first or left first:

((0,1),A,a,A), ((1,2), A, a, B), ((2,0),B,a,A), ((2,3),B,a,A).
((0,1),A,a,A), ((1,2), A, a, B), ((2,3),B,a,A), ((2,0),B,a,A).

Then we use the lexicographical order of nodes (and the same labels for edges), and still we have two possible DFS codes for the same graph. As such, by the lexicographical order of nodes only, we can see that we can have not only one but also multiple DFS codes for one graph.

However if we have some complete rules over the priority, like the lexicographical order over both nodes and edges, we can fix the most prioritized DFS code. For example, in (b) of Fig. 7.1, the node labels and edge labels are all same, under which the order in node labels and edge labels cannot work. In fact there are two possible DFS codes for the right and left sides of (b):

((0,1), A, a, A), ((0,2), A, a, A).
((0,1), A, a, A), ((1,2), A, a, A).

Even in this case, for example, we can prioritize the DFS code, in which smaller numbers for nodes are used, and also smaller appear earlier. Then the first code (left graph) would be selected. Thus entirely we can see that the most prioritized code can be decided by using complete rules.

If the most prioritized DFS code is decided for each graph, we can compare different graphs each other through the most prioritized ones. We call this most prioritized code as the *minimum DFS code*, which should have the following property:

Proposition 7.2 (Minimum DFS Code). *Minimum DFS code is unique and complete. These two properties mean:*

Uniqueness: *Among numerous DFS codes for a particular graph, the minimum DFS code is the only one DFS code.*

Completeness: *For any graph, there must be the minimum DFS code, meaning that all graphs have the minimum DFS codes. Thus the space is completely covered by the minimum DFS codes only.*

These properties can be understood intuitively. The next property is further useful:

Proposition 7.3. *The prefix subcode of the minimum DFS code is always the minimum DFS code of the corresponding graph.*

As mentioned by toy examples, a DFS code shows a process of scanning over a graph. In other words, a DFS code shows a process of growing the graph from just one node to complete the corresponding graph. Thus the above property says that the minimum DFS code of a graph can be generated from only the minimum DFS code of a smaller subgraph on the graph growing process. This reduces the data space of graphs greatly.

Algorithm Framework

From these observations, we always need to decide not only the order of nodes but also the order of edge labels, and also additional detailed rules over priority. We call these criteria *DFS lexicographical order*. That is, under DFS lexicographical order, the minimum DFS code is already decided for any graph.

Thus from the above consideration, the following algorithm framework can be generated:

1. **Generate one-edge graphs:** We generate one-edge graphs, in the DFS lexicographical order, and compute the support of each of all one-edge graphs.

2. **Extend graphs by one-edge:** We extend each of generated graphs by one-edge, and this procedure is repeated. However when we add one edge, the generated graph can be discarded (pruned) if either of the following two conditions is not satisfied:

   Condition 1: Minimum DFS code: Adding one edge to the current graph means adding 4-tuple to the last of the (minimum) DFS code of the current graph. Then we can check whether the resultant DFS code is the minimum DFS code. If the minimum DFS code, we can keep the new graph; otherwise it is discarded.

   Condition 2: support $\geq$ min_sup: If the support of the new graph obtained by adding one edge to the current graph is larger than or equal to min_sup, we can keep the new graph.

Please note that in this algorithm framework, **Condition 1** is from **Proposition 7.3**, while **Condition 2** is from **Proposition 7.1**.

Algorithm 7.1: `Mine_Subgraph`: Recursive searching frequent subgraphs.

```
Function Mining_frequent_subgraphs(G, D, S)
    Data: graph: G, label set: D, frequent subgraphs: S
    Result: frequent subgraphs
    if G is not a minimum DFS code then /* Condition 1 */
        return ;
    Add G to S ;
    foreach G', one edge extension of G by D do
        if support of G' is not less than Min_Sup then /* Condition 2 */
            Mining_frequent_subgraphs(G', D, S) ;
```

Checking Condition 1

For **Condition 1**, we need to check if the DFS code of the graph generated by adding one edge is the minimum DFS code. We can use the following procedure to check this. That is, first if DFS code c of graph G is given, G can be easily generated from c:

$$G = \mathrm{graph}(c). \tag{7.3}$$

Also in the above algorithm, G is generated from F with a one smaller number of edges and DFS code b of graph F which is already known. So c is also easily generated by adding the 4-tuple of one edge to b.

$$c = \mathrm{code}(G). \tag{7.4}$$

If c is the minimum DFS code, the following is correct.

$$c = \min\{\mathrm{code}(\mathrm{graph}(c))\}. \tag{7.5}$$

Thus, in order to check if the newly generated code c is the minimum DFS code of graph G, we can check if (7.5) is true for them.

Algorithm

Algorithm 7.2 shows a pseudocode obtained by implementing the idea of the above algorithm framework. This algorithm recursively call the function for exploring frequent subgraph: `Mining_Frequent_Subgraph` to obtain the minimum DFS code and enumerate frequent subgraphs.

In Section 6.3, we showed two definitions of subtrees, and for the flexible definition of subtrees, the solution was to apply the frequent subgraph mining algorithm to the problem of frequent subtree mining. In fact, **Algorithm 7.2** is for graphs, while this algorithm can be applied to the problem of frequent subtree mining, where subtrees are defined by the flexible definition in Section 6.3.

Algorithm 7.2: Main part of the gSpan algorithm.

Function `Main_part_of_gSpan`($\mathcal{G}$, $\boldsymbol{D}$, Min_Sup)
  Data: graph set: $\mathcal{G}$, label set: $\boldsymbol{D}$, minimum support: Min_Sup
  Result: all frequent subgraphs
  Sort labels in $\boldsymbol{D}$ by their frequency.;
  Remove infrequent labels and corresponding nodes (also and edges for which connected nodes are removed).;
  $\boldsymbol{S}^1 \leftarrow$ all frequent 1-edge graphs.;
  Sort 1-edge graphs in $\boldsymbol{S}^1$ in DFS lexicographical order.;
  $\boldsymbol{S} \leftarrow \boldsymbol{S}^1$.;
  foreach *edge in* $\boldsymbol{S}^1$ **do**
    Generate graph G as one edge graph.;
    `Mining_frequent_subgraphs`(G, $\boldsymbol{D}$, $\boldsymbol{S}$).;
  Output all $G \in \boldsymbol{S}$.;

7.1.3 Reverse Search

Criterion to Satisfy the Completeness and Uniqueness

We have shown the frequent subgraph mining algorithm, which focuses only on the minimum DFS codes to reduce the originally huge search space, to make the algorithm correct and efficient. In fact, to guarantee that the algorithm is correct and efficient, the following two points must be satisfied:

Completeness (exhaustive enumeration): All frequent subgraphs are enumerated.

Uniqueness: The same frequent subgraph is not counted more than once.

Then let us consider a general setting of *searching*, being away from frequent subgraph mining. In fact, an abstract and general concept to guarantee the above *completeness* and *uniqueness* of the algorithm for searching in a space is *reverse search* [6]. The completeness means covering the entire possible search space corresponds to the above *correctness*, while the uniqueness means no redundancy, connecting to the above *efficiency*.

In reverse search, a *rooted spanning tree*, called an *enumeration tree*, is defined:

Definition 7.6 (Enumeration tree). *Enumeration tree is a rooted spanning tree, which is the search space. For example, given graphs, to enumerate subgraphs,*

1. **Search space:** *Each node corresponds to a graph to be examined in the search space.*

2. **Root:** *The root corresponds to a graph with no edges.*

3. **Nodes in the same layer:** *Nodes in the 1st layer, which is directly connected to the root, correspond to all one-edge graphs. As such, nodes in*

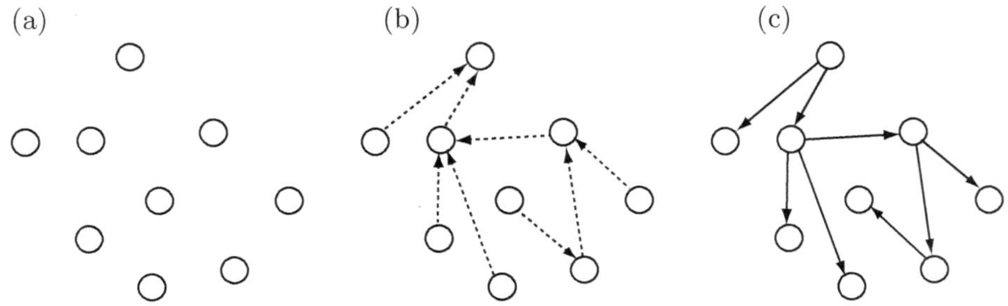

Figure 7.2: Schematic diagram of "reverse search": (a) nodes to be enumerated, (b) unique parents of nodes decided and (c) reverse search: enumeration tree over (a).

the k-th layer correspond to all graphs with k-edges and all these graphs in the same layer are different.

Again the enumeration tree is exactly the search space. If the algorithm is designed so that each of all nodes should be visited only once, the algorithm must satisfy the completeness and uniqueness.

When we generate the enumeration tree (rooted spanning tree), to satisfy the above two points, i.e. completeness and uniqueness, the important criterion is as follows:

Proposition 7.4 (Enumeration Tree Satisfies Completeness and Uniqueness). *For an enumeration tree with nodes covering the search space entirely, if each node has only one single parent, an algorithm which can search all nodes along with the edges from the root can satisfy the completeness and uniqueness.*

In the enumeration tree, starting from the root, children are generated from the parent. Each time when a child is generated from the parent, we need to check not only the direction from the parent to its children, but each child to the parent (i.e. a *reverse way*) so that a child has to be only one parent. Thus as the above proposition says, the criterion that each node always has only one parent satisfies the above completeness and uniqueness. Fig. 7.2 shows a schematic diagram of reverse search. In this figure, (a) just shows a set of nodes, which must be searched. In (b), when we generate parent-child relations, we need to check them so that each child has only one parent. Then in (c), the search from the parent to children is so efficient as to satisfy the completeness and uniqueness.

We have explained a rather general concept. Let us go back to the problem of frequent subgraph mining. For frequent subgraph mining, first each node of the enumeration tree corresponds to the minimum DFS code. Each graph has only one minimum DFS code to satisfy the uniqueness. Also we can use **Proposition 7.3** as a criterion for a node (child) to select its parent uniquely. That is, to decide the parent of one node with the minimum DFS code $c = (a_1, a_2, \dots, a_m)$, we can first remove the last part of the minimum DFS code, i.e. $c = (a_1, a_2, \dots, a_{m-1})$, and decide the node corresponding to this code, as its parent. In this way, by

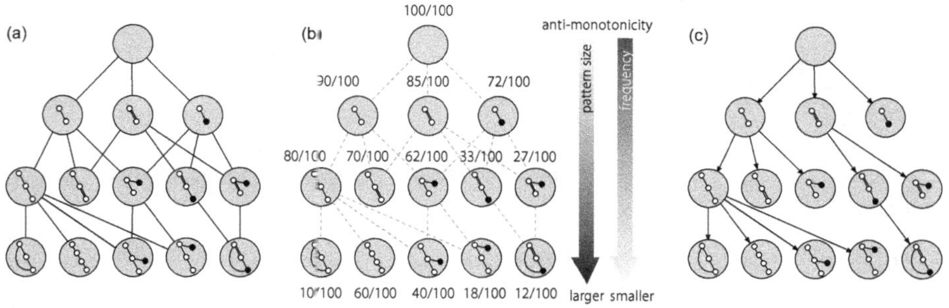

Figure 7.3: Schematic diagram of an enumeration tree for frequent subgraph mining (given 100 graphs): (a) a so-called Hasse diagram from subgraph isomorphism, (b) the number of support, where 90/100 indicates the corresponding subgraph appears in 90 cases out of all given 100 graphs and (c) an enumeration tree, generated from (a).

deciding the parent of each node, we can organize the enumeration tree, allows us to perform reverse search.

Thus, in **Algorithm 7.2**, the procedure uses the minimum DFS code only for search, by checking all possibilities of extending one-edge from the current minimum DFS code, and keeps only the minimum DFS code among them. This procedure covers the search space, achieving the above completeness and uniqueness. Fig. 7.3 shows the real procedure of **Algorithm 7.2**: (a) Each node corresponds to a graph, and two nodes are connected by an edge if a graph corresponding to one of the two nodes can be generated from the graph corresponding to the other node. This diagram is called a *Hasse diagram*. The Hasse diagram is a graph, while keeping layers as a tree, and the size (the number of edges) of graphs at each layer is kept as the same. (b) Note that the support is monotonically decreased as the layer goes to deeper, meaning that the number of edges in a graph increases. (c) In the real algorithm, we just choose the minimum DFS code only, resulting in the enumeration tree (redundant edges in (a) are not used).

We explain a general concept: reverse search for frequent subgraph mining, while this idea can be applied to itemsets, subsequences, subtrees and subgraphs for frequent pattern mining. This concept guarantees the validity of the algorithms for all these patterns, i.e. itemsets, subsequences, subtrees and subgraphs.

7.1.4 Removing Redundant Frequent Subgraphs

In frequent pattern mining, smaller patterns included in a frequent pattern (its support is equal to or larger than the minimum support) are all frequent. This property makes outputs redundant.

To reduce this redundancy, two concepts of frequent patterns are defined. Let $\mathcal{P}$ be a pattern, support($\mathcal{P}$) be the support of $\mathcal{P}$ and min_sup is minimum support. Given two patterns $\mathcal{P}$ and $\mathcal{P}'$, let $\mathcal{P} \subset \mathcal{P}'$ be the situation that pattern $\mathcal{P}'$ is larger than $\mathcal{P}$ and also included in $\mathcal{P}$.

Definition 7.7 (Maximal frequent pattern).

$$\mathcal{F}_M = \{\mathcal{P} \mid (\mathit{support}(\mathcal{P}) \geq \mathit{min\_sup}) \ \& \ (\mathit{support}(\mathcal{P}') < \mathit{min\_sup})\}. \ (\mathcal{P} \subset \mathcal{P}') \tag{7.6}$$

That is, maximal frequent pattern is a frequent pattern, where all patterns larger than this pattern are infrequent.

Definition 7.8 (Closed frequent pattern $\mathcal{P}$).

$$\mathcal{F}_C = \{\mathcal{P} \mid (\mathit{support}(\mathcal{P}) \geq \mathit{min\_sup}) \ \& \ (\mathit{support}(\mathcal{P}') \neq \mathit{support}(\mathcal{P}'))\}. \ (\mathcal{P} \subset \mathcal{P}') \tag{7.7}$$

That is, closed frequent pattern is a frequent pattern with support S, where the support of any larger pattern is smaller than S.

Thus, the maximal frequent pattern is the largest frequent pattern, while the set of only closed frequent patterns means that among frequent patterns with the same support, all frequent patterns are removed except that with the largest support. Thus, for the set of regular, closed and maximal frequent patterns, $\mathcal{F}$, $\mathcal{F}_c$ and $\mathcal{F}_M$, respectively,

$$\mathcal{F} \supseteq \mathcal{F}_c \supseteq \mathcal{F}_M. \tag{7.8}$$

If we use the set of maximal frequent patterns, we keep only the largest frequent patterns, by which frequent patterns are expected to be reduced a lot. On the other hand, if we use the set of closed frequent patterns, only frequent patterns with the same support are reduced, and so, it is expected that as increasing the size of given data, different frequent patterns are not more likely to have the same support, making it harder to reduce the redundant patterns. Thus, we have a definition, which connects these two concepts:

δ-tolerance closed frequent pattern

$$\mathcal{F}_{\delta C} =$$
$$\{\mathcal{P} \mid (\text{support}(\mathcal{P}) \geq \text{min\_sup}) \ \& \ (\text{support}(\mathcal{P}') < (1-\delta)\text{support}(\mathcal{P}))\}. \tag{7.9}$$

$$(\mathcal{P} \subset \mathcal{P}', 0 \leq \delta \leq 1) \tag{7.10}$$

That is, δ-tolerance closed frequent pattern is a frequent pattern with support S, where the support of any larger pattern is less than (1 - δ) $\times S$.

The support has the monotone property. That is, the support always decreases (or keeps the same) as the size of patterns increases. Thus starting with the smallest pattern, when we increase the size of patterns, the pattern is always frequent until the support comes down to the minimum support. In this process of increasing the pattern, maybe the support of a pattern $\mathcal{P}'$ (generated from $\mathcal{P}$ by one-edge extension) decreases suddenly a lot. In this case, we can regard pattern

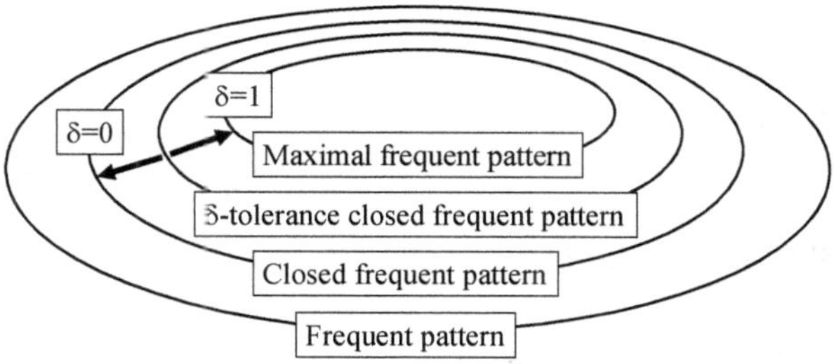

Figure 7.4: Hierarchy of frequent patterns.

$\mathcal{P}$ as an important pattern and this pattern can be a δ-tolerance closed frequent pattern, since support($\mathcal{P}'$) can be smaller than (1 - δ) $\times$ support($\mathcal{P}$) (7.9).

In particular, δ-tolerance closed frequent pattern is a good property that this pattern is the same as the closed frequent pattern when δ=0, and the same as the maximal frequent pattern when δ=1, as shown below:

Proposition 7.5.

$$\mathcal{F} \supseteq \mathcal{F}_c \supseteq \mathcal{F}_{\delta C} \supseteq \mathcal{F}_M. \quad (0 \leq \delta \leq 1) \tag{7.11}$$

Fig. 7.4 shows the above property as the hierarchical structure of frequent patterns. We note that using this property, we can reduce the redundant frequent patterns, particularly controlling the number of frequent patterns by changing δ.

One approach of generating δ-tolerance closed frequent patterns is to keep such patterns after generating all frequent patterns. However, we can generate δ-tolerance closed frequent patterns in the algorithm enumerating frequent patterns [88]. In particular, the number of $\mathcal{F}_C$ can be smaller than the number of $\mathcal{F}$ in the algorithm already, which contributes to that the size of the search space in the enumeration tree is pruned, reducing the computation cost. That is, the advantage is in not only reducing the redundancy but also fast computation.

7.2 Kernel Learning

If we define the kernel function between instances, i.e. graphs, we can perform kernel learning regardless of the data types of instances. Thus we consider kernel functions for graphs, i.e. two given graphs.

Any polynomial time algorithm is still unknown for the graph isomorphism problem, which checks if given two graphs G_1 and G_2 are isomorphic. This means that we do not have any efficient (polynomial-time) procedure to compute the distance of two given graphs. This is true of computing the kernel function of two given graphs. Then a lot of (heuristics-based) methods have been proposed to design the graph kernel functions, by using various properties of graphs. Thus we here briefly show these methods to compute the kernel function $K(G_1, G_2)$ for two graphs G_1 and G_2 which we call *graph kernel*. Below we will describe the methods

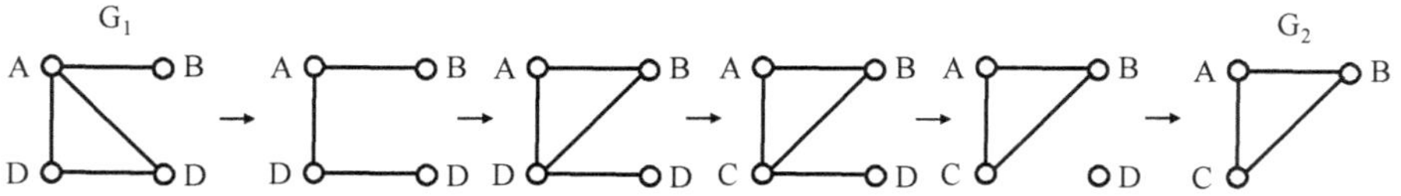

Figure 7.5: Example of editing G_1 to transform into G_2.

and ideas, from simpler ones to more complex ones, which are summarized into five types:

1. **Edit distance kernel [65]:**
   This kernel can be computed from the number of steps to modify from G_1 to G_2 by using prefixed operations, such as removing an edge, adding an edge and an node, etc.,

   For example, edit distance $d(G_1, G_2)$ can be computed as follows:

$$d(G_1, G_2) = \min_{\pi \in \Pi(G_1, G_2)} \sum_{q \in \pi} c(q), \tag{7.12}$$

   where π is a set of operations to edit from G_1 to G_2, and Fig. 7.5 is an example of π. $\Pi(G_1, G_2)$ is a set of π, and q is a step (operation) of π, and $c(q)$ is the cost of operation q.

2. **Path or shortest path kernel [13]:**
   For each graph, we can sample two nodes and then check the path between the two nodes. Enumerating all paths of all node pairs is a NP-hard problem, while computing the shortest path (distance) is tractable in polynomial time. By using this property, the shortest path kernel of node pairs can be defined as follows:

$$K(G_1, G_2) = \sum_{v_i, v_j \in V_1} \sum_{v_k, v_l \in V_2} K_{\text{len}}(d(v_i, v_j), d(v_k, v_l)), \tag{7.13}$$

   where K_{len} is a kernel function, comparing two shortest distances $d(v_i, v_j)$ and $d(v_k, v_l))$, which can use a variety of kernels, such as linear kernel:

$$K_{\text{len}} = d_{v_i, v_j} \cdot d_{v_k, v_l}. \tag{7.14}$$

3. **Random walk kernel [93]:**
   The idea behind this kernel is when random walk over each graph is repeated, the same walk shared by two graphs appears more, the distance between two graphs should be closer. This kernel needs random walk many times, which rises the computational load, while this kernel is most well-studied and computational efficiency is being improved. This can be regarded as a general case of the above kernels using the distance between two graphs.

4. **Subtree kernel [69]:**
   We repeat sampling a subtree from each graph randomly, where the subtree is by the flexible definition in Section 6.3. This is sampling with replacement, meaning that any overlap is allowed (random sampling is regarded as the same process of sampling a path randomly, instead of the shortest distance path). Then multiple subtrees from each graph are compared in a many-to-many way. We can think that a subtree has structure information, comparing with the distance-based kernel described above, while the problem is computational load.

5. **Subgraph kernel [79]:**
   We first fix several small graphs, called *graphlets*, and compute the distance between two graphs, if each of the graphlets is shared by the two graphs. Again solving the graph isomorphism problem has no efficient algorithms, which makes computation heavy, even with small graphlets only.

Below rather than going into the detail of these approaches, we consider to generate kernel functions which are similar to each of the above five categories, by using the techniques we have studied on graphs. In particular, we can focus on the two following practical situations.

1. **Learning from multiple graphs:** Graph kernel functions are considered only for two input graphs while regular machine learning settings using graph kernels assume a larger number of graphs as input. In fact usually above kernel functions just focus on two input graphs, without considering the multiple graphs behind the two graphs. Kernel functions are used for machine learning settings, and so it would be more reasonable to consider the kernel functions under the situation that multiple graphs are given.

2. **Prior knowledge of enumeration tree:** In frequent subgraph mining, we define the minimum DFS code, through which we can check the graph isomorphism among multiple graphs easily. Then the enumeration tree over the minimum DFS codes can be built. The enumeration tree is a tree, which makes the computation highly efficient. Furthermore, similar to the point already mentioned in the above #1, the enumeration tree is generated by using given multiple input graphs and pruned over the patterns (subgraphs), which are not frequent, because the enumeration tree is built for frequent subgraph mining. Of course, if computation time is allowed, the minimum support can be set at 1, by which all appeared subgraphs can be enumerated. Also another advantage is, by increasing the minimum support, we can reduce the number of appearing frequent subgraphs.

From these consideration, we can use the enumeration tree for computing the kernel function. Before doing this, we first check how one graph can be represented by the enumeration tree. Fig. 7.6 (a) is a sample graph G_1 with four nodes and three edges. Similarly Fig. 7.6 (c) is another sample graph G_2 with the same structure G_1, while one edge label is different. Suppose that the enumeration tree is Fig. 7.3 (c), Figs. 7.6 (b) and (d) show where the graphs (and

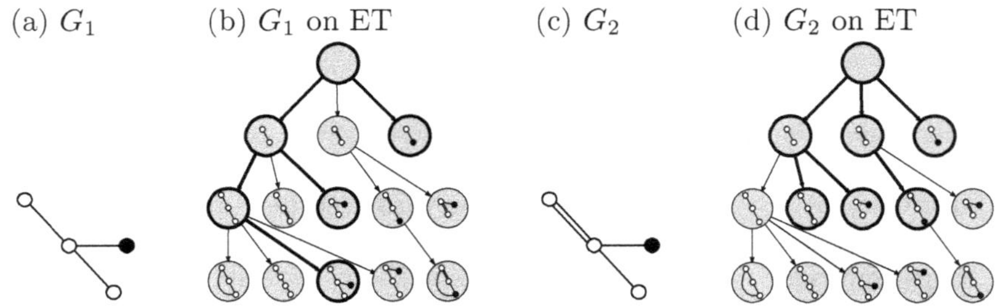

Figure 7.6: Two example graphs (a) G_1 and (b) G_2, and also (c) shows enumeration tree (ET) where all subgraphs includded in G_1 are in boldface and also (d) shows that for G_2.

subgraphs) of (a) G_1 and (c) G_2, respectively, are located. Due to the property of the frequent patterns in **Proposition 7.1**, as shown in these figures, if the subgraph corresponding to some node appears in the graph in question, ancestors from the node to the root all appear in that graph.

If we implement the idea of each of the above five types of kernel functions, under the constraint of the enumeration tree, the possible kernel function can be given as follows (explanation is done more weighed on intuitive understanding):

1. **Edit distance:**
   When we use the enumeration tree, which has both G_1 and G_2, we can move from the node corresponding to G_1 to the node corresponding to G_2 using the edges of the enumeration tree. Then, the number of edges needed for this moving corresponds to the edit distance between G_1 and G_2 on the enumeration tree, and so this can be used for the kernel function. This number of edges is not necessarily consistent with the regular edit distance, because

   1) This distance is the distance under the constraint of the enumeration tree.

   2) The enumeration tree itself can be changed by the given data and the minimum support.

   If there are no nodes corresponding to G_1 (or G_2), we can use the node corresponding to a larger, closest graph.

2. **Shortest path:**
   Each subgraph corresponding to a node in the enumeration tree is obtained by adding one edge to the subgraph corresponding to the parent of the node. Thus, the shortest distance between two nodes of G_1 can be regarded as the number of edges between two nodes (in the enumeration tree) corresponding to two subgraphs of G_1. Thus, we can repeat choosing such two nodes and computing the number of edges between them, for each of G_1 and G_2, and then compare the series of distances from G_1 with those from G_2.

Algorithm 7.3: Generate path kernel (for graphs) over enumeration tree. lay (n) is a function, which returns the layer depth (root is zero) of node n.

Function `Path_Kernel_over_Enumeration_Tree(`$\mathcal{T}$, G_1, G_2, C`)`

  Data: enumeration tree: $\mathcal{T}$, two graphs: G_1 and G_2, #random trials: C

  Result: kernel: $K(G_1, G_2)$

  Generate set $\mathcal{S}_1$ of nodes corresponding to subgraphs in G_1.;

  Generate set $\mathcal{S}_2$ of nodes corresponding to subgraphs in G_2.;

  repeat

    for $i = 1$ **to** 2 **do**

      Pick up two nodes n_1 and n_2 out of S_i randomly.;

      Find the closest common ancestor a of n_1 and n_2.;

      Compute the distance of two nodes: $l = (\text{lay}(n_1)$ - $\text{lay}(a)) + (\text{lay}(n_2)$ - $\text{lay}(a))$.;

      $d_i[l] + +$.;

  until C *times*;

  Compute the inner product of d_1 and d_2, distributions over the distance between two nodes: $K(G_1, G_2) = \sum_{l=0}^{|T|} d_1[l] \cdot d_2[l]$.;

  Output the above.;

---

3. Random walk:

In fact the random walk means choosing two nodes in a graph, which can be again considered as choosing two nodes (in the enumeration tree) corresponding to two subgraphs. Thus this can be the same as the above #2. However in this case, another lazy way is to choose only one node in the enumeration tree and compute only the distance between the selected node and the root. For all above 1 to 3, these operations are over the tree, and can be processed very fast. For example, in the above lazy case, we can just check the node layer selected and this is the distance. We can save the depth of layer at each node, and so the complexity is $O(1)$ for each iteration. **Algorithm 7.3** shows a pseudocode of this algorithm, where the distribution of the distance (number of edges) between two randomly selected nodes of G_1 and that of G_2 are computed and their inner product is computed as the kernel function.

Also the above lazy computation makes things more intuitive, because the kernel is the comparison of the distribution of layers randomly sampled for G_1 with that of randomly sampled for G_2, which would reflect (the topology of) the distribution of subgraphs of G_1 (and G_2) over the enumeration tree.

4, 5. Subgraphs:

The enumeration tree has a subgraph for each node, and so we consider subgraphs already, which are more complex than subtrees. We can compare the nodes corresponding to subgraphs of G_1 and those of G_2. If the overlap is larger, the distance between G_1 and G_2 is closer. Simply we can generate a binary vector with the length of $|T|$ (the number of nodes of the enumeration tree T) for G_1, where if the element vector is 1, the subgraph of the corresponding node is a part of G_1; otherwise zero. We can generate the vector with the same length for G_2 as well. Using the inner product of these two vectors, we can generate kernel function $K(G_1, G_2)$:

$$K(G_1, G_2) = \boldsymbol{u}^\top \cdot \boldsymbol{v}, \tag{7.15}$$

where $\boldsymbol{u}^\top = (u_1, \ldots, u_{|T|})$, $\boldsymbol{v}^\top = (v_1, \ldots, v_{|T|})$,

$$u_i = \begin{cases} 1 & \text{if } G_1 \text{ has the subgraph corresponding to node } i, \\ 0 & \text{otherwise.} \end{cases} \tag{7.16}$$

and

$$v_i = \begin{cases} 1 & \text{if } G_2 \text{ has the subgraph corresponding to node } i, \\ 0 & \text{otherwise.} \end{cases} \tag{7.17}$$

The vector length is longer as the size (number of nodes) of the enumeration tree is larger, but subgraphs of G_1 and G_2 would not cover the entire enumeration tree. Also the following proposition assists to prune the enumeration tree and narrow the search space, which contributes to fast generation of vectors.

Proposition 7.6. *If graph G does not have a subgraph which corresponds to a node in the enumeration tree, G does not have any subgraphs corresponding to all nodes on the paths from the node to leaves.*

If graph G has subgraph g corresponding to a node in the enumeration tree, G has all subgraphs corresponding to nodes on the path from the node to the root. Thus, in the above vector, all elements corresponding to the above nodes have 1. We can regard this as assigning 1 to each edge of g, while a smaller subgraph than g is assigned by 1 more than one times, which may look redundant. In this case, one approach might set weights over layers, to emphasize a bigger graph, like that a weight closer to the root is smaller and the weight is larger for more far from the root. An extreme case would assign 1 for the largest subgraph and zero for the other subgraphs.

7.2.1 Notes on Kernel Learning

Once an appropriate kernel function is built, we can use various kernel learning algorithms in Section 3.4. On the other hand, as we mentioned in Section 3.4, missing values are not allowed in kernel functions, making kernel functions unscalable.

We have showed a lot of kernels which can be generated by using prior knowledge, i.e. the enumeration tree, in which we did not use the support (the number of appearances in given data) of each subgraph, while one idea is to consider the supports of subgraphs to compute the kernels.

Table 7.1: Top 10 frequent subgraphs from CPDB (Sup. = Support).

	Frequent mining (Freq)		δ=0 (Closed)		δ=0.2 (δ20)		δ=1 (Maximal)	
	Sup.	Graph	Sup.	Graph	Sup.	Graph	Sup.	Graph
1	618		618		618		97	
2	468		468		401		95	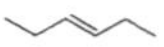
3	456		456		369		92	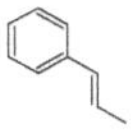
4	429		429		354		92	
5	410		410		309		88	
6	401	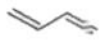	401	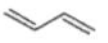	265		87	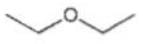
7	401		401		242	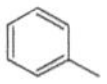	86	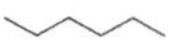
8	398	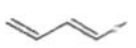	398	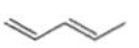	205	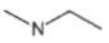	83	
9	369	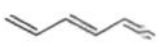	369	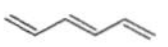	196	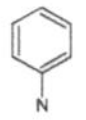	83	
10	369		369		196		82	

7.3 Applications to Life Sciences

As mentioned in the beginning of this chapter, typical graph data in life sciences is chemical structures of chemical compounds which are called molecular graphs. Already as shown in Fig. 3.36 of Section 3.5, an application of learning vectors showed that molecular graphs have various substructures (subgraphs). Thus frequent subgraph mining can be applied to any database with chemical compounds as records to understand chemical substructures (subgraphs). Table 7.1 shows results obtained by applying δ-tolerance closed frequent mining to chemical compound database CPDB (Carcinogenic Potency Database). This is a typical result of frequent subgraph mining for chemical compound data. The most left-hand side column shows the frequent subgraphs with the largest 10 supports. It can be easily seen that frequent subgraphs are all rather small and many of them are redundant, resulting in that this column is not so useful for knowledge discovery. The next column from the left shows the results of closed frequent subgraphs, while this column is totally the same as that of the regular frequent subgraphs, indicating that mining closed subgraphs does not work so well. In fact, as mentioned in Section 7.1.4, for the large database, the supports of frequent patterns are not easily tied. On the other hand, the most right column shows maximal frequent subgraphs, which are the largest frequent patterns. Indeed the subgraphs in this column are very different from each other, while the supports themselves are entirely rather not big. Finally the next from the right is the results of δ-tolerance closed frequent subgraphs when δ = 0.2. This column shows larger supports than those of maximal frequent subgraphs, and also graphs in this column are moderately different from each other, implying that entirely the balance was kept.

Chapter 8

Learning Nodes in a Graph

As written in the beginning of the preceding chapter, in machine learning, graphs are used as data in two different ways: 1) one graph is one instance, and 2) one node in each graph is an instance. In this chapter, we focus on the second usage, where we call learning nodes in a graph *graph learning*: Input data has two cases: 1) one graph or 2) multiple graphs, sharing the same node set. In both cases, nodes are unique, since they are instances. First we show several practical examples, being divided into two areas:

Internet: For example, world wide web (WWW) is a huge graph, where nodes are homepages and edges are hyperlinks. Also social network services (SNS) are of course a graph, where each node is an individual and an edge is the connection between individuals. Note that we can think this network not only in Internet but also in real-world.

Life science data: Interactions of biological molecules. For example, protein-interactions, gene networks (particularly gene regulatory network, signal transduction network, gene expression network) [7].

In fact, originally the finding of nodes and edges themselves were biological discovery. However, recently, due to the advancement of the so-called high-throughput techniques, a large amount of experimental results can be obtained simultanelusly, resulting in generating a large number of networks. However, high-throughput data are noisy, which also generate many problems to be solved over the networks.

In science and engineering, graphs are rather new data to be dealt with broadly. For example, a research field called *network sciences*, related with traditional *complex systems* in physics and/or economics but more strongly focused on research in graphs, has emerged and advanced in the 21st century. Thus again graphs have become a major topic in science, and this is true of machine learning, data mining and bioinformatics, which has made techniques in machine learning and data mining highly advanced.

In a graph where nodes are instances, a feature of each instance is edges, i.e. the information is only the connectivity among nodes. Nodes are unique, and

(a) Graph

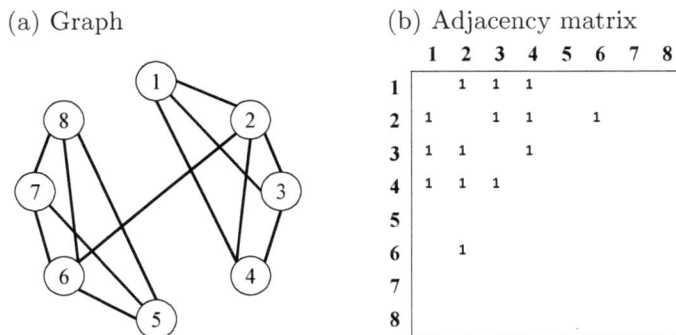

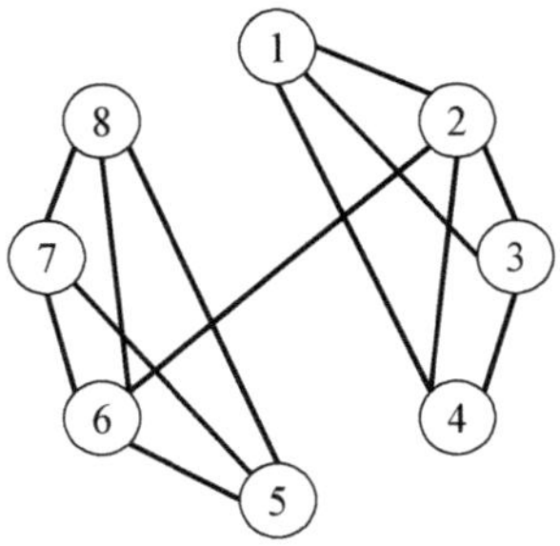

(b) Adjacency matrix

	1	2	3	4	5	6	7	8
1		1	1	1				
2	1		1	1		1		
3	1	1		1				
4	1	1	1					
5								
6		1						
7								
8								

Figure 8.1: A toy example of (a) graph and (b) adjacency matrix. These two show the same topology.

so a graph can be a matrix, where nodes can be both rows and columns, and elements are 1 if there are edges corresponding node pairs; otherwise zero. This matrix is called an *adjacency matrix*, which is equivalent to a graph. Fig. 8.1 shows an example adjacency matrix (over nodes), and a corresponding graph. As such, again a graph can be represented by an adjacency matrix. If we regard rows and columns in the adjacency matrix as instances and features, respectively, each instance is a vector, and approaches of learning vectors can be applied to this matrix. Thus, in this chapter, we do not regard this input matrix as such a matrix, and instead introduce the methods, which regard this matrix as a graph and describe the comparison between the way of dealing with the matrix and that with the graph.

Problem settings can be separated into two situations, with or without labels:

With labels: Instances are nodes, and so labels are assigned to nodes. If all nodes have labels, we have no room to predict the labels. Thus the problem setting is *semi-supervised learning*, where only part of nodes have labels. In fact, we can say that semi-supervised learning is a typical problem setting for supervised learning of graphs.

Without labels: In this case, we do not have any labels, and the only information are edges to connect nodes, i.e. instances. Then a typical problem setting is clustering nodes in a graph. This problem is often called *graph partitioning*.

Also regardless of labels, we can have two input settings: 1) one graph, and 2) multiple graphs. When we have multiple graphs as input, they share the node set, in which each node is an individual. For example, individuals (or users, i.e. nodes) use different social network services (or graphs), which have some connections (edges) between users. The node set is shared among these graphs, while edges are different, depending on graphs, because for the case of social network services, individuals use different graphs, depending on situations, such as in public and in private. Thus multiple graph input is important, and might be more meaningful than the input of only one graph. Also learning from multiple graphs connects to the subsequent chapter on data integrative learning.

Finally we mention the comparison between learning nodes in a graph (graph learning) and kernel learning, although this is mentioned already in the last of kernel learning of Section 3.4. Kernel functions are a matrix representing similarity between nodes, and can be regarded as a graph (weighted graph) with nodes as instances. However, a clear difference between kernel learning with kernel functions and graph learning with nodes as instances is as follows:

Missing values: Kernel functions in kernel learning do not allow missing values, while a graph has edges if the corresponding nodes have high similarity, and graph learning has nothing to do with missing values.

In particularly for multiple graphs as input, this difference is more sizable. Kernel function needs each graph to be complete, while graph learning just uses only edges already obtained.

Sparse learning: The above point on the missing values allows graph learning to be highly useful for sparse data, which are usual in the real world, while kernel functions cannot be directly used for sparse data.

Entirely graph learning is much more flexible than kernel learning, and thanks to the sparsity, graph learning can keep high scalability.

Notation in This Chapter

Let $\boldsymbol{W}$ be an adjacency matrix for graph $G = (V, E)$:

$$\boldsymbol{W}_{ij} = \begin{cases} 1 & \text{if there is an edge between node } i \text{ and node } j, \\ 0 & \text{otherwise.} \end{cases} \tag{8.1}$$

Let N be the number of nodes of graph G, and $\boldsymbol{W}$ is a $N \times N$-matrix:$\{0,1\}^{N \times N}$, where $\boldsymbol{W}_{ij}$ cannot be necessarily binaries but can be real-values between 0 and 1 (*weighted graph*). That is, the discussion in this chapter can be applied to a weighted graph.

The number of edges of each node is called a *degree* d_i of node i which can be computed from $\boldsymbol{W}$:

$$d_i = \sum_j \boldsymbol{W}_{ij}. \tag{8.2}$$

Also a diagonal matrix showing degrees on the diagonal is $\boldsymbol{D} = \text{diag}(d_1, \ldots, d_N)$.

8.1 Unsupervised Learning (Single Graph)

8.1.1 Spectral Clustering

Two Clusters

A representative approach in unsupervised learning is clustering. First we consider that nodes in graph G are clustered into two clusters (generally it is called *graph partitioning*). First we have *cluster assignment vector* $\boldsymbol{z} = (z_1, \ldots, z_N)$ ($z_i = \{+1, -1\}$), which assigns N nodes into one of the two clusters, i.e. if $z_i = 1$, cluster 1; otherwise cluster -1. Here, to check if two nodes are in the same cluster or not, let $\boldsymbol{B}_{ij} = ||z_i - z_j||^2$, i.e. the squared loss on the difference of $\boldsymbol{z}$. Obviously $\boldsymbol{B}$ follows:

$$\boldsymbol{B}_{ij} = \begin{cases} 0 & \text{if two nodes } i \text{ and } j \text{ are in the same cluster,} \\ 4 & \text{otherwise.} \end{cases} \tag{8.3}$$

If we do not think about any constraint on $\boldsymbol{z}$, minimizing $\sum_{i,j} \boldsymbol{B}_{ij}$ means making all node pairs zero, i.e. that all nodes are in the same cluster, which is not clustering at all, while we have constraints or data, i.e. edges of the given graph. Then, clustering nodes in the given graph means nodes are clustered to be consistent with edges of the graph. In fact, it would be reasonable to think clustering according to the existence of edges. That is, two nodes are in the same cluster if they are connected by an edge. The existence of the edge between nodes i and j are given by $\boldsymbol{W}_{ij}$, which is one if there is the edge; otherwise zero. On the other hand, if two nodes i and j are in the same cluster or not is given by $\boldsymbol{B}_{ij}$, which shows zero in the same cluster; otherwise 4, as shown above. That is, if $\boldsymbol{W}_{ij} = 1$, $\boldsymbol{B}_{ij}$ should be zero, and should not be 4.

That is, we can set up the following minimization problem, for clustering nodes in the two clusters:

$$\min \sum_{i,j} \boldsymbol{W}_{ij} \boldsymbol{B}_{ij}. \tag{8.4}$$

Multiplying the sum by $\frac{1}{2}$,

$$\begin{aligned}
\frac{1}{2}\sum_{i,j} \boldsymbol{W}_{ij}\boldsymbol{B}_{ij} &= \frac{1}{2}(\sum_{i,j} \boldsymbol{W}_{ij}||z_i - z_j||^2) = \frac{1}{2}(\sum_{i,j} \boldsymbol{W}_{ij}(z_i^2 - 2z_iz_j + z_j^2)) && (8.5)\\
&= \frac{1}{2}(\sum_{i,j} \boldsymbol{W}_{ij}z_i^2 - 2\sum_{i,j} \boldsymbol{W}_{ij}z_iz_j + \sum_{i,j} \boldsymbol{W}_{ij}z_j^2) && (8.6)\\
&= \frac{1}{2}(2\sum_{i} z_i^2 \sum_{j} \boldsymbol{W}_{ij} - 2\sum_{i,j} \boldsymbol{W}_{ij}z_iz_j) = \sum_{i} z_i^2 d_i - \sum_{i,j} \boldsymbol{W}_{ij}z_iz_j && (8.7)\\
&= \boldsymbol{z}^\top(\boldsymbol{D} - \boldsymbol{W})\boldsymbol{z}. && (8.8)
\end{aligned}$$

(8.4) becomes the following minimization problem:

$$\min_{\boldsymbol{z}} \boldsymbol{z}^\top \boldsymbol{L} \boldsymbol{z}, \tag{8.9}$$

where $\boldsymbol{L}$ ($= \boldsymbol{D} - \boldsymbol{W}$) is *Graph Laplacian*, which is defined and explained more in Section A.7.1. The key point of graph Laplacian is as follows:

Proposition 8.1 (Graph Laplacian). *Graph Laplacian is a positive semidefinite matrix.*

Proof. In (8.5), $\boldsymbol{W}_{ij}$ is non-negative for all i and j, and also $||z_i - z_j||^2$ are also non-negative for all i and j, then (8.5) are both non-negative always, and

$$\boldsymbol{z}^\top \boldsymbol{L} \boldsymbol{z} = \frac{1}{2} \sum_{i,j} \boldsymbol{W}_{ij} ||z_i - z_j||^2 \geq 0. \tag{8.10}$$

□

Again $\min_{\boldsymbol{z}} \boldsymbol{z}^\top \boldsymbol{L} \boldsymbol{z}$ is the objective function to be minimized.

$\boldsymbol{z}$ is a cluster assignment matrix, and its elements take +1 or -1, and so

$$z_i \cdot z_i = 1. \tag{8.11}$$

That is,

$$\boldsymbol{z}^\top \boldsymbol{z} = N. \tag{8.12}$$

This is the constraint of the minimization problem.

In summary the problem can be formulated as follows:

$$\min_{\boldsymbol{z}} \boldsymbol{z}^\top \boldsymbol{L} \boldsymbol{z} \text{ subject to } \boldsymbol{z}^\top \boldsymbol{z} = N. \tag{8.13}$$

We can relax z to take real values, and due to only the equality condition, using Lagrange multipliers (see Section A.8.5), this problem can be an eigenvalue problem:

$$\boldsymbol{L} \boldsymbol{z} = \lambda \boldsymbol{z}, \tag{8.14}$$

where $\boldsymbol{z}^\top \boldsymbol{L} \boldsymbol{z} \geq 0$:

$$\boldsymbol{z}^\top \boldsymbol{L} \boldsymbol{z} = \lambda \boldsymbol{z}^\top \boldsymbol{z} = \lambda \geq 0. \tag{8.15}$$

From this, we can show another property of graph Laplacian:

Proposition 8.2. *The eigenvalue of graph Laplacian is equal to or larger than zero.*

More Than Two Clusters

Now we consider a more general setting, where the number of clusters, K is more than two. We first expand the cluster assignment vector like this way: $\boldsymbol{z}^k = (z_1^k, \ldots, z_N^k)$ for the k-th cluster, which satisfies the following:

$$z_i^k = \begin{cases} 1 & \text{if node } i \text{ is in cluster } k, \\ 0 & \text{otherwise.} \end{cases} \tag{8.16}$$

Then, (8.4) for two clusters can hold for k-th cluster. Then, we use $\boldsymbol{z}^k$ as k-th column vector to generate *cluster assignment matrix* $\boldsymbol{Z}$. Now using this matrix $\boldsymbol{Z} = (\boldsymbol{z}^1, \ldots, \boldsymbol{z}^K)$, the objective function (8.4) for two clusters can be written for more than two clusters as follows:

$$\min_{\boldsymbol{Z}} \operatorname{trace}(\boldsymbol{Z}^\top \boldsymbol{L} \boldsymbol{Z}). \tag{8.17}$$

For example, constraints over $\boldsymbol{Z}$ can be written as follows:

$$\boldsymbol{Z}^\top \boldsymbol{Z} \sim \frac{N}{K} \boldsymbol{I}_K. \tag{8.18}$$

The right-hand side is the diagonal matrix and each of the K diagonals corresponds to the number of nodes in each cluster, which is kept at the same value, i.e. $\frac{N}{K}$. In fact, the above equation (8.18) is totally the same as (3.5) of Section 3.1.3. Thus we can make the same discussion over this matrix as in (3.5). That is, as explained in the subsequent section further, the clusters should be equal each other, and then as in (8.18), the right-hand side of $\boldsymbol{I}_K$ should be treated as a constraint.

Thus, the optimization (minimization) problem is given as follows:

$$\min \operatorname{trace}(\boldsymbol{Z}^\top \boldsymbol{L} \boldsymbol{Z}) \text{ subject to } \boldsymbol{Z}^\top \boldsymbol{Z} = \boldsymbol{I}_K. \tag{8.19}$$

Similar to the case that the number of clusters is two, we can relax the values taken by $\boldsymbol{Z}$ as real values, and due to only the equality constraint, by using Lagrange multipliers (see Section A.8.5), we have the following eigenvalue problem:

$$\boldsymbol{L}\boldsymbol{Z} = \lambda \boldsymbol{Z}. \tag{8.20}$$

By solving this eigenvalue problem, we can have $\boldsymbol{Z}$.

However, now $K > 2$, and so assigning nodes (instances) to clusters by using the resultant $\boldsymbol{Z}$ is not easy. Then, usually, we regard $\boldsymbol{Z}$ as a matrix (or each instance is a vector), and apply some clustering method for vectors to the matrix and have the final clusters. Usually K-means is used as a clustering method for vectors.

Various Graph Cuts

Now again let us think about clustering nodes in a graph, i.e. graph partitioning. Originally there is one single connected component in a graph, and graph partitioning means that the number of connected components is increased. Thus in order to increase the number of connected components, we need to remove edges to connect the components. This is called *graph cut*. As such, when we increase the connected components, selecting edges to be cut for having good clustering results is the main issue of clustering. The first, reasonable criterion we can have regarding graph cut is that the number of edges to be removed should be smallest. We can think about this by examples: Fig. 8.2 (a) is the original given graph. We can think about dividing this single connected component into two separate

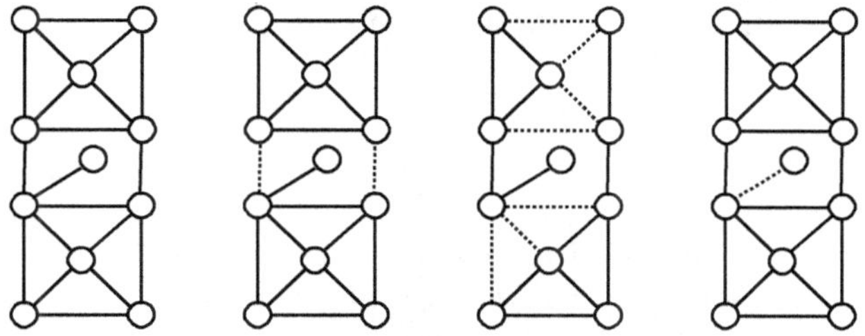

Figure 8.2: (a) original graph and two clusters by cutting (b) two edges, (c) eight edges and (d) only one edge.

connected components by cutting/removing edges. Then we have two examples: (b) and (c) in Fig. 8.2, where we remove only two edges in (b), but we have to remove eight edges in (c). Thus comparing with these two graph cuts, clearly (b) would be more reasonable clustering. That is, simply speaking, if the smaller number of edges are cut, that would be better. This concept is called *minimum cut*.

When we considered clustering given adjacency matrix $\boldsymbol{W}$ into two clusters, we set up $\boldsymbol{B}$ by using cluster assignment vector $\boldsymbol{z}$. In this case, the idea of formulating the problem is if the edge exists ($\boldsymbol{W}_{ij}$ is one), the cluster assignment $\boldsymbol{B}_{ij}$ is zero, meaning that nodes i and j are the same cluster. Then the objective function was set to minimize that $\boldsymbol{B}_{ij}$ is nonzero for $\boldsymbol{W}_{ij}$ with one:

$$\min \sum_{i,j} \boldsymbol{W}_{ij} \boldsymbol{B}_{ij}. \tag{8.21}$$

This is exactly the idea of the minimum cut.

Thus the idea of the minimum cut is already formulated and set as the objective function. Indeed the minimum cut seems the most appropriate for the objective function, while the minimum cut cannot be the best criterion always.

For example, Fig. 8.2 (d) shows another example that the original graph (a) is partitioned into two clusters, and the number of edges removed in (d) is only one, which is smaller than (b) of two. However the clusters in (d) is very biased in the sense that one cluster has only one node, and the other cluster is almost the same as the input. This would not be the two clusters we would like to have. That is, to fix the problem of minimum cut, we need to have a set of clusters, where the cluster sizes are as equal as possible.

In detail, for clustering the original graph into three or more clusters, again the left-hand side of (8.18) becomes a diagonal matrix, and each element of the diagonals corresponds to the number of nodes in each cluster (see Section 3.5):

$$\boldsymbol{Z}^{\mathsf{T}} \boldsymbol{Z} = \boldsymbol{F}, \tag{8.22}$$

where

$$\boldsymbol{F} = \mathrm{diag}(e_1, \ldots, e_K). \tag{8.23}$$

and e_k is the number of 1s in vector $\boldsymbol{z}^k$, i.e. the number of nodes in k-th cluster. Then to make the number of nodes in clusters constant,

$$e_1 = \cdots = e_K. \tag{8.24}$$

and so $\boldsymbol{F}$ should be the constant times an identity matrix, and we can use the identity matrix in (8.18). Also this is equivalent to use $\boldsymbol{D}$ in constrained K-means clustering, when diagonals of $\boldsymbol{D}$ are equal values, like (3.15) and (3.14). Among the criteria of graph cut, this constraint is called *ratio cut* [35]. That is, (8.18) follows the idea of ratio cut. In other words, (8.18) realized the idea of ratio cut already.

The idea of ratio cut is to keep the number of nodes in each cluster balanced. However to keep the balance of clusters, we can think about not only the number of nodes but also various aspects of a graph. For example, even if the number of nodes is balanced for each cluster, the number of outgoing edges from each cluster may be totally different, which might cause a problem. For this purpose, instead of the number of nodes, we can use the number of edges to be balanced, and the constraints can be given as follows:

$$\boldsymbol{Z}^\top \boldsymbol{D} \boldsymbol{Z} = \boldsymbol{I}_K, \tag{8.25}$$

where $\boldsymbol{D}$ is given as a diagonal matrix, such as (3.15), where the diagonals are cluster sizes.

This constraint tries to keep the number of edges in each cluster balanced, and called *normalized cut* [23]. Of course we can further think about other criteria, while the most well-accepted criteria are these two, i.e. ratio cut and normalized cut.

We can formulate the problem of using normalized cut:

$$\min_{\boldsymbol{Z}} \mathrm{trace}(\boldsymbol{Z}^\top \boldsymbol{L} \boldsymbol{Z}) \text{ subject to } \boldsymbol{Z}^\top \boldsymbol{D} \boldsymbol{Z} = \boldsymbol{I}_K. \tag{8.26}$$

As well as ratio cut, we can relax the values taken by $\boldsymbol{Z}$ into real values, and due to the equality constraints only, using the Lagrange multiplier (Section A.8.5), the following generalized eigenvalue problem must be solved [80]:

$$\boldsymbol{L}\boldsymbol{Z} = \lambda \boldsymbol{D}\boldsymbol{Z}. \tag{8.27}$$

On the other hand, instead of solving the generalized eigenvalue problem, we can first normalize graph Laplacian and solve a regular eigenvalue problem [66]:

$$\tilde{\boldsymbol{L}} = \boldsymbol{I} - \boldsymbol{D}^{-1/2} \boldsymbol{W} \boldsymbol{D}^{-1/2}. \tag{8.28}$$

This graph Laplacian is called *normalized graph Laplacian.*

Algorithm 8.1 shows a pseudocode of a representative algorithm on this type of problems. In this algorithm, for ratio cut, graph Laplacian is used without normalization.

As we have explained so far, approaches for solving the problem of graph partitioning as an eigenvalue problem or a generalized eigenvalue problem are

Algorithm 8.1: Spectral clustering algorithm.

```
1 Function Spectral_clustering(G, K, M)
    Data: graph: G(= W), #clusters: K, #eigenvectors used: M
    Result: node clusters: C_1, ..., C_K
2   Compute D (from W) and then graph Laplacian L.;
3   Compute normalized graph Laplacian L~ from L.;
4   Run Eigen_Decomposition(L~, M) of Algorithm 3.2 to have first M
    eigenvectors to generate (N × M)-matrix Z, by using the M
    eigenvalues.;
5   Run K-means(Z, K) (see Algorithm 3.1) over rows of Z to have
    clusters: C_1, ..., C_K.;
```

entirely called *spectral clustering* [61]. In spectral clustering, as mentioned earlier, graph Laplacian is a positive semidefinite matrix, and also the identity matrix and diagonal matrices for constraints are all positive semidefinite matrices. Thus, spectral clustering is also an algorithm/method, which takes the form of Rayleigh quotient (Section A.8.7).

Connection to Clustering Vectors

If we regard graph $\boldsymbol{W}$ as a matrix (instances are vectors), we can apply a various clustering algorithms over the matrix.

K-means The objective function of K-means (Section 3.1.2) is (3.10), which is again shown as follows:

$$\text{trace}(\boldsymbol{Z}^\top \boldsymbol{E} \boldsymbol{Z}), \tag{8.29}$$

where $\boldsymbol{Z}$ is the cluster assignment matrix, which is the same as that in spectral clustering, and $\boldsymbol{E}$ is given by (3.11) and again shown as follows:

$$\boldsymbol{E}_{ij} = \frac{1}{2N}(\boldsymbol{x}_i - \boldsymbol{x}_j)^2. \tag{8.30}$$

Each element of $\boldsymbol{E}$ is the squared **distance** between the two corresponding instances. That is, if instances are more similar, the distance is small; otherwise, the distance is larger. Such $\boldsymbol{E}$ is sandwiched by $\boldsymbol{Z}$. On the other hand, the input in this section is a graph (adjacency matrix) $\boldsymbol{W}$, which shows the **similarity** between instances and so reverse to the distance, while in the objective function, graph Laplacian is sandwiched by $\boldsymbol{Z}$, as follows:

$$\begin{aligned}
\text{trace}(\boldsymbol{Z}^\top \boldsymbol{L} \boldsymbol{Z}) &= \text{trace}(\sum_k (\boldsymbol{z}^k)^\top \boldsymbol{L} \boldsymbol{z}^k) && (8.31)\\
&= \text{trace}(\sum_k (\boldsymbol{z}^k)^\top (\boldsymbol{D} - \boldsymbol{W}) \boldsymbol{z}^k) && (8.32)\\
&= \text{trace} \sum_k \sum_{ij} \boldsymbol{W}_{ij} (z_i^k - z_j^k)^2. && (8.33)
\end{aligned}$$

As such, eventually adjacency matrix $\boldsymbol{W}$ weights over the **distance** between the two corresponding cluster assignments. Then both K-means and spectral clustering are minimization problems.

Constrained K-means We explained constrained K-means clustering in Section 3.1.3. Constrained K-means keeps the same objective function as K-means, while the difference from K-means is to keep the size of clusters equal over all clusters. To satisfy this point, for example, we can use the following constraint over $\boldsymbol{Z}$:

$$\boldsymbol{Z}^{\top}\boldsymbol{Z} = \boldsymbol{D}, \tag{8.34}$$

where $\boldsymbol{D}$ can be given as a diagonal matrix with any prespecified size of clusters for diagonal elements, like (3.15) and a special case with equal values, like (3.15).

Then entirely the optimization problem is given as follows:

$$\min_{\boldsymbol{Z}} \quad \text{trace}(\boldsymbol{Z}^{\top}\boldsymbol{E}\boldsymbol{Z}) \text{ subject to } \boldsymbol{Z}^{\top}\boldsymbol{Z} = \boldsymbol{D}. \tag{8.35}$$

This optimization problem is equivalent to the problem of ratio cut, for the constraint with equal sizes of clusters.

Kernel K-means Kernel K-means introduced in Section 3.4.1 has kernel function $\boldsymbol{K}$, which is sandwiched by $\boldsymbol{Z}$. Here graph Laplacian is a positive semidefinite matrix, and so kernel functions $\boldsymbol{K}$ and graph Laplacian $\boldsymbol{L}$ are changeable, and so kernel K-means and spectral clustering are equivalent. Kernel K-means allows to use any arbitrary kernel, and in this sense spectral clustering can be a special case of kernel K-means. However, practically kernel functions do not allow missing values, while the input graph and graph Laplacian allow missing values as long as the algorithm for solving the eigenvalue problem allows missing values. In this sense graph learning is more flexible and scalable.

Connection to Feature Learning

PCA A typical method for feature learning or dimensionality reduction is principal component analysis (PCA), where input vectors are projected on one-dimensional space, and the variance of the projected vectors is the objective function to be maximized. In fact this problem was formulated as in (3.237), and if the coefficient vector $\boldsymbol{w}$ in (3.237) is denoted by $\boldsymbol{z}$, and (3.237) can be written as follows:

$$\max_{\boldsymbol{z}} \boldsymbol{z}^{\top}\boldsymbol{X}^{\top}\boldsymbol{X}\boldsymbol{z} \quad \text{subject to} \quad \boldsymbol{z}^{\top}\boldsymbol{z} = \text{Constant}. \tag{8.36}$$

Then in PCA, this problem was eventually solved as an eigenvalue problem. Thus this procedure looks totally the same as K-means and also spectral clustering (minimization). However, there are two clear differences between PCA and K-means (spectral clustering).

Dimension of matrices: In K-means, the objective function has the matrix sandwiched by cluster assignment matrices. The dimension of the sandwiched matrix is N, which is the number of instances. Similarly, graph Laplacian in spectra learning is also $N \times N$-matrix.

However in PCA, for example, the above (8.36) shows $\boldsymbol{X}^\top \boldsymbol{X}$ is the $M \times M$ matrix, where M is the number of features.

Thus formulation looks the same but the dimension is totally different. This is due to the difference in focus: clustering is on instances while feature selection/dimensionality reduction focuses on features.

Only K eigenvalues used Both K-means and spectral clustering solve the eigenvalue problems, and all eigenvectors are used to do further clustering. However in PCA, only the top K eigenvectors are used as principal components.

Laplacian Eigenmaps: Also another method of feature learning in Section 3.3.3 is Laplacian eigenmaps. Laplacian eigenmaps transforms the input matrix into a graph. After transformation into a graph, the procedure of Laplacian eigenmaps is totally the same as spectral clustering. Only one difference is that spectral clustering runs K-means at the final step, while instead Laplacian eigenmaps just outputs the top eigenvectors as feature vectors. This is because Laplacian eigenmaps was motivated by manifold learning[1], including graph learning, and so this consistency is more than coincidence.

Connection to Matrix Factorization

Clustering methods for vectors, i.e. K-means and kernel K-means, solve eigenvalue problems, and also PCA and Laplacian Eigenmaps use eigenvalue problems, and so they use the same procedure as spectral clustering. In fact matrix factorization also follows this idea.

In matrix factorization, input matrix $\boldsymbol{X}$ can be factorized into two low-rank matrices:

$$\boldsymbol{X} \sim \boldsymbol{U}\boldsymbol{V}^\top. \tag{8.37}$$

Then, $\boldsymbol{U}$ and $\boldsymbol{V}$ are estimated to minimize the following:

$$\min_{\boldsymbol{U},\boldsymbol{V}} ||\boldsymbol{X} - \boldsymbol{U}\boldsymbol{V}^\top||_2^2. \tag{8.38}$$

When we think about the graph input, input $\boldsymbol{X}$ is adjacency matrix $\boldsymbol{W}$ or graph Laplacian $\boldsymbol{L}$, which are both symmetric matrices, and two low-rank matrices must be the same matrix. That is, $\boldsymbol{U} = \boldsymbol{V}$. Thus, for example, if we use graph Laplacian, instead of $\boldsymbol{X}$, and the formulation can be written as follows:

$$\min_{\boldsymbol{U}} ||\boldsymbol{L} - \boldsymbol{U}\boldsymbol{U}^\top||_2^2. \tag{8.39}$$

[1]Learning the problem of dimensionality reduction is said by several ways. In particular non-linear dimensionality reduction is called *manifold learning*

Algorithm 8.2: Spectral clustering algorithm for multiple graphs.

Function `Spectral_clustering_for_multiple_graphs`($G^{(i)}(i = 1, \ldots, T)$, K, M)
  Data: graph: $G^{(i)}(= \boldsymbol{W}^{(i)})(i = 1, \ldots, T)$, #clusters: K, #eigenvectors used: M
  Result: node clusters: $C_1, \ldots, C_K$
  Compute $\boldsymbol{L}$ and $\boldsymbol{D}$ by (8.42) (and let G be the corresponding graph).;
  Run `Spectral_clustering`(G, K, M) (**Algorithm 8.1**) to output K clusters.;

This is obtained by changing vector $\boldsymbol{u}$ of (3.57), which appeared in the last of Section 3.1.8 on the connection to the Rayleigh quotient, by matrix $\boldsymbol{U}$. This formulation also results in (3.60) of eigenvalue decomposition. Thus we can interpret that spectral clustering is a matrix factorization, resulting in eigenvalue decomposition.

From these observation, we can see that spectra clustering is connected to a variety of unsupervised learning approaches.

8.2 Unsupervised Learning (Multiple Graphs)

The input is T graphs, $G^{(1)}, \ldots, G^{(T)}$, which share a set of nodes, while edges are different depending on graphs: $G^{(i)} = (V, E^{(i)})$ $(i = 1, \ldots, T)$, where i-th adjacency matrix $\boldsymbol{W}^{(i)}$ corresponds to the i-th graph $G^{(i)}$.

8.2.1 Spectral Clustering

We can first compute each graph Laplacian $\boldsymbol{L}_i$ from adjacency matrix $\boldsymbol{W}_i$:

$$\boldsymbol{L}^{(i)} = \boldsymbol{D}^{(i)} - \boldsymbol{W}^{(i)}, \tag{8.40}$$

where $\boldsymbol{D}^{(i)}$ is the diagonal matrix for the i-th graph, showing the degree of nodes.

The problem setting is to estimate cluster assignment matrix, which can be the most consistent with all given graphs (matrices). However, there are no information on which graphs are most reliable. Thus, we can use a hyperparameter which weighs graphs, turning into weighing graph Laplacians. We then learn the cluster assignment matrix so that this matrix is smooth over the weighted mean of the graph Laplacians:

$$\min_{\boldsymbol{Z}} \boldsymbol{Z}^\top \boldsymbol{L} \boldsymbol{Z} \text{ subject to } \boldsymbol{Z}^\top \boldsymbol{Z} = \boldsymbol{I}_K, \tag{8.41}$$

where

$$\boldsymbol{L} = \sum_i w_i \boldsymbol{L}^{(i)}, \quad \boldsymbol{D} = \sum_i w_i \boldsymbol{D}^{(i)}. \quad (0 \leq w_i \leq 1, \ \sum_i w_i = 1) \tag{8.42}$$

The weights are tentatively hyperparameters, and so just we can use a uniform distribution if there are no prior information.

The (8.42) can be treated in the same way as spectral clustering over a single graph. **Algorithm 8.2** shows a pseudocode of the above algorithm.

8.2.2 Matrix Factorization

For matrix factorization with multiple graphs, again if there are no information on what graphs are most reliable, the main idea would be to estimate low-rank matrices which can be shared by the given graphs (matrices) and the low-rank matrices should approximate to the entire graphs as consistently as possible. We can first start with bi-matrix factorization, which generates two low-rank matrices. If both matrices are shared over all input graphs, the problem is very simple: this would be the same as that we can just combine all graphs beforehand and then run spectral clustering over the combined graph (matrix). Thus one interesting idea is that out of the two matrices, we make one the matrix shared by given graphs, and the other is the matrix peculiar to each graph. That is, each graph Laplacian is factorized into low rank matrices as follows:

$$\boldsymbol{L}^{(i)} \sim \boldsymbol{U}\boldsymbol{V}_i^{\top}. \tag{8.43}$$

In this factorization, $\boldsymbol{U}$ is common to given all graphs. On the other hand, $\boldsymbol{V}_i$ is a low-rank matrix special for the i-th graph[2]. We note that the input graph (matrix) is symmetric, and so $\boldsymbol{U}$ and $\boldsymbol{V}_i$ are the matrices with the same size. Also we note that $\boldsymbol{U}$ and $\boldsymbol{V}_i$ are changeable in the sense that either of the two low-rank matrices is a common matrix, and the other is a matrix peculiar to each graph. We can formulate this problem into the following optimization problem:

$$\min_{\boldsymbol{U},\boldsymbol{V}} \sum_i ||\boldsymbol{L}^{(i)} - \boldsymbol{U}\boldsymbol{V}_i^{\top}||_2^2 + \lambda(||\boldsymbol{U}||_2^2 + \sum_i ||\boldsymbol{V}_i||_2^2). \tag{8.44}$$

Lagrange function can be written as follows:

$$L(\boldsymbol{U},\boldsymbol{V}) = \sum_i | \boldsymbol{L}^{(i)} - \boldsymbol{U}\boldsymbol{V}_i^{\top}||_2^2 + \lambda(||\boldsymbol{U}||_2^2 + \sum_i ||\boldsymbol{V}_i||_2^2). \tag{8.45}$$

We can estimate these parameters by alternating least square (see (A.8.3)).

We then set the gradient (partial derivative) of the Lagrangian with respect to $\boldsymbol{U}$ (or $\boldsymbol{V}_i$ alternately) equal to zero, where $\boldsymbol{U}$ and $\boldsymbol{V}_i$ are done alternately. This results in the following equations:

$$\frac{\partial L(\boldsymbol{U},\boldsymbol{V})}{\partial \boldsymbol{U}} = -2\sum_i \boldsymbol{L}^{(i)}\boldsymbol{V}_i + 2\boldsymbol{U}\sum_i \boldsymbol{V}_i^{\top}\boldsymbol{V}_i + 2\lambda\boldsymbol{U}. \tag{8.46}$$

$$\frac{\partial L(\boldsymbol{U},\boldsymbol{W})}{\partial \boldsymbol{V}_i} = -2(\boldsymbol{L}^{(i)})^{\top}\boldsymbol{U} + 2\boldsymbol{V}_i(\boldsymbol{U}^{\top}\boldsymbol{U}) + 2\lambda\boldsymbol{V}_i. \tag{8.47}$$

[2]Entirely matrix $\boldsymbol{V}_i$ can be written as $\boldsymbol{V}$, which is a tensor, by using $\boldsymbol{V}_i (i = 1, \ldots, T)$.

Algorithm 8.3: Alternating least squares (ALS) for solving (8.44).

Function `ALS_for_factorizing_multiple_graphs`($G^{(i)}(i = 1, \ldots, T)$, K)
  Data: graphs: $G^{(i)}(= \boldsymbol{W}^{(i)})$, rank: K (low-rank matrices)
  Result: $\boldsymbol{U}, \boldsymbol{V}_i(i = 1, \ldots, T)$
  Compute graph Laplacian $\boldsymbol{L}^{(i)}, i = 1, \ldots, T$;
  Initialize $\boldsymbol{U}$ and $\boldsymbol{V}_i, i = 1, \ldots, T$.;
  repeat
    Update $\boldsymbol{U}$ by (8.48);
    for $i = 1$ **to** T **do**
      Update $\boldsymbol{V}_i^\top$ by (8.49);
  until *convergence*;
  Output $\boldsymbol{U}$ and $\boldsymbol{V}(i = 1, \ldots, T)$.;

Here $\boldsymbol{U}$ are used for all graphs, while $\boldsymbol{V}_i$ is for only the i-th graph, which makes derivation easy (the same as only single graph Laplacian).

We then have the following update rules:

$$\boldsymbol{U} \leftarrow (\sum_i \boldsymbol{L}^{(i)} \boldsymbol{V}_i)(\sum_i \boldsymbol{V}_i^\top \boldsymbol{V}_i + \lambda \boldsymbol{I}_K)^{-1}. \tag{8.48}$$

$$\boldsymbol{V}_i^\top \leftarrow (\boldsymbol{U}^\top \boldsymbol{U} + \lambda \boldsymbol{I}_K)^{-1} \boldsymbol{U}^\top \boldsymbol{L}^{(i)}. \tag{8.49}$$

From the above, we can see that $\boldsymbol{U}$ is updated by using all graphs, while $\boldsymbol{V}_i$ is updated by using $\boldsymbol{U}$ and only $\boldsymbol{L}_i$. **Algorithm 8.3** shows a pseudocode of estimating low-rank matrices over multiple graphs using the above update rules.

From the results, factors common to the given graphs can be found from low-rank matrix $\boldsymbol{U}$, while factors peculiar to each graph can be checked by $\boldsymbol{V}_i$. Thus we can check the difference between $\boldsymbol{V}_i$ of different i.

8.3 Label Propagation: Semi-Supervised Learning (Single Graph)

8.3.1 Problem Setting

Given a graph in which several nodes are labeled already and the other nodes are unlabeled, then the problem is to predict the labels of unknown nodes. This problem is called *label propagation*. Fig. 8.3 shows a schematic picture of label propagation. In this figure, we have two labels: dark gray and light gray, and not all but part of nodes have these labels. Again the problem is to predict labels of unknown nodes by using nodes already known.

Out of N nodes, if the first L nodes have labels, and the rest $N - L$ nodes are unlabeled:

$$\{y_1, \ldots, y_L\}, \quad y_i \in \{-1, +1\}. \tag{8.50}$$

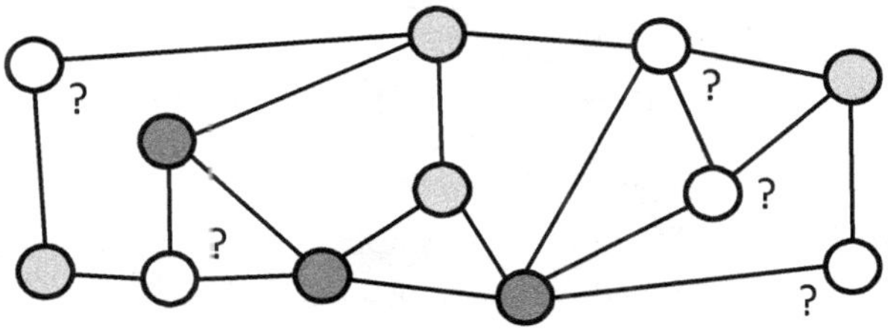

Figure 8.3: Schematic picture of label propagation.

Then the objective is to estimate labels of the rest of the nodes:

$$\{y_{L+1}, \ldots, y_N\}. \tag{8.51}$$

We write all node labels $\boldsymbol{y} = (y_1, \ldots, y_N)^\top$. Here, we set up label assignment function $\boldsymbol{f} = (f_1, \ldots, f_N)^\top (f_i \in \{-1, +1\})$. That is, we estimate f_i $(i = 1, \ldots, N)$, and then our objective will be done:

$$y_i \leftarrow f_i \ (i = L+1, \ldots, N). \tag{8.52}$$

Information for deciding f_i are two parts: 1) graph information and label information. Thus, $\boldsymbol{f}$ should be consistent with both graph edge connectivity and given label information. We here first explain a basic method of label propagation [106] and then next, a more flexible formulation [67].

8.3.2 Basic Label Propagation

1. **Edge Connectivity:** In graph partitioning in Section 8.1, we assign nodes to clusters, along with edge connectivity. In more detail, nodes connected by edges should be in the same cluster. Similarly we can consider here that nodes connected by edges can have the same label; otherwise not necessarily. Thus, we can consider the same formulation as (8.5) for $\boldsymbol{f}$:

$$\frac{1}{2}(\sum_{i,j} \boldsymbol{W}_{ij} ||f_i - f_j||^2) = \boldsymbol{f}^\top (\boldsymbol{D} - \boldsymbol{W}) \boldsymbol{f} \tag{8.53}$$

$$= \boldsymbol{f}^\top \boldsymbol{L} \boldsymbol{f}, \tag{8.54}$$

where $f_{L+1}, \ldots, f_N$ are unknown, and so they are set at zero so that $||f_i - f_j||^2$ should be the same for both +1 and -1.

2. **Consistency with labels:** The loss function can be given by the squared error against labels:

$$\sum_i ||f_i - y_i||_2^2 = ||\boldsymbol{f} - \boldsymbol{y}||_2^2. \tag{8.55}$$

Algorithm 8.4: Standard label propagation algorithm.

```
Function Basic_label_propagation(W, y_1, ..., y_L)
    Data: Graph: G(= W), Labels: y_1, ..., y_L
    Result: node labels: y_{L+1}, ..., y_N
    Compute D (from W) and then graph Laplacian L.;
    Compute the normalized graph Laplacian L̃ from L.;
    Compute (8.59) from L̃ and y to obtain f̂, which is the output.;
```

3. **Formulation:** We formulate the problem so that both (8.54) and (8.55) are minimized. Either of them can be the objective function or the regularization term. For example, we can formulate the problem as follows:

$$\arg\min_{\boldsymbol{f}} (\boldsymbol{f}-\boldsymbol{y})^\top(\boldsymbol{f}-\boldsymbol{y}) \text{ subject to } \boldsymbol{f}^\top \boldsymbol{L}\boldsymbol{f} \leq \gamma. \tag{8.56}$$

   Note that as shown in Section 8.1, we can use normalized graph Laplacian for the graph Laplacian in the above formulation:

$$\tilde{\boldsymbol{L}} = \boldsymbol{I} - \boldsymbol{D}^{-1/2}\boldsymbol{W}\boldsymbol{D}^{-1/2}. \tag{8.57}$$

4. **Solution:** The error function over the entire formulation can be given as follows:

$$\epsilon = (\boldsymbol{f}-\boldsymbol{y})^\top(\boldsymbol{f}-\boldsymbol{y}) + \lambda(\boldsymbol{f}^\top \boldsymbol{L}\boldsymbol{f} - \gamma). \tag{8.58}$$

   We can then set the partial derivative of the error function with respect to $\boldsymbol{f}$ equal to zero, i.e. $\frac{\partial \epsilon}{\partial \boldsymbol{f}} = 0$. Then this problem can be directly solved analytically:

$$\hat{\boldsymbol{f}} = (\boldsymbol{I} + \lambda \boldsymbol{L})^{-1}\boldsymbol{y}. \tag{8.59}$$

   We estimate $\hat{f}_{L+1}, \ldots, \hat{f}_N$ and assign them to $y_{L+1}, \ldots, y_N$. We call this entire procedure *Basic Label Propagation.*

Algorithm 8.4 is a pseudocode of the above procedure, in which normalized graph Laplacian is used.

8.3.3 Flexible Label Propagation

The problem setting of label propagation has two different types of information: graphs and labels. The basic label propagation formulate this information by using the two errors, each corresponding to graphs and labels, i.e. (8.54) and (8.55), respectively. Both are simple, and there are no problems in formulation, while either of the two sides (i.e. graphs and labels) cannot have their weights, and both edge connectivity and label consistency allow only one type of form, which cannot be modified at all.

Then, to combine the two sides more flexibly to optimize, we use the same form for the two objective functions:

Graph: Objective function is the same as basic label propagation:

$$\frac{\boldsymbol{f}^\top \boldsymbol{L}\boldsymbol{f}}{\boldsymbol{f}^\top \boldsymbol{f}}. \tag{8.60}$$

Label: We can adopt the form of Rayleigh quotient (Section A.8.7) for the objective function of label consistency, making easier to integrate the objective function of graphs.

$$\frac{\boldsymbol{f}^\top \boldsymbol{f}}{\boldsymbol{f}^\top \boldsymbol{Y}\boldsymbol{f}} \text{ or } -\frac{\boldsymbol{f}^\top \boldsymbol{Y}\boldsymbol{f}}{\boldsymbol{f}^\top \boldsymbol{f}}, \tag{8.61}$$

where $\boldsymbol{Y}_{ij} = y_i y_j$.

Thus the both objective functions take the form of Rayleigh quotient, making the integration of the two objective functions easier. In particular, we can flexibly integrate the two objective function, by using two types of integration: additive and subtractive models. Also we will present a multiplicative model just for reference. Below c is a real-value between zero to one.

Additive:

$$\frac{c\boldsymbol{f}^\top \boldsymbol{L}\boldsymbol{f} + (1-c)\boldsymbol{f}^\top \boldsymbol{f}}{c\boldsymbol{f}^\top \boldsymbol{f} + (1-c)\boldsymbol{f}^\top \boldsymbol{Y}\boldsymbol{f}} \tag{8.62}$$

$$= \frac{\boldsymbol{f}^\top (c\boldsymbol{L} + (1-c)\boldsymbol{I})\boldsymbol{f}^\top}{\boldsymbol{f}^\top (c\cdot \boldsymbol{I} + (1-c)\boldsymbol{Y})\boldsymbol{f}}. \tag{8.63}$$

The optimization problem can be written as follows:

$$\arg\min_{\boldsymbol{f}} \frac{\boldsymbol{f}^\top (c\boldsymbol{L} + (1-c)\boldsymbol{I})\boldsymbol{f}^\top}{\boldsymbol{f}^\top (c\cdot \boldsymbol{I} + (1-c)\boldsymbol{Y})\boldsymbol{f}}. \tag{8.64}$$

Subtractive:

$$c\cdot \frac{\boldsymbol{f}^\top \boldsymbol{L}\boldsymbol{f}}{\boldsymbol{f}^\top \boldsymbol{f}} - (1-c)\cdot \frac{\boldsymbol{f}^\top \boldsymbol{Y}\boldsymbol{f}}{\boldsymbol{f}^\top \boldsymbol{f}} \tag{8.65}$$

$$= \frac{\boldsymbol{f}^\top (c\cdot \boldsymbol{L} - (1-c)\cdot \boldsymbol{Y})\boldsymbol{f}}{\boldsymbol{f}^\top \boldsymbol{f}}. \tag{8.66}$$

The optimization problem can be written as follows:

$$\arg\min_{\boldsymbol{f}} \frac{\boldsymbol{f}^\top (c\cdot \boldsymbol{L} - (1-c)\cdot \boldsymbol{Y})\boldsymbol{f}}{\boldsymbol{f}^\top \boldsymbol{f}}. \tag{8.67}$$

Multiplicative:

$$\left(\frac{\boldsymbol{f}^\top \boldsymbol{f}}{\boldsymbol{f}^\top \boldsymbol{Y}\boldsymbol{f}}\right)^c \cdot \left(\frac{\boldsymbol{f}^\top \boldsymbol{L}\boldsymbol{f}}{\boldsymbol{f}^\top \boldsymbol{f}}\right)^{(1-c)}. \tag{8.68}$$

This optimization problem is not so easy to solve, and so we take $c = 0.5$, and the objective function of this case is as follows:

$$\arg\min_{\boldsymbol{f}} \frac{\boldsymbol{f}^\top \boldsymbol{f}}{\boldsymbol{f}^\top \boldsymbol{Y} \boldsymbol{f}} \cdot \frac{\boldsymbol{f}^\top \boldsymbol{L} \boldsymbol{f}}{\boldsymbol{f}^\top \boldsymbol{f}} \tag{8.69}$$

$$= \arg\min_{\boldsymbol{f}} \frac{\boldsymbol{f}^\top \boldsymbol{L} \boldsymbol{f}}{\boldsymbol{f}^\top \boldsymbol{Y} \boldsymbol{f}}. \tag{8.70}$$

where c is a hyperparameter to control the weights over the two types of information. All above three objective functions keep the form of Rayleigh quotient. Thus all formulations can be solved by, for example, an eigenvalue problem. However, we will explain more efficient solution to this problem in the next chapter.

8.4 Label Propagation: Semi-Supervised Learning (Multiple Graphs)

8.4.1 Input Data: Multiple Graphs

We have multiple graphs (not a single graph) under the same setting as unsupervised learning, and the objective is to perform label propagation over these graphs [49]. Now let us remember the multiple graph setting:

Unique node shared: Multiple graphs share the same set of nodes. Thus each node in a graph is unique, and then this unique node appears only one time in a graph and appears the number of graphs times.

Edge connectivity different: Edges are different depending on graphs. That is, for the same node pair, an edge might exist in a graph but may not in another graph.

8.4.2 Problem Setting 1: Weights over Graphs

Then now we assume that like edges, graph reliability can be different, depending on graphs. Thus the objective of label propagation over multiple graphs is to find more reliable graphs and select them.

In more detail, given K graphs, let $\boldsymbol{\mu} = (\mu_1, \cdots, \mu_K)^\top$ be K weights over K graphs, and the objective is to learn these weights, where a graph with a larger weight should be more reliable.

Direct Extension from Basic Label Propagation: Excess Sparse Solution

First we attempt to extend the "Basic Label Propagation" we introduced in the beginning of Section 8.3 for a single graph to multiple graphs. Then the straightforward expansion can be given as follows:

$$\min_{\boldsymbol{f},\boldsymbol{\mu}} \sum_k \frac{\mu_k}{Z_k} \boldsymbol{f}^\top \boldsymbol{L}_k \boldsymbol{f} + \lambda ||\boldsymbol{y} - \boldsymbol{f}||_2^2 \text{ subject to } \boldsymbol{\mu}^\top \boldsymbol{1} = 1, \boldsymbol{\mu} \geq 0. \tag{8.71}$$

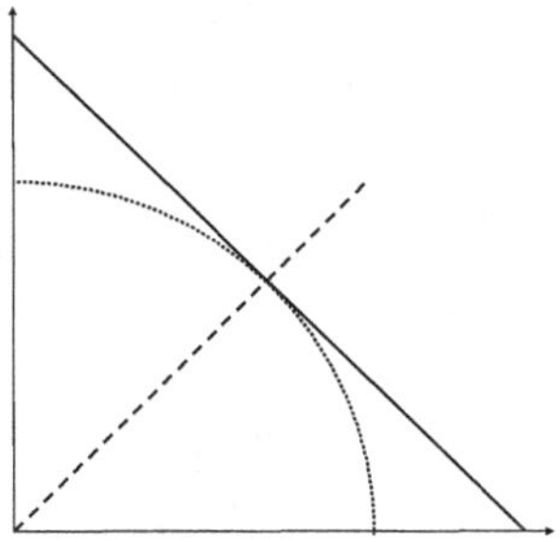

Figure 8.4: Regularization of $\boldsymbol{\mu}$, x-axis: μ_i, y-axis: μ_j.

where Z_k is a constant for normalization[3], and $\mathbf{1}$ is a vector, in which all elements are 1.

In this formulation, we keep the same term of label information as that of a single graph, and use weights $\boldsymbol{\mu}$ for the graph information of multiple graphs. However, $\boldsymbol{f}^\top \boldsymbol{L}_k \boldsymbol{f}$ is a scalar, and so the minimization problem of (8.71) can be solved by selecting the smallest among $\boldsymbol{f}^\top \boldsymbol{L}_k \boldsymbol{f} (k = 1, \ldots, K)$ and letting μ_{k^*} be 1, where k^* is selected to minimize the above term as follows, and all other k be zero:

$$k^* = \arg\min_k \boldsymbol{f}^\top \boldsymbol{L}_k \boldsymbol{f}. \tag{8.72}$$

That is, this problem can be solved by just choosing one graph. Then let $\boldsymbol{e}_i$ be a unit vector, in which i-th element is one; other elements are all zero. The solution can be given as follows:

$$\boldsymbol{\mu} \leftarrow \boldsymbol{e}_{k_{min}} \text{ where } k_{min} = \arg\min_k \frac{1}{Z_k} \boldsymbol{f}^\top \boldsymbol{L}_k \boldsymbol{f}. \tag{8.73}$$

Formulation for Obtaining Moderately Sparse Solution

The objective of this problem is to choose several reliable graphs, while (8.71) selects only one graph and this solution is too sparse and too simple. In fact, given multiple graphs, it would be rather easy to understand that the performance of using multiple (hopefully a few) graphs would be better than using only one graph. Then to select a moderate number of graphs, we formulate the problem as follows [49].

$$\min_{\boldsymbol{f},\boldsymbol{\mu}} \sum_k \frac{\mu_k}{Z_k} \boldsymbol{f}^\top \boldsymbol{L}_k \boldsymbol{f} + \lambda_1 ||\boldsymbol{y} - \boldsymbol{f}||_2^2 + \lambda_2 ||\boldsymbol{\mu}||_2^2 \text{ subject to } \boldsymbol{\mu}^\top \mathbf{1} = 1, \boldsymbol{\mu} \geq 0. \tag{8.74}$$

The two formulation, i.e. (8.71) and (8.74), are the same, only except the last term $||\boldsymbol{\mu}||_2^2$ of (8.74). This term is L^2 norm, and plays a role of relaxing the excess

[3] In fact $Z_k = ||\boldsymbol{L}_k||_F$, because $\boldsymbol{f}^\top \boldsymbol{L}_k \boldsymbol{f} = \langle \boldsymbol{L}_k, \boldsymbol{f}\boldsymbol{f}^\top \rangle_F$.

sparseness. This is similar to the situation that LASSO (3.247) of linear regression is too sparse and so the L^2 norm is also introduced to use both L^1 and L^2 terms, which is called *elastic net* (3.248). In fact the elastic net can relax the excessive sparseness caused when using only the L_1 norm. Our formulation has no L^1 terms and so not necessarily totally the same as the elastic net, while a common point is that originally the formulation was excessively sparse, and so we added L^2 norm to the formulation, in order to relax the excessive sparsity.

We here examine the change of $\boldsymbol{\mu}$ when λ_2 is varied. When $\lambda_2 = 0$, (8.74) is equivalent to (8.71), and only one graph is selected at $\boldsymbol{\mu} = \boldsymbol{e}_{k_{min}}$. On the other hand, when $\lambda_2 \to \infty$, $\mu_k = \frac{1}{K}$ $(k = 1, \dots K)$, i.e. μ_k becomes a uniform distribution. Thus our formulation estimates the optimal distribution of $\boldsymbol{\mu}$, between the excessive sparse solution of choosing only one graph and a uniform distribution, which is totally reverse to the idea of sparse solution.

Deriving Moderately Sparse Solution

To solve (8.74), we adopt an alternating least square algorithm, in which two parameters $\boldsymbol{f}$ and $\boldsymbol{\mu}$ are estimated alternately. That is, we repeat two steps alternately: 1) fixing $\boldsymbol{\mu}$, we estimate $\boldsymbol{f}$, and 2) fixing $\boldsymbol{f}$, we estimate $\boldsymbol{\mu}$.

Estimating $\boldsymbol{f}$ When we fix $\boldsymbol{\mu}$, the new term added to (8.74) is a constant, and so solving (8.74) is equal to solving (8.71). Setting the derivative of the left of (8.71) with respect to $\boldsymbol{f}$ equal to zero, we can obtain the update equation to estimate $\boldsymbol{f}$ analytically:

$$\hat{\boldsymbol{f}} = \lambda_1(\lambda_1 \boldsymbol{I} + \sum_k \frac{\mu_k}{Z_k}\boldsymbol{L}_k)^{-1}\boldsymbol{y}. \tag{8.75}$$

This (8.75) is equivalent to the update rule of $\boldsymbol{f}$ in the "Basic Label Propagation" of a single graph, while multiple graphs are incorporated.

Estimating $\boldsymbol{\mu}$ When we fix $\boldsymbol{f}$, (8.74) is the following optimization problem:

$$\min_{\boldsymbol{\mu}} \boldsymbol{v}^\top \boldsymbol{\mu} + \frac{\lambda_2}{2}||\boldsymbol{\mu}||_2^2 \text{ subject to } \boldsymbol{\mu}^\top \boldsymbol{1} = 1, \boldsymbol{\mu} \geq 0. \tag{8.76}$$

where $\boldsymbol{v} = (v_1, \dots, v_K)^\top, v_k = \frac{1}{Z_k}\boldsymbol{f}^\top \boldsymbol{L}_k \boldsymbol{f}$.

This is a quadratic programming problem and can be solved in that way. We can however show a faster solution of using Karush-Kuhn-Tucker (KKT) conditions (see Section A.8.6) to solve the optimization problem with inequality conditions.

First we rank the term v_k on graph Laplacians in the ascending order:

$$v_1 \leq \cdots \leq v_K. \tag{8.77}$$

Graph Laplacians are not equal each other, and so we can assume that they have to be ordered, without loss of generality.

Proposition 8.3. *The optimum solution of* (8.76) *is given as follows:*

$$\mu_k = \frac{\eta - v_k}{\lambda_2}, \quad k = 1, \ldots, m. \tag{8.78}$$

$$\mu_k = 0, \quad k = M + 1 \ldots, K. \tag{8.79}$$

where

$$\eta = \frac{\lambda_2 + \sum_k v_k}{m}. \tag{8.80}$$

and

$$m = |\{k | \eta - v_k > 0, k = 1, \ldots, K\}|. \tag{8.81}$$

Proof. We prove this by using the KKT conditions (see Section A.8.6). First the Lagrange function of (8.76) is given as follows:

$$L = \boldsymbol{v}^{\mathsf{T}} \boldsymbol{\mu} + \frac{\lambda_2}{2} \boldsymbol{\mu}^{\mathsf{T}} \boldsymbol{\mu} + \eta(1 - \boldsymbol{\mu}^T \mathbf{1}) - \boldsymbol{\xi}^{\mathsf{T}} \boldsymbol{\mu}. \tag{8.82}$$

where $\eta \geq 0$, $\boldsymbol{\xi} = (\xi_1, \ldots, \xi_K)^{\mathsf{T}}$ and $\xi_k \geq 0, k = 1, \ldots, K$.

First from the Stationary condition of the KKT conditions, we can set the derivative of the Lagrange function (8.82) with respect to u_k equal to zero, which gives the following:

$$\mu_k = \frac{\eta - v_k + \xi_k}{\lambda_2}. \tag{8.83}$$

Next, from the Complementary slackness of the KKT conditions, we can obtain below:

$$\xi_k \mu_k = 0 \quad (k = 1, \ldots, K), \tag{8.84}$$

which means that either of ξ_k or μ_k must be zero always.

Now we can estimate ξ_k and μ_k by depending on if $\eta - v_k$ in (8.83) is positive, negative or zero, as follows:

$\eta - v_k > 0$**:** Since $\xi_k \geq 0$, $\mu_k > 0$ by using (8.83). Then from (8.84), $\xi_k = 0$.

$\eta - v_k < 0$**:** μ_k must be positive, and so $\xi_k > 0$ by using (8.83). Then $\mu_k = 0$, by using (8.84).

$\eta - v_k = 0$**:** $\mu_k = 0$ due to the same reason as $\eta - v_k < 0$.

We can summarize the above discussion as follows:

$$\mu_k = \frac{\eta - v_k}{\lambda_2}, \quad k \in \{k | \eta - v_k > 0\}. \tag{8.85}$$

$$\mu_k = 0, \quad k \in \{k | \eta - v_k \leq 0\}. \tag{8.86}$$

Algorithm 8.5: Obtain $\boldsymbol{\mu}$ by KKT conditions of (8.82).

Function `Estimate_`$\boldsymbol{\mu}(v_1, \ldots, v_K)$
 Data: graph regularizers: $v_1, \ldots, v_K$
 Result: $\boldsymbol{\mu}$
 for $m \leftarrow 1$ **to** K **do**
  $\eta \leftarrow \frac{\lambda_2 + \sum_k v_k}{m}$.;
  if $m = |\{k | \eta - v_k > 0\}, k = 1, \ldots, K|$ **then**
   break
 $\mu_k = \frac{\eta - v_k}{\lambda_2},\ k \in \{k | \eta - v_k > 0\}$.;
 $\mu_k = 0,\ k \in \{k | \eta - v_k \leq 0\}$.;
 Output $\boldsymbol{\mu}$.;

Algorithm 8.6: Label propagation algorithm for multiple graphs.

Function `Label_propagation_for_multiple_graphs`$(G_1, \ldots, G_K, y_1, \ldots, y_L, \lambda_1, \lambda_2)$
 Data: Graphs: $G_1, \ldots, G_K$, Labels: $y_1, \ldots, y_L$, λ_1, λ_2
 Result: $\boldsymbol{f}$, $\boldsymbol{\mu}$
 Compute graph Laplacian L_k for all $k, k = 1, \ldots, K$.;
 Initialize $\boldsymbol{\mu}$.;
 repeat
  1. Optimize $\boldsymbol{f}$ by (8.75).;
  2. Optimize $\boldsymbol{\mu}$ by `Estimate_`$\boldsymbol{\mu}(v_1, \ldots, v_K)$.;
 until *convergence*;
 Output both $\boldsymbol{f}$ and $\boldsymbol{\mu}$.;

They correspond to (8.78) and (8.79), respectively.

Since v_k $(k = 1, \ldots, K)$ are ranked in the ascending order in (8.85) (see (8.77)), when v_k is small, $\eta - v_k$ can be positive, but when v_k is getting larger, $\eta - v_k$ cannot be positive. Thus we can obtain m by k which keeps $\eta - v_k$ positive, as follows:

$$m = |\{k | \eta - v_k > 0\}|. \tag{8.87}$$

This corresponds to (8.81). We now have m, and so (8.78) and (8.79) are substituted into $\boldsymbol{\mu}^\mathsf{T}\mathbf{1} = 1$ to have η. This corresponds to (8.80). □

Algorithm 8.6 and **Algorithm 8.5** show a pseudocode of the above entire procedure.

It would be easily understood by the algorithm derivation and also the algorithm pseudocode that this algorithm is very fast, and the time complexity for learning is only $O(K)$.

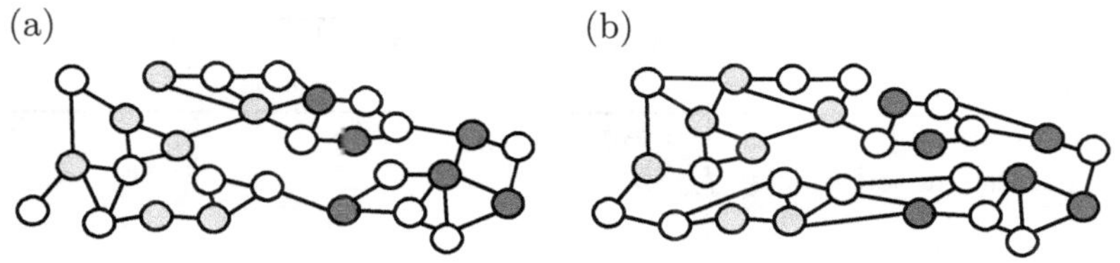

Figure 8.5: Two sample graphs.

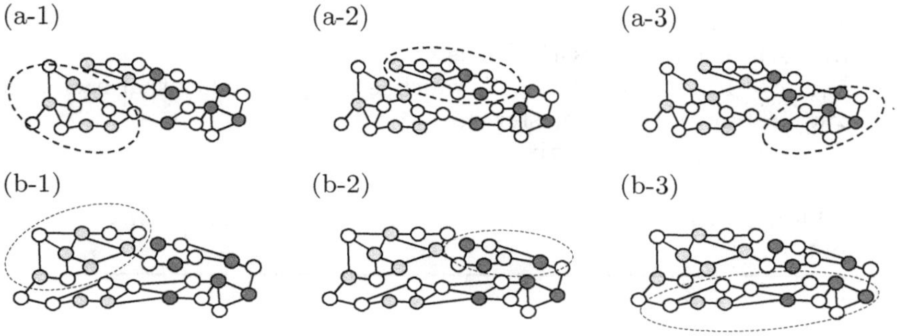

Figure 8.6: Three clusters from graphs (a) and (b) of 8.5.

8.4.3 Problem Setting 2: Weights over Localized Information

In Problem 1, when we integrate multiple graphs, we have considered what graphs should be integrated (selected to integrate) among given multiple graphs. That is, given multiple graphs include reliable and unreliable graphs, and the objective is to select the reliable graphs. However, even in one single graph, there might be reliable parts and unreliable parts. For example, Fig. 8.5 shows two graph examples, (a) and (b), both share the same set of nodes with the same labels, while edges are different. This is exactly the same data setting of label propagation of multiple graphs (semi-supervised learning of graphs).

Now let us consider which of (a) and (b) in Fig. 8.5 is more reliable. There are three portions in (a), in which edge connectivity is dense. They are all surrounded by dotted line in Fig. 8.6 (a-1)-(a-3).

From these three, we can see that in (a-1) and (a-3), labels surrounded by the dotted line are consistent, while in (a-2), labels inside the dotted circle are inconsistent. Similarly Fig. 8.5 (b) shows three densely connected portions, corresponding to three parts specified by the dotted circles of (b-1)-(b-3) of Fig. 8.6, and in (b-1) and (b-2), labels and edges are consistent, while in (b-3) they are inconsistent. That is, both Figs. 8.5 (a) and (b), we can see the inconsistent parts (i.e. (a-2) and (b-3)), by which we are unable to decide which graph of (a) and (b) is more reliable. On the other hand, in (a), (a-1) and (a-3) and in (b), (b-1) and (b-2) are consistent with labels and edges, implying that these parts are reliable

---

Algorithm 8.7: Label propagation algorithm for multiple graphs, considering local information.

Data: Graph:$G_1, \ldots, G_M$, labels:$y_1, \ldots, y_L$, λ_1, λ_2
Result: estimated labels: $\boldsymbol{f}$

1 **foreach** G_m **do**
2   Compute normalized graph Laplacian $\tilde{\boldsymbol{L}}_m$ from $\boldsymbol{W}_m$ and $\boldsymbol{D}_m$.;
3   Run `Eigen_Decomposition`($\tilde{\boldsymbol{L}}_m$,) of **Algorithm 3.2** to have first K eigenvectors to generate $(N \times K)$-matrix $\boldsymbol{Z}$, by using the K eigenvalues.;
4   Run **Algorithm 3.4** with C clusters to have $\hat{p}_{c|\boldsymbol{z}_i}, c = 1, \ldots, C, i = 1, \ldots, N$, which is the probability that each i-th row (instance) $\boldsymbol{z}_i$ is in cluster c.;
5   Generate adjacency matrix $\boldsymbol{W}_{i,j}^{c,m} = \hat{p}_{c|\boldsymbol{z}_i}\hat{p}_{c|\boldsymbol{z}_j}\boldsymbol{W}_{ij}(i = 1, \ldots, N, j = 1, \ldots, N)$.;
6 Run **Algorithm 8.6** over all $\boldsymbol{W}^{c,m}, c = 1, \ldots, C, m = 1, \ldots, M$;
7 Output $\boldsymbol{f}$.;

---

and should be used for integrating these graphs. Therefore, the problem is not to select graphs but parts of the graphs.

Combination of Clustering with Label Propagation over Weighted Graphs for Localized Information

Thus we can focus on parts on graphs more than graphs themselves. A possible solution is rather straight-forward on the observations we had so far. That is, the method has the following three step procedure [82]:

1. **Graph partitioning:** We first perform graph partitioning over each of given M graphs into C subgraphs.

2. **Subgraph to graph:** We then regard each subgraph obtained by graph partitioning as a graph. Note that the number of graphs is now $M \times C$.

3. **Label propagation over multiple graphs:** We run label propagation over $M \times C$ graphs. In this procedure, we select reliable graphs so that we can assign labels to unknown nodes.

Algorithm 8.7 shows a pseudocode of the above procedure, which follow the above ideas and procedure. In more detail, in this algorithm, we first compute the normalized graph Laplacian of each of the given graphs (line 2), then compute K eigenvectors from eigen decomposition (line 3) and run clustering by finite mixture model by using the top K eigenvectors (line 4). We generate a graph (adjacency matrix) corresponding to each cluster of the clustering results (line 5). Finally we run label propagation over multiple graphs on all graphs generated and assign labels from the label propagation result (line 6). This is a heuristics but a reasonable approach.

Chapter 9

Data Integrative Learning

We have seen machine learning and data mining methods for different types of data sets.

In this chapter, we always have more than one datasets, for example two datasets, meaning that for example one instance has two (different) vectors, or one vector and one graph. In fact, this is a more real-world and practical setting than that only one dataset is given. For example demographic data (such as gender, age, occupation, etc.) is usually a vector, and the vector of individuals is saved in a lot of different places in not the same but slightly different contents, depending on the requirements. That is, people subscribe more than one E-commerce sites, and these sites save demographic data of individuals in different formats and contents. Simply speaking, different information sources can give different vectors for each individual. Also as we have seen in this book, the information is not necessarily a vector, and that can be a set, a sequence, a graph or a node in one or more graphs. For example each individual has demographic data as a vector and at the same time is also a member of social network services (SNS), which can be a graph in which each individual is a node. Furthermore, it would be possible to have different demographic data, which can be multiple vectors for each individual, and also each individual can join multiple SNS, which makes an individual a node shared with multiple graphs. Machine learning methods in this chapter are for learning this type of inputs.

As we mentioned in Section 2.3.7, sets, sequences, trees and graphs have inclusion relations, and we can regard sets, sequences and trees as all special cases of graphs. Thus algorithms for graphs can be applied to sequences and trees. Then when we think about the combination of six data types (vectors, sets, sequences, trees, graphs and nodes in a graph), we can skip sets, sequences and trees, and instead focus on graphs. In other words, we can just focus on vectors, graphs and nodes in a graph and consider the learning method which can learn the combination of them as input, integrating the input during learning. In more detail, sections in this chapter are as follows:

Section 9.1: Vectors and vectors (multiple sets of vectors)

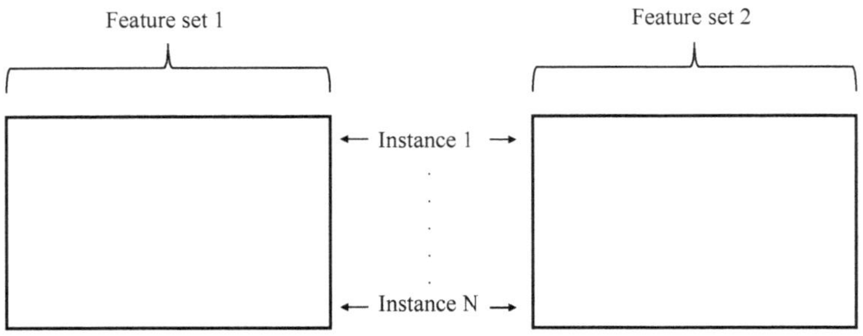

Figure 9.1: Multiview learning: schematic diagram of two views.

Section 9.2: Vectors and graphs

Section 9.3: Vectors and nodes in a graph

Section 9.4: Nodes in one graph and nodes in the other graph

9.1 Integrative Learning for Multiple Vectors

Each instance has two different vectors. Fig. 9.1 shows a schematic picture of this input, where one instance has two different feature sets, each row being a vector for one instance. This way of learning multiple vectors at the same time is called in many ways, where one relative reasonable way is *multiview learning*, where one view means one vector.

The constraint over these two matrices of Fig. 9.1 is that the number of rows should be equal each other, though the size of feature sets can be different. Thus the input is two matrices which can have the different sizes of columns but share the size of rows between two matrices.

A typical sample of this matrix which is often introduced in machine learning is that one vector is a natural language sentence (features are words, under either a bag of words assumption or keeping the word order). That is, two vectors are written in different languages, say English and Spanish, but these two sentences share the totally same meaning. Indeed such sentences would numerously exist definitely, while the objective of this setting of data would be rather unclear, and it might be too simple for any problem setting we can make. This is because there would be almost one-to-one correspondence in features (words) between the two sentences of one instance, such as English and Spanish, if the languages are derived from the Latin language.

Then probably the following sample would be more practical and also often the case:

Items purchased in different E-commerce sites For example, one side is a book store, and the other side is a sports item store, where features are items. Each instance is a customer, and an element of the matrix is 1 if the corresponding item is purchased by the corresponding customer; otherwise

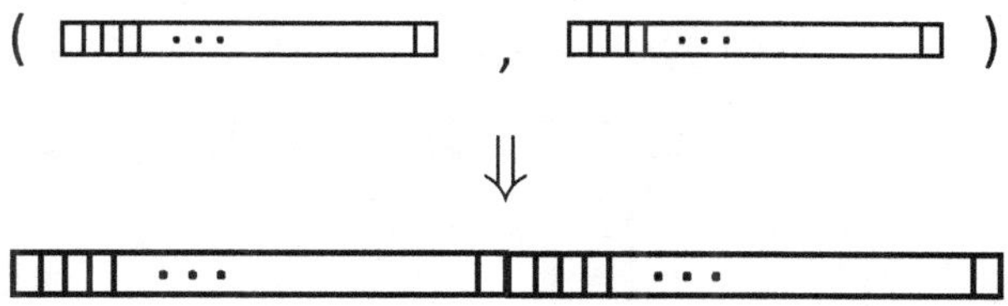

Figure 9.2: Two vectors, given as an instance, can be simply concatenated to be used for regular machine learning methods for vectors.

zero. There would be some connections in the purchased items between books and sports items (the idea behind this is more like market basket analysis, expecting that for example simply travel books and car magazines are purchased at the same time, etc.), and the found rules might characterize customers (customer categorization) by the data eventually.

As such, when we can have multiple vectors for each instance, the simplest or baseline approach for using this type of data is we just concatenate multiple vectors of each instance. Fig. 9.2 shows a schematic procedure of concatenating given multiple vectors. By doing this, as mentioned in Chapter 3, we can use a variety of approaches for dealing with vectors in various problem settings, such as unsupervised and supervised learning, and also feature learning and kernel learning. However, this uses all input views just equally and does not work at all to distinguish different vectors (views), which might be fundamentally different in importance. For example, in the above example of E-commerce, the original objective would be to find the connection between the bookstore purchase and sports item purchase. However we may be unable to focus on this connection between the two different vectors. For example, on average, books might be purchased ten times than sports items, and then the combinations of purchased items are just only from the book side if the two given vectors are just concatenated.

Thus we need consider learning methods more than just concatenating given two vectors for this situation. The first step for our data of two or more vectors is kernel learning.

9.1.1 Kernel learning

Even when we use kernel learning, as mentioned in the above, we can first concatenate the two given vectors into one vector, and generate a kernel from the long vector. However we already mentioned the problem of this approach which cannot distinguish the importance of the original vectors. Thus an alternative, better idea is to generate a kernel from each of the given two vectors and combine these kernels. Fig. 9.3 shows a schematic picture of this procedure. First, we write that the i-th instance $\boldsymbol{x}_i$ consists of the first vector (view) $\boldsymbol{x}_i^{(1)}$ and the second vector (view) $\boldsymbol{x}_i^{(2)}$. Thus, the kernel function between the i-th instance $\boldsymbol{x}_i$

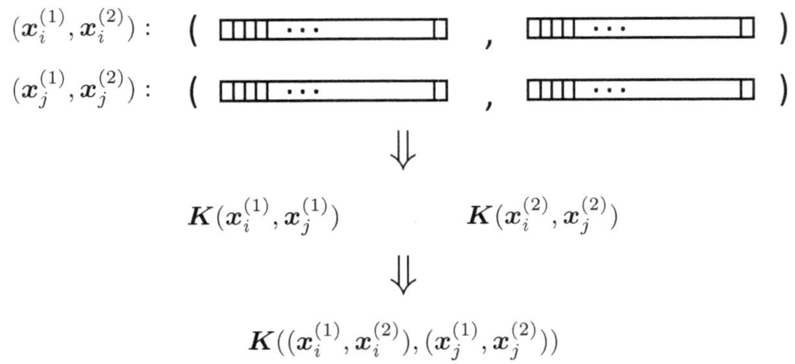

Figure 9.3: Two views, given as an instance, can be transformed into two kernels, and then finally a kernel between two instances is generated (where one instance is made from two views). This can be extended to K views. $\boldsymbol{x}_i^{(j)}$ means the j-th view of the i-th instance.

and the j-th instance $\boldsymbol{x}_j$ can be written as follows:

$$\boldsymbol{K}(\boldsymbol{x}_i, \boldsymbol{x}_j) = \boldsymbol{K}((\boldsymbol{x}_i^{(1)}, \boldsymbol{x}_i^{(2)}), (\boldsymbol{x}_j^{(1)}, \boldsymbol{x}_j^{(2)})). \tag{9.1}$$

The two kernel functions of the first and second vectors (views) are given as follows:

$$\boldsymbol{K}(\boldsymbol{x}_i^{(1)}, \boldsymbol{x}_j^{(1)}), \qquad \boldsymbol{K}(\boldsymbol{x}_i^{(2)}, \boldsymbol{x}_j^{(2)}). \tag{9.2}$$

Then the entire kernel function can be computed from them, for example by using their product, as follows:

$$\boldsymbol{K}((\boldsymbol{x}_i^{(1)}, \boldsymbol{x}_i^{(2)}), (\boldsymbol{x}_j^{(1)}, \boldsymbol{x}_j^{(2)})) = \boldsymbol{K}(\boldsymbol{x}_i^{(1)}, \boldsymbol{x}_j^{(1)}) \times \boldsymbol{K}(\boldsymbol{x}_i^{(2)}, \boldsymbol{x}_j^{(2)}). \tag{9.3}$$

Along with this procedure, we can consider not only two but also any arbitrary number of vectors (views):

$$\boldsymbol{K}((\boldsymbol{x}_i^{(1)}, \ldots, \boldsymbol{x}_i^{(K)}), (\boldsymbol{x}_j^{(1)}, \ldots, \boldsymbol{x}_j^{(K)})) \quad = \quad \boldsymbol{K}(\boldsymbol{x}_i^{(1)}, \boldsymbol{x}_j^{(1)}) \times \cdots \times \boldsymbol{K}(\boldsymbol{x}_i^{(K)}, \boldsymbol{x}_j^{(K)}). \tag{9.4}$$

However, the product is the simplest method, which are rather close to the same level as just concatenating given vectors. It would be more reasonable to consider the importance of each vector (view), meaning that we can assign some weights over given vectors. This way of using kernel functions over multiple vectors (views) is called *multiple kernel learning* (MKL), which is not necessarily for given vectors, but for any data types, which can be kernels. Thus MKL is a method for learning given multiple kernels with weights over them. We describe MKL in Section 9.5, in more detail.

However more importantly, one obvious drawback of kernel learning is that the original feature values are not kept, meaning that we are unable to capture

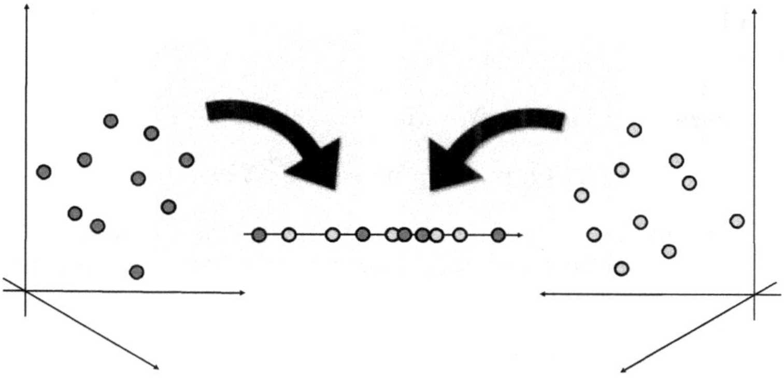

Figure 9.4: Concept of canonical correlation analysis.

the relations between features in different vectors. However, there must be a situation, in which given multiple vectors (views) should be considered differently, and also the relations between features in different vectors (views) should be checked. Below, we introduce the methods, which allow to solve that problem.

9.1.2 Canonical Correlation Analysis

Given two matrices, keeping the constraints in Fig. 9.1, possible questions we have would be mainly the following two:

Relations between features in different vectors As mentioned in the above, one interesting point would be to find the relations between features in different vectors. A brute force manner would be to check all possible one-to-one pairs between one of the features in one vector and one in the other. A better approach would be to find one-to-many or many-to-many relations distributed in different vectors automatically.

Feature reduction in one vector If we are interested in the above relations between different vectors, maybe features not involved with such relations would be out of our interests and can be removed.

One approach which can realize these two at the same time is *canonical correlation analysis* (CCA). Below we will explain CCA [78]. First let $\boldsymbol{X}$ be one of the given two matrices and also $\boldsymbol{Y}$ be the other of the given two matrices. To compare these two, we need to project each of them onto the same space.

Principal component analysis also projects the input $\boldsymbol{X}$ on one dimensional space (see Fig. 3.31) so that the variance of the projected instances should be large and so the given instances are well distributed and understandable well. Similarly CCA projects $\boldsymbol{X}$ onto one dimensional space (Fig. 9.4) as done by principal component analysis (PCA). At the same time CCA projects $\boldsymbol{Y}$ onto the same one-dimensional space on which $\boldsymbol{X}$ is projected already (Fig. 9.4). By doing this, we can compare the two inputs, $\boldsymbol{X}$ and $\boldsymbol{Y}$.

Back to PCA, letting $\boldsymbol{w}$ be the projection weight of $\boldsymbol{X}$, PCA projects $\boldsymbol{X}$ so that the variance of the projected distribution ($\boldsymbol{W}\boldsymbol{w}$) should be the largest. In other words, $\boldsymbol{w}$ is optimized so that the variance of the distribution ($\boldsymbol{X}\boldsymbol{w}$) should be the maximized (see (3.236)). We write (3.236) here again:

$$\max(\boldsymbol{X}\boldsymbol{w})^\top \boldsymbol{X}\boldsymbol{w} = \max \boldsymbol{w}^\top \boldsymbol{X}^\top \boldsymbol{X}\boldsymbol{w}. \tag{9.5}$$

Then CCA can also follow this idea, while instead of variance, we use covariance of $\boldsymbol{X}$ and $\boldsymbol{Y}$, as below. Let $\boldsymbol{w}_X$ and $\boldsymbol{w}_Y$ be weights for the projection of $\boldsymbol{X}$ and those of $\boldsymbol{Y}$, respectively.

$$\max(\boldsymbol{X}\boldsymbol{w}_X)^\top \boldsymbol{Y}\boldsymbol{w}_Y = \max \boldsymbol{w}_X^\top \boldsymbol{X}^\top \boldsymbol{Y}\boldsymbol{w}_Y. \tag{9.6}$$

The constraint of PCA was that the size of weights (L^2 norm) keeps the constant ($\boldsymbol{w}^\top\boldsymbol{w} = 1$). Here similarly, we can place the same L^2 norm constraints for $\boldsymbol{w}_X$ and $\boldsymbol{w}_Y$ of $\boldsymbol{X}$ and $\boldsymbol{Y}$, respectively:

$$\boldsymbol{w}_X^\top \boldsymbol{w}_X = 1. \tag{9.7}$$
$$\boldsymbol{w}_Y^\top \boldsymbol{w}_Y = 1. \tag{9.8}$$

However in CCA, we focus on the correlation between $\boldsymbol{X}$ and $\boldsymbol{Y}$ more, and attempt to remove correlations within $\boldsymbol{X}$ and those within $\boldsymbol{Y}$. That is, the objective function of PCA is now used as the constraints for $\boldsymbol{X}$ and $\boldsymbol{Y}$:

$$(\boldsymbol{X}\boldsymbol{w}_X)^\top \boldsymbol{X}\boldsymbol{w}_X = \boldsymbol{w}_X^\top \boldsymbol{X}^\top \boldsymbol{X}\boldsymbol{w}_X = 1. \tag{9.9}$$
$$(\boldsymbol{Y}\boldsymbol{w}_Y)^\top \boldsymbol{Y}\boldsymbol{w}_Y = \boldsymbol{w}_Y^\top \boldsymbol{Y}^\top \boldsymbol{Y}\boldsymbol{w}_Y = 1. \tag{9.10}$$

Thus we can summarize the formulation of CCA as follows:

$$\max_{\boldsymbol{w}_X, \boldsymbol{w}_Y} \quad \boldsymbol{w}_X^\top \boldsymbol{C}_{XY} \boldsymbol{w}_Y \tag{9.11}$$
$$\text{subject to} \quad \boldsymbol{w}_X^\top \boldsymbol{C}_{XX} \boldsymbol{w}_X = 1 \text{ and } \boldsymbol{w}_Y^\top \boldsymbol{C}_{YY} \boldsymbol{w}_Y = 1. \tag{9.12}$$

Here $\boldsymbol{C}_{XX} = \boldsymbol{X}^\top\boldsymbol{X}$, $\boldsymbol{C}_{XY} = \boldsymbol{Y}^\top\boldsymbol{X}$ and $\boldsymbol{C}_{YY} = \boldsymbol{Y}^\top\boldsymbol{Y}$. These can be computed from data beforehand (before starting the optimization), and so they are the input.

Then we can write the Lagrange function as follows:

$$L(\boldsymbol{w}_X, \boldsymbol{w}_Y) = \boldsymbol{w}_X^\top \boldsymbol{C}_{XY} \boldsymbol{w}_Y - \lambda_X(\boldsymbol{w}_X^\top \boldsymbol{C}_{XX} \boldsymbol{w}_X - 1) - \lambda_Y(\boldsymbol{w}_Y^\top \boldsymbol{C}_{YY} \boldsymbol{w}_Y - 1), \tag{9.13}$$

where λ_X and λ_Y are regularization coefficients.

We then set the derivative of the Lagrange function with respect to $\boldsymbol{w}_X$ (and $\boldsymbol{w}_Y$) equal to zero: That is, $\frac{\partial L(\boldsymbol{w}_X, \boldsymbol{w}_Y)}{\partial \boldsymbol{w}_X} = 0$ and $\frac{\partial L(\boldsymbol{w}_X, \boldsymbol{w}_Y)}{\partial \boldsymbol{w}_Y} = 0$, and then the optimum $\boldsymbol{w}_X$ and $\boldsymbol{w}_Y$ have to keep the following two equations:

$$\boldsymbol{C}_{XY}\boldsymbol{w}_Y - \lambda_X \boldsymbol{C}_{XX}\boldsymbol{w}_X = 0. \tag{9.14}$$
$$\boldsymbol{C}_{XY}^\top\boldsymbol{w}_X - \lambda_Y \boldsymbol{C}_{YY}\boldsymbol{w}_Y = 0. \tag{9.15}$$

Algorithm 9.1: Canonical correlation analysis algorithm.

Function `Canonical_correlation_analysis(`$\boldsymbol{X}, \boldsymbol{Y}$`)`
  Data: $\boldsymbol{X}$, $\boldsymbol{Y}$
  Result: $\boldsymbol{w}_X$, $\boldsymbol{w}_Y$
  Compute $\boldsymbol{C}_{XX}$, $\boldsymbol{C}_{XY}$ and $\boldsymbol{C}_{XY}$ from $\boldsymbol{X}$ and $\boldsymbol{Y}$;
  Solve the general eigenvalue problem of CCA, i.e. (9.17), to output $\boldsymbol{w}_X$ and $\boldsymbol{w}_Y$;

From these two equations, we remove the term of $\boldsymbol{C}_{XY}$ and then we have the following:

$$\lambda_X \boldsymbol{w}_X^\top \boldsymbol{C}_{XX} \boldsymbol{w}_X = \lambda_Y \boldsymbol{w}_Y^\top \boldsymbol{C}_{YY} \boldsymbol{w}_Y. \tag{9.16}$$

From this, we can see $\lambda_X = \lambda_Y$. Then from (9.14) and (9.15), we can obtain the optimum $\boldsymbol{w}_X$ and $\boldsymbol{w}_Y$ by solving the following:

$$\begin{pmatrix} \mathbf{0} & \boldsymbol{C}_{XY} \\ \boldsymbol{C}_{YX} & \mathbf{0} \end{pmatrix} \begin{pmatrix} \boldsymbol{w}_X \\ \boldsymbol{w}_Y \end{pmatrix} = \lambda \begin{pmatrix} \boldsymbol{C}_{XX} & \mathbf{0} \\ \mathbf{0} & \boldsymbol{C}_{YY} \end{pmatrix} \begin{pmatrix} \boldsymbol{w}_X \\ \boldsymbol{w}_Y \end{pmatrix}. \tag{9.17}$$

where $\boldsymbol{C}_{XY}^\top = (\boldsymbol{X}^\top \boldsymbol{Y})^\top = \boldsymbol{Y}^\top \boldsymbol{X} = \boldsymbol{C}_{YX}$.

This is a generalized eigenvalue problem, but this is easily reduced to a regular eigenvalue problem (see Section A.4.4).

The point of the computation of CCA is time-efficiency, because the matrices of (9.17) have nothing to do with the number of instances. This is because all matrices, i.e. $\boldsymbol{C}_{XY}$, $\boldsymbol{C}_{XX}$ and $\boldsymbol{C}_{YY}$, have the size of (number of features in $\boldsymbol{X}$)$\times$(number of features in $\boldsymbol{Y}$). So the generalized eigenvalue problem has nothing to do with the number of instances, but with the number (size) of features. The size of instances affects the computation of $\boldsymbol{C}_{XY}$, $\boldsymbol{C}_{XX}$ and $\boldsymbol{C}_{YY}$, but multiplying $M \times N$-matrix by $N \times M$-matrix needs the computational complexity of $O(M^2 \cdot N)$ only, meaning that the above computation of $\boldsymbol{C}_{XY}$, $\boldsymbol{C}_{XX}$ and $\boldsymbol{C}_{YY}$ needs only the linear computation time with respect to the number of instances.

Also we can see that (9.17) has matrices with zeros in diagonals of the right-hand side and in non-diagonals of the left-hand side. Thus if eigenvalue λ and eigenvector

$$\begin{pmatrix} \boldsymbol{w}_X \\ \boldsymbol{w}_Y \end{pmatrix} \tag{9.18}$$

are the optimum solution, eigenvalue $-\lambda$ and eigenvector

$$\begin{pmatrix} \boldsymbol{w}_X \\ -\boldsymbol{w}_Y \end{pmatrix}. \tag{9.19}$$

are also the optimum solution. This means that we can first focus on the optimum solutions with positive eigenvalues only, by which solving the problem can be more efficient further.

Algorithm 9.2: Generalized canonical correlation analysis algorithm.

Function `Generalized_canonical_correlation_analysis(`$\boldsymbol{X}_1, \ldots, \boldsymbol{X}_K$`)`
  Data: $\boldsymbol{X}_1, \ldots, \boldsymbol{X}_K$
  Result: $\boldsymbol{w}_1, \ldots, \boldsymbol{w}_K$
  for $i \leftarrow 1$ **to** K **do**
    for $j \leftarrow 1$ **to** K **do**
      Compute $\boldsymbol{C}_{X_i X_j} = \boldsymbol{X}_i^\top \boldsymbol{X}_j$.;
  Solve the general eigenvalue problem of (9.23), to output $\boldsymbol{w}_1, \ldots, \boldsymbol{w}_K$.;

Also as well as PCA, larger eigenvalues and their eigenvectors (weights) between $\boldsymbol{X}$ and $\boldsymbol{Y}$ are correlated more. **Algorithm 9.1** shows a pseudocode of the above procedure.

Here let us think the data we have more than two different vectors (views). First (9.14) and (9.15) can be transformed into the following:

$$\boldsymbol{C}_{XX}\boldsymbol{w}_X + \boldsymbol{C}_{XY}\boldsymbol{w}_Y - (1+\lambda_X)\boldsymbol{C}_{XX}\boldsymbol{w}_X = 0. \tag{9.20}$$

$$\boldsymbol{C}_{XY}^\top\boldsymbol{w}_X + \boldsymbol{C}_{YY}\boldsymbol{w}_Y - (1+\lambda_Y)\boldsymbol{C}_{YY}\boldsymbol{w}_Y = 0. \tag{9.21}$$

Then the generalized eigenvalue problem (9.17) also can be written as follows:

$$\begin{pmatrix} \boldsymbol{C}_{XX} & \boldsymbol{C}_{XY} \\ \boldsymbol{C}_{YX} & \boldsymbol{C}_{YY} \end{pmatrix} \begin{pmatrix} \boldsymbol{w}_X \\ \boldsymbol{w}_Y \end{pmatrix} = (1+\lambda) \begin{pmatrix} \boldsymbol{C}_{XX} & \boldsymbol{0} \\ \boldsymbol{0} & \boldsymbol{C}_{YY} \end{pmatrix} \begin{pmatrix} \boldsymbol{w}_X \\ \boldsymbol{w}_Y \end{pmatrix}. \tag{9.22}$$

From these observations, if K different vectors (views) are given, we can write the equation which the optimum weights have to keep, generally, as follows:

$$\begin{pmatrix} \boldsymbol{C}_{11} & \boldsymbol{C}_{12} & \ldots & \boldsymbol{C}_{1K} \\ \boldsymbol{C}_{21} & \boldsymbol{C}_{22} & \ldots & \boldsymbol{C}_{2K} \\ \vdots & \vdots & \ddots & \vdots \\ \boldsymbol{C}_{K1} & \boldsymbol{C}_{K2} & \ldots & \boldsymbol{C}_{KK} \end{pmatrix} \begin{pmatrix} \boldsymbol{w}_1 \\ \boldsymbol{w}_2 \\ \vdots \\ \boldsymbol{w}_K \end{pmatrix} = \lambda \begin{pmatrix} \boldsymbol{C}_{11} & \boldsymbol{0} & \ldots & \boldsymbol{0} \\ \boldsymbol{0} & \boldsymbol{C}_{22} & \ldots & \boldsymbol{0} \\ \vdots & \vdots & \ddots & \vdots \\ \boldsymbol{0} & \boldsymbol{0} & \ldots & \boldsymbol{C}_{KK} \end{pmatrix} \begin{pmatrix} \boldsymbol{w}_1 \\ \boldsymbol{w}_2 \\ \vdots \\ \boldsymbol{w}_K \end{pmatrix}. \tag{9.23}$$

where $\boldsymbol{C}_{ij}$ is the covariance matrix between i-th vector (view) and j-th vector (view).

Overall **Algorithm 9.2** is a pseudocode of the algorithm of generalized CCA for K vectors.

9.1.3 Kernel Canonical Correlation Analysis

The simpler kernel learning explained in Section 9.1.1 cannot keep the features given in each vector. However, by extending CCA in the framework of kernel learning, we can capture the relations (particularly non-linear relations) between features in different vectors.

First, as was done by kernel learning of Section 3.4, we represent the weights $\boldsymbol{w}_X$ and $\boldsymbol{w}_Y$ of input $\boldsymbol{X}$ and $\boldsymbol{Y}$, respectively, by a linear combination of input $\boldsymbol{X}$ and $\boldsymbol{Y}$, respectively, as follows:

$$\boldsymbol{w}_X = \boldsymbol{X}^\top \boldsymbol{\beta}_X. \tag{9.24}$$

$$\boldsymbol{w}_Y = \boldsymbol{Y}^\top \boldsymbol{\beta}_Y. \tag{9.25}$$

This means that instead of $\boldsymbol{w}_X$ and $\boldsymbol{w}_Y$, we estimate $\boldsymbol{\beta}_X$ and $\boldsymbol{\beta}_Y$ for solving this problem of CCA in kernel learning.

We then directly substitute them into (9.11) and (9.12), resulting in the following optimization problem:

$$\max_{\boldsymbol{\beta}_X, \boldsymbol{\beta}_Y} \quad \boldsymbol{\beta}_X^\top \boldsymbol{X}\boldsymbol{X}^\top \boldsymbol{Y}\boldsymbol{Y}^\top \boldsymbol{\beta}_Y. \tag{9.26}$$

$$\text{subject to} \quad \boldsymbol{\beta}_X^\top \boldsymbol{X}\boldsymbol{X}^\top \boldsymbol{X}\boldsymbol{X}^\top \boldsymbol{\beta}_X = 1 \text{ and } \boldsymbol{\beta}_Y^\top \boldsymbol{Y}\boldsymbol{Y}^\top \boldsymbol{Y}\boldsymbol{Y}^\top \boldsymbol{\beta}_Y = 1. \tag{9.27}$$

As we did in kernel learning, covariance matrices $\boldsymbol{X}\boldsymbol{X}^\top$ and $\boldsymbol{Y}\boldsymbol{Y}^\top$ can be replaced with kernel functions $\boldsymbol{K}_X$ and $\boldsymbol{K}_Y$, respectively. Then we can have the following formulation of *kernel canonical correlation analysis* (KCCA) [78]:

$$\max_{\boldsymbol{\beta}_X, \boldsymbol{\beta}_Y} \quad \boldsymbol{\beta}_X^\top \boldsymbol{K}_X \boldsymbol{K}_Y \boldsymbol{\beta}_Y. \tag{9.28}$$

$$\text{subject to} \quad \boldsymbol{\beta}_X^\top \boldsymbol{K}_X^2 \boldsymbol{\beta}_X = 1 \text{ and } \boldsymbol{\beta}_Y^\top \boldsymbol{K}_Y^2 \boldsymbol{\beta}_Y = 1. \tag{9.29}$$

We can write the Lagrange function as follows:

$$L(\boldsymbol{\beta}_X, \boldsymbol{\beta}_Y) = \boldsymbol{\beta}_X^\top \boldsymbol{K}_X \boldsymbol{K}_Y \boldsymbol{\beta}_Y - \lambda_X(\boldsymbol{\beta}_X^\top \boldsymbol{K}_X^2 \boldsymbol{\beta}_X - 1) - \lambda_Y(\boldsymbol{\beta}_Y^\top \boldsymbol{K}_Y^2 \boldsymbol{\beta}_Y - 1). \tag{9.30}$$

where λ_X and λ_Y are regularization coefficients.

We set the derivative of the Lagrange function with respect to $\boldsymbol{\beta}_X$ (and $\boldsymbol{\beta}_Y$) equal to zero: $\frac{\partial L(\boldsymbol{\beta}_X, \boldsymbol{\beta}_Y)}{\partial \boldsymbol{\beta}_X} = 0$ and $\frac{\partial L(\boldsymbol{\beta}_X, \boldsymbol{\beta}_Y)}{\partial \boldsymbol{\beta}_Y} = 0$. From them we can have the following equations which both weights $\boldsymbol{\beta}_X$ and $\boldsymbol{\beta}_Y$ have to keep:

$$\boldsymbol{K}_X \boldsymbol{K}_Y \boldsymbol{\beta}_Y - \lambda_X \boldsymbol{K}_X^2 \boldsymbol{\beta}_X = 0. \tag{9.31}$$

$$\boldsymbol{K}_Y \boldsymbol{K}_X \boldsymbol{\beta}_X - \lambda_Y \boldsymbol{K}_Y^2 \boldsymbol{\beta}_Y = 0. \tag{9.32}$$

Kernel functions $\boldsymbol{K}_X$ and $\boldsymbol{K}_Y$ are originally covariance matrices of $\boldsymbol{X}$ and $\boldsymbol{Y}$, respectively. So if given instances are independent of each other, the covariance matrices are full-rank matrices (regular matrices). Suppose that kernel functions also follow the covariance matrices and then regular matrices, so they have the inverse matrices. Then by using (9.31), we can estimate $\boldsymbol{\beta}_Y$:

$$\boldsymbol{\beta}_Y = \lambda_X \boldsymbol{K}_Y^{-1} \boldsymbol{K}_X \boldsymbol{\beta}_X. \tag{9.33}$$

From these observations, regarding the regularization coefficients, we can expect $\lambda_X = \lambda_Y$ again. Then the following equation can be generated:

$$\boldsymbol{K}_X \boldsymbol{\beta}_X - \lambda^2 \boldsymbol{K}_X \boldsymbol{\beta}_X = 0. \tag{9.34}$$

This equation means that for $\lambda = 1$, any arbitrary vector $\boldsymbol{\beta}_X$ can be allowed, meaning that we cannot obtain any solution. This is originally because we are now using highly flexible kernel functions which can make the formulation (9.30) overfitted to the given data, making us hard to have any solutions.

Then we now go back to CCA, and to reduce the above flexibility, we attempt to add more regularization terms. In fact, the regularization terms of CCA was (9.9) and (9.10). We here write them again, as follows:

$$\boldsymbol{w}_X^\top \boldsymbol{X}^\top \boldsymbol{X} \boldsymbol{w}_X = 1, \quad \boldsymbol{w}_Y^\top \boldsymbol{Y}^\top \boldsymbol{Y} \boldsymbol{w}_Y = 1. \tag{9.35}$$

KCCA follows these regularization terms, which can be written in the kernel forms, as follows:

$$\boldsymbol{\beta}_X^\top \boldsymbol{K}_X^2 \boldsymbol{\beta}_X = 1, \quad \boldsymbol{\beta}_Y^\top \boldsymbol{K}_Y^2 \boldsymbol{\beta}_Y = 1. \tag{9.36}$$

However, as we wrote when we derived (9.9) and (9.10), another possibility was to use the terms equivalent to the original PCA, i.e. (9.7) and (9.8). We write them again as follows:

$$\boldsymbol{w}_X^\top \boldsymbol{w}_X = 1, \quad \boldsymbol{w}_Y^\top \boldsymbol{w}_Y = 1. \tag{9.37}$$

In fact what we need now is to restrict the high flexibility and so adding regularization terms more would be useful for this objective. Thus we use these two types of regularization terms, weighing over the two terms, which is like elastic net, explained in Section 3.3.4[1]. That is, we can formulate the problem, using the both two regularization terms, as follows:

$$\max_{\boldsymbol{w}_X, \boldsymbol{w}_Y} \quad \boldsymbol{w}_X^\top \boldsymbol{C}_{XY} \boldsymbol{w}_Y \tag{9.38}$$

$$\text{subject to} \quad (1-\tau_X) \boldsymbol{w}_X^\top \boldsymbol{C}_{XX} \boldsymbol{w}_X + \tau_X \boldsymbol{w}_X^\top \boldsymbol{w}_X = 1 \text{ and} \tag{9.39}$$

$$(1-\tau_Y) \boldsymbol{w}_Y^\top \boldsymbol{C}_{YY} \boldsymbol{w}_Y + \tau_Y \boldsymbol{w}_Y^\top \boldsymbol{w}_Y = 1. \tag{9.40}$$

where τ_X and τ_Y are hyperparameters, weighing over two regularization terms and taking a value between zero and one, as follows:

$$0 < \tau_X < 1, \quad 0 < \tau_Y < 1. \tag{9.41}$$

This means that for $\tau_X = 0$, only the term with $\boldsymbol{C}_{XX}$ is used, and the term derived here is not used. On the contrary, for $\tau_X = 1$, the term with $\boldsymbol{C}_{XX}$ is not used, and only the term derived here is used. Thus, τ_X should be a value between zero and one. This is true of τ_Y.

Again, using (9.24) and (9.25), we transform the above formulation into kernel canonical correlation analysis (KCCA), as follows:

$$\max_{\boldsymbol{\beta}_X, \boldsymbol{\beta}_Y} \quad \boldsymbol{\beta}_X^\top \boldsymbol{K}_X \boldsymbol{K}_Y \boldsymbol{\beta}_Y \tag{9.42}$$

$$\text{subject to} \quad (1-\tau_X) \boldsymbol{\beta}_X^\top \boldsymbol{K}_X^2 \boldsymbol{\beta}_X + \tau_X \boldsymbol{\beta}_X^\top \boldsymbol{K}_X \boldsymbol{\beta}_X = 1 \text{ and} \tag{9.43}$$

$$(1-\tau_Y) \boldsymbol{\beta}_Y^\top \boldsymbol{K}_Y^2 \boldsymbol{\beta}_Y + \tau_Y \boldsymbol{\beta}_Y^\top \boldsymbol{K}_Y \boldsymbol{\beta}_Y . = 1 \tag{9.44}$$

[1] A clear difference between this case and elastic net is that elastic net used the L^1 and L^2 norms, while in this case both are squared term of the parameters.

We can write the Lagrange function as follows:

$$
\begin{aligned}
L(\boldsymbol{\beta}_X, \boldsymbol{\beta}_Y) &= \boldsymbol{\beta}_X^\top \boldsymbol{K}_X \boldsymbol{K}_Y \boldsymbol{\beta}_Y && (9.45)\\
&- \lambda_X((1-\tau_X)\boldsymbol{\beta}_X^\top \boldsymbol{K}_X^2 \boldsymbol{\beta}_X + \tau_X \boldsymbol{\beta}_X^\top \boldsymbol{K}_X \boldsymbol{\beta}_X - 1) && (9.46)\\
&- \lambda_Y((1-\tau_Y)\boldsymbol{\beta}_Y^\top \boldsymbol{K}_Y^2 \boldsymbol{\beta}_Y + \tau_Y \boldsymbol{\beta}_Y^\top \boldsymbol{K}_Y \boldsymbol{\beta}_Y - 1). && (9.47)
\end{aligned}
$$

We then set the derivative of the Lagrange function with respect to $\boldsymbol{\beta}_X$ (and $\boldsymbol{\beta}_Y$) equal to zero: That is, $\frac{\partial L(\boldsymbol{\beta}_X,\boldsymbol{\beta}_Y)}{\partial \boldsymbol{\beta}_X} = 0$ and $\frac{\partial L(\boldsymbol{\beta}_X,\boldsymbol{\beta}_Y)}{\partial \boldsymbol{\beta}_Y} = 0$. Again, due to symmetricity, we can see $\lambda_X = \lambda_Y = \lambda$, and then the weights to be estimated have to keep the following equations:

$$
\begin{aligned}
\boldsymbol{K}_X \boldsymbol{K}_Y \boldsymbol{\beta}_Y - \lambda((1-\tau_X)\boldsymbol{K}_X^2 \boldsymbol{\beta}_X + \tau_X \boldsymbol{K}_X \boldsymbol{\beta}_X) &= 0. && (9.48)\\
\boldsymbol{K}_Y \boldsymbol{K}_X \boldsymbol{\beta}_X - \lambda((1-\tau_Y)\boldsymbol{K}_Y^2 \boldsymbol{\beta}_Y + \tau_Y \boldsymbol{K}_Y \boldsymbol{\beta}_Y) &= 0. && (9.49)
\end{aligned}
$$

where first kernel functions are positive semidefinite matrices and so are regular matrices, meaning that they have inverse matrices (see Section A.4.6). Thus, by multiplying from the left-hand side of (9.48) and (9.49) by the inverse matrices of kernel functions, i.e. $\boldsymbol{K}_X^{-1}$ and $\boldsymbol{K}_Y^{-1}$, respectively, we can have an equivalent formulation to (9.17) of CCA. That is, we can see that we should solve the following generalized eigenvalue problem:

$$
\begin{pmatrix} \mathbf{0} & \boldsymbol{K}_Y \\ \boldsymbol{K}_X & \mathbf{0} \end{pmatrix} \begin{pmatrix} \boldsymbol{\beta}_X \\ \boldsymbol{\beta}_Y \end{pmatrix} = \lambda \begin{pmatrix} (1-\tau_X)\boldsymbol{K}_X + \tau_X \boldsymbol{I} & \mathbf{0} \\ \mathbf{0} & (1-\tau_Y)\boldsymbol{K}_Y + \tau_Y \boldsymbol{I} \end{pmatrix} \begin{pmatrix} \boldsymbol{\beta}_X \\ \boldsymbol{\beta}_Y \end{pmatrix}. \quad (9.50)
$$

This formulation looks very equivalent to (9.17), while the decisive difference is that the matrices of (9.50) have much larger sizes. That is, although this problem is peculiar to kernel learning, both $\boldsymbol{K}_X$ and $\boldsymbol{K}_Y$ are $(N \times N)$-matrices, resulting in that the size of the matrices in (9.50) is double the number of instances in both rows and columns. This is totally opposite to that the size of the matrices in (9.50) of CCA has nothing to do with the number of instances. In fact kernel learning always suffers from the size of kernel functions, which are always $(N \times N)$ matrices.

However a good news of KCCA is that we can reduce the size of matrices in the main generalized eigenvalue problem, i.e. (9.50), by using the product of kernel functions in (9.48) and (9.49). First, kernel functions are positive semi-definite matrices, and so Cholesky decomposition is possible over kernels (see Section A.4.6):

$$
\boldsymbol{K}_X = \boldsymbol{R}_X^\top \boldsymbol{R}_X, \qquad \boldsymbol{K}_Y = \boldsymbol{R}_Y^\top \boldsymbol{R}_Y. \quad (9.51)
$$

where you can see that the size of columns of $\boldsymbol{R}_X$ and $\boldsymbol{R}_Y$ has to be the number of instances, while the size of rows can be far reduced from the number of instances.

Thus first we replace kernel functions with $\boldsymbol{R}_X$ and $\boldsymbol{R}_Y$ by using (9.48) and (9.49) and then multiply from the left-hand side by the inverse matrices of $\boldsymbol{R}_X^\top$

Algorithm 9.3: Kernel canonical correlation analysis algorithm.

Function `Kernel_canonical_correlation_analysis(`$\boldsymbol{X}$, $\boldsymbol{Y}$, τ_X, τ_Y`)`
  Data: $\boldsymbol{X}$, $\boldsymbol{Y}$, τ_X, τ_Y
  Result: $\boldsymbol{\beta}_X$, $\boldsymbol{\beta}_Y$
  1. Compute kernels: $\boldsymbol{K}_X$ and $\boldsymbol{K}_Y$.;
  2. Decompose kernels into $\boldsymbol{R}_X$ and $\boldsymbol{R}_Y$, due to (9.51).;
  3. Solve the eigenvalue problem (9.55) to obtain $\boldsymbol{\alpha}_X$ and $\boldsymbol{\alpha}_Y$.;
  4. Output $\boldsymbol{\beta}_X$ and $\boldsymbol{\beta}_Y$, which are obtained from $\boldsymbol{\alpha}_X$ and $\boldsymbol{\alpha}_Y$ by using (9.52).;

and $\boldsymbol{R}_Y^\top$. We then switch the parameters we should obtain from $\boldsymbol{\beta}_X$ and $\boldsymbol{\beta}_Y$ to $\boldsymbol{\alpha}_X$ and $\boldsymbol{\alpha}_Y$, respectively, by using the following:

$$\boldsymbol{\alpha}_X = \boldsymbol{R}_X \boldsymbol{\beta}_X, \qquad \boldsymbol{\alpha}_Y = \boldsymbol{R}_Y \boldsymbol{\beta}_Y. \tag{9.52}$$

Finally (9.48) and (9.49) can be given as follows:

$$\boldsymbol{R}_X \boldsymbol{R}_Y^\top \boldsymbol{\alpha}_Y - \lambda(1-\tau_X)\boldsymbol{R}_X \boldsymbol{R}_X^\top \boldsymbol{\alpha}_X - \lambda \tau_X \boldsymbol{\alpha}_X = 0. \tag{9.53}$$

$$\boldsymbol{R}_Y \boldsymbol{R}_X^\top \boldsymbol{\alpha}_X - \lambda(1-\tau_Y)\boldsymbol{R}_Y \boldsymbol{R}_Y^\top \boldsymbol{\alpha}_Y - \lambda \tau_Y \boldsymbol{\alpha}_Y = 0. \tag{9.54}$$

From these two equations, $\boldsymbol{\alpha}_X$ and $\boldsymbol{\alpha}_Y$ can be estimated by solving the following generalized eigenvalue problem:

$$\begin{pmatrix} \mathbf{0} & \boldsymbol{R}_X \boldsymbol{R}_Y^\top \\ \boldsymbol{R}_Y \boldsymbol{R}_X^\top & \mathbf{0} \end{pmatrix} \begin{pmatrix} \boldsymbol{\alpha}_X \\ \boldsymbol{\alpha}_Y \end{pmatrix} = \lambda \begin{pmatrix} (1-\tau_X)\boldsymbol{R}_X \boldsymbol{R}_X^\top + \tau_X \boldsymbol{I} & \mathbf{0} \\ \mathbf{0} & (1-\tau_Y)\boldsymbol{R}_Y \boldsymbol{R}_Y^\top + \tau_Y \boldsymbol{I} \end{pmatrix} \begin{pmatrix} \boldsymbol{\alpha}_X \\ \boldsymbol{\alpha}_Y \end{pmatrix}. \tag{9.55}$$

In the above generalized eigenvalue problem, the size of matrices $\boldsymbol{R}_X \boldsymbol{R}_X^\top$, $\boldsymbol{R}_X \boldsymbol{R}_Y^\top$ and $\boldsymbol{R}_Y \boldsymbol{R}_Y^\top$ can be much smaller than the number of instances. Also due to symmetricity, all we have to do is just to obtain only the half of all eigenvalues and eigenvectors. **Algorithm 9.3** shows a pseudocode of the above procedure.

9.2 Integrative Learning for Vectors and Graphs

We have explained integrative learning for multiple vectors, particularly two different types of vectors. We now consider that one of the two vectors is a more complex data type, particularly a graph (as mentioned in the beginning of this chapter, sets, sequences and trees can be special cases of graphs, and so we focus on graphs as a more complex structure of vectors). Thus each instance has one vector and one graph.

In fact, the most typical example of graphs as instances are chemical compounds, i.e. their chemical structures or molecular graphs. The following would be the most typical example of the instance with one vector and one graph.

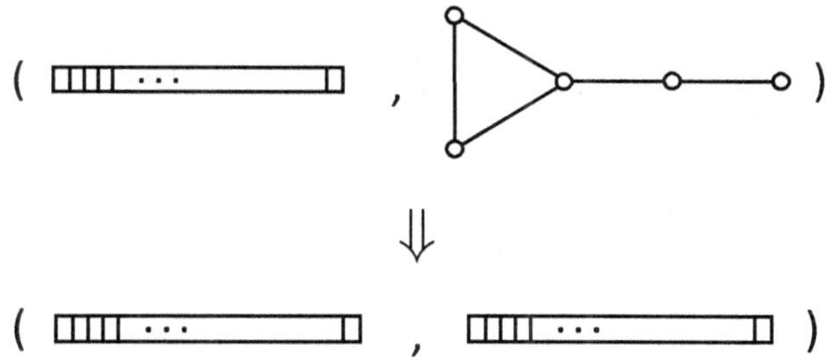

Figure 9.5: A vector and a graph, given as an instance, can be simply two vectors by transforming the graph into a vector.

Chemical compound database Each instance is a chemical compound, consisting of two parts:

1. **physico-chemical properties:**
   Hydrophobicity, electricity, etc., and each of them is numerical or discrete values. Entirely these properties are represented by a vector.
2. **molecular graph:**
   Chemical structures, called molecular graphs, have elements for nodes and chemical bonds between elements for edges.

The former is a vector, and the latter is a graph, by which each instance has a vector and a graph.

We consider machine learning methods for instances with these two different data types.

Again a simple method for this dataset would be first to extract features from each graph and then each graph is transformed into a vector. By doing this, we have two different types of vectors for each instance, and we can apply CCA and KCCA that we have just explained in Section 9.1 to these instances, each being with two different types of vectors. Fig. 9.5 shows a schematic picture of changing one graph into a vector, resulting in two different types of vectors, which can be used for integrative learning of two different vectors. In fact, by using CCA and KCCA, we can see the relations between features in different vectors. However, the vector derived from a graph cannot necessarily be directly connected to particular nodes, edges and subgraphs. This means that even if we have some relations between vectors, these relations may not be connected to the original graphs. This is because transforming graphs into vectors does not necessarily keep the original graphs or transforming graphs into vectors might lose the information in original graphs. Thus, we consider machine learning methods which can keep the information of given vectors and graphs as much as possible.

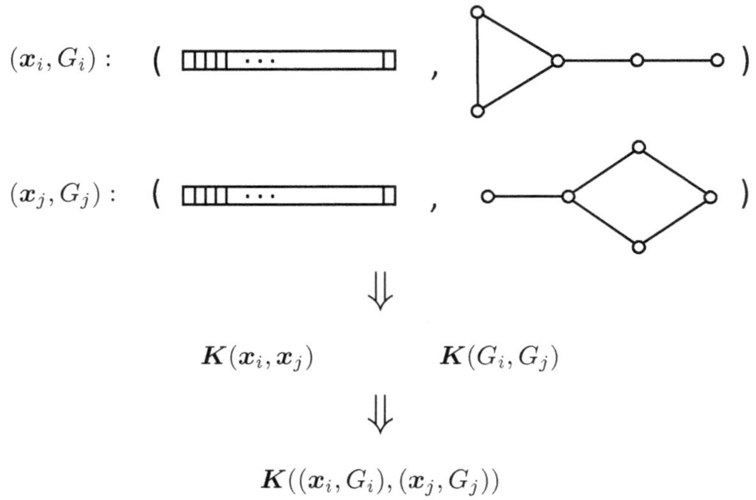

Figure 9.6: A vector and a graph, given as an instance, can be transformed into two kernels, and then finally a kernel between two instances (where one instance is made from one vector and one graph).

9.2.1 Kernel Learning

Again one possible approach would be kernel learning. In 9.1.1, we already explained kernel learning for multiple vectors (views), and we keep the same procedure, except that we use graphs instead of vectors for one of the two sides. That is, we generate a kernel function from vectors and also another kernel functions from graphs, and then we combine these two kernels for kernel learning. Fig. 9.6 shows a schematic picture of this procedure. First, from two vectors $\boldsymbol{x}_i$, $\boldsymbol{x}_j$, kernel function $K(\boldsymbol{x}_i, \boldsymbol{x}_j)$ is generated, and from two graphs G_i and G_j, kernel function $K(G_i, G_j)$ is generated (see Section 7.2 for graph kernels). We then compute the final kernel by combining these two kernels to generate kernel function between two instances, each being with a vector and a graph ($K((\boldsymbol{x}_i, G_i), (\boldsymbol{x}_j, G_j))$).

For example, as we explained in Section 9.1.1, a simple kernel would be just the product of the kernel from vectors and the kernel from graphs:

$$\boldsymbol{K}((\boldsymbol{x}_i, G_i), (\boldsymbol{x}_j, G_j)) = \boldsymbol{K}(\boldsymbol{x}_i, \boldsymbol{x}_j) \times \boldsymbol{K}(G_i, G_j). \tag{9.56}$$

However, as explained in Section 9.1.1, this "product" approach is too simple to integrate the kernels generated from different part of data. For this problem, possible approaches are the following two:

Multiple kernel learning (MKL) which allows to learn weights between vectors and graphs, considering their significance. MKL will be explained more in Section 9.5.

Prior knowledge Also another possible approach is to combine kernels using some prior knowledge. If we generate an integrative kernel by using proper knowledge on the combination, we can use a lot of approaches in kernel learning by using the generated kernel.

In summary, kernel learning can be summarized as follows:

Advantage: We do not have to transform data (graphs) into vectors, and so original graph information is not lost like the data transformation.

Disadvantage: However, the original graphs themselves are not kept. Thus we cannot see what portion of the original graph is working to obtain the result. This is a drawback shared by all kernel learning.

9.2.2 Frequent Pattern Mining

Another approach to avoid transforming graphs into vectors and keep the original graph and vector information is frequent pattern mining. We have seen frequent pattern mining for different data types: itemsets, sequences, trees and graphs. Thus instead of vectors, we now think itemsets, the most basic data type for frequent pattern mining, which are equivalent to vectors. Thus each instance has an itemset and also a graph.

The purpose of frequent pattern mining is to enumerate all frequent patterns in given datasets. For example, for given itemsets, if the support of an itemset is equal to or larger than the prespecified minimum support, this itemset is called a frequent itemset. For graphs, if the support of a subgraph is no less than the prespecified minimum support, this subgraph is called a frequent subgraph. Then now for pairs of itemsets and graphs, the support of a pair is no less than the prespecified minimum support, this pair can be a frequent pattern:

Definition 9.1 (Frequent pair mining). *For a dataset, in which each instance is a pair of an itemset and a graph, if both an itemset and a subgraph appear at the same time no less than the minimum support times, this pair is frequent.*

The point is a frequent pair must be frequent as a *pair*. In other words, even if both its itemset and its subgraph are frequent independently, a pair of an itemset and a graph cannot necessarily be frequent. Thus we just need to check if a given pair is frequent.

The following point can be just noted:

Proposition 9.1. *If a pair (of an itemset and a subgraph) is frequent (equal to or larger than the prespecified minimum support), both its itemset and its subgraph must be frequent.*

However again even if an itemset and a subgraph are both frequent independently, their pair cannot be necessarily frequent. This indicates that we should examine if each pair is frequent directly[2]. The idea of the algorithm for frequent pair mining follows that of other frequent pattern mining: we can first start with

[2]This is because the pair has a more wide data (search) space than the space by itemsets (or subgraphs).

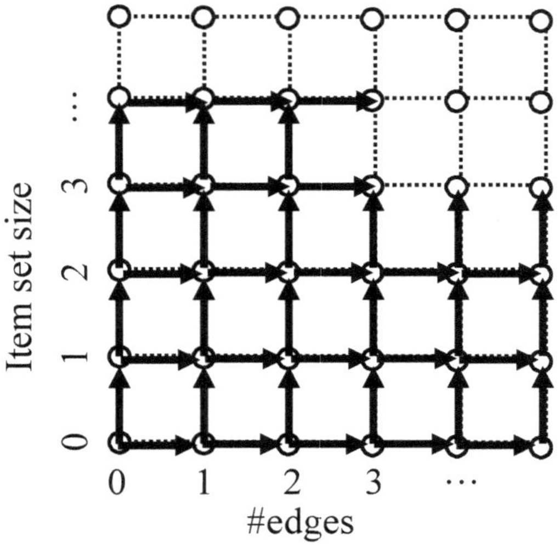

Figure 9.7: Pairwise alignment is equivalent to the problem of finding the least cost path on the two-dimensional grid.

one item or one edge, then make the frequent pair larger by adding one item or one edge to the current frequent pair and check the larger pairs to keep each of them if it is frequent. This algorithm can use the basic idea of frequent pattern mining: **downward closure property** as follows:

Proposition 9.2 (Downward Closure Property (pair)). *If the support of a pair (of an itemset and a subgraph) is smaller than the minimum support, any pair including the itemset or the subgraph cannot be frequent.*

Thus again when we increase the itemset or the subgraph of a pair, if the pair is no longer frequent, any pair including the itemset or the subgraph cannot be frequent and so we do not have to search along with that direction any more.

We write this more mathematically. First we define notations: Let $\mathcal{P}(i, e)$ be a set of pairs, where each pair has an itemset of i items and a subgraph of e edges. Let $\mathcal{F}(i, e)$ be a set of frequent pairs, where each pair has an itemset of i items and a subgraph of e edges:

$$\mathcal{P}(i, e) \supseteq \mathcal{F}(i, e). \tag{9.57}$$

Note that a set of all frequent pairs $\mathcal{F}$ is the union of $\mathcal{F}(i, e)$ over all possible i and e:

$$\mathcal{F} \leftarrow \cup_{i,e} \mathcal{F}(i, e). \tag{9.58}$$

Here let us consider an algorithm for enumerating all possible frequent pairs. First, $\mathcal{F}(i, e)$ are grids of the two-dimensional lattice with one axis for the number

Algorithm 9.4: Checking if a given pair is frequent and new.

```
Function Checking_if_given_pair_is_frequent((I, G), Min_Sup)
    Data: Pair (set + graph): (I, G), Minimum support: Min_Sup
    if G is not a minimum DFS code then
        return.;
    if Support of (I, G) is not less than Min_Sup and (I, G) is new to F_{|I|+|G|} then
        Add (I, G) to F_{|I|+|G|}.;
```

Algorithm 9.5: Recursively obtaining a complete set of frequent pairs, which are one letter/edge extension of a given set of frequent pairs.

```
Function Generate_frequent_pairs(F_t, Min_Sup)
    Data: Complete set of frequent pairs with the size of t: F_t, minimum
          support: Min_Sup
    Result: F_{t+1}
    foreach (I, G) ∈ F_t do
        foreach (I^+, G^+), where (I^+ is one item extension of I and G^+ is
        G) or (G^+ is one edge extension of G and I^+ is I) do
            Checking_if_given_pair_is_frequent((I^+, G^+), Min_Sup).;
    if F_{t+1} is not empty then
        Run Generate_frequent_pairs(F_{t+1}, Min_Sup).;
```

of itemsets, i, and the other axis for the number of edges, e. Fig. 9.7 shows the two dimensional lattice. The objective of the algorithm is to enumerate all possible frequent grids, i.e. $\mathcal{F}(i,e)$. The key point is that instances with a pair (of an itemset I with i items and a subgraph G with e edges) in $\mathcal{F}(i,e)$ are all in instances with the pair (itemset I^-, one item smaller than I and G), and those with the pair (I and G^-, one edge smaller than G). In other words, the pair of I and G can be obtained from both (I^- and G) and (I and G^-). Then, reversely, if a pair P with an itemset of $i-1$ items and a graph of e edges is already not in $\mathcal{F}(i-1,e)$, a new pair obtained by adding one item to P cannot be in $\mathcal{F}(i,e)$. This is true of the graph side.

Thus, to generate $F(i,e)$, we increase one item or one edge, respectively, of pairs in $F(i-1,e)$ or $F(i,e-1)$, and check if new pairs are frequent and also new to $F(i,e)$ (There are two ways to reach $F(i,e)$, i.e. from $F(i-1,e)$ or $F(i,e-1)$, and so which one reaches $F(i,e)$ earlier depends on the algorithm), and then $F(i,e)$ generated in this manner are saved. Thus in general, as shown by the arrows in Fig. 9.7, we repeat increasing one item or one edge, in a dynamic programming manner, keeping that a smaller is earlier in the size of $(|I|+|G|)$.

Algorithms 9.4–9.6 show a series of pseudocodes, which implement the above procedure and ideas. The algorithms are separated into three parts: The main algorithm is **Algorithm 9.6**, where first enumerate all frequent pairs of

Algorithm 9.6: Main part of mining frequent pairs (sets + graphs).

Function `Start_mining_frequent_pairs(`$\boldsymbol{P}$, Min_Sup`)`
  Data: Pairs (set + graph): $\boldsymbol{P}$, minimum support: Min_Sup
  Result: freqent pair set: $\mathcal{F}$
  $\mathcal{L} \leftarrow$ all letters appearing in $\boldsymbol{P}$.;
  $\mathcal{E} \leftarrow$ all edges appearing in $\boldsymbol{P}$.;
  foreach $a \in \mathcal{L}$ **do**
    if *support of* $(a, \varnothing)$ *is not less than* Min_Sup **then**
      Add $(a, \varnothing)$ to $\mathcal{F}_1$.;
  foreach $e \in \mathcal{E}$ **do**
    if *support of* $(\varnothing, e)$ *is not less than* Min_Sup **then**
      Add $(\varnothing, e)$ to $\mathcal{F}_1$.;
  Run `Generate_frequent_pairs(`$\mathcal{F}_1$, Min_Sup`)`.;
  Output $\mathcal{F}$ $(= \cup_t \mathcal{F}_t)$.;

(one item and null set) and (null set and one edge), which is $\mathcal{F}_1$. **Algorithm 9.5** has a set of frequent pairs $\mathcal{F}_t$ with the size (the number of items + the number of edges) of t as input and generate $\mathcal{F}_{t+1}$. In this step, we increase the size of the input by one and save the frequent patterns generated if the newly generated pairs are new in $\mathcal{F}_{t+1}$. This process is recursive in the sense that $\mathcal{F}_{t+1}$ becomes the next input. In this procedure, **Algorithm 9.4** examines if the pair is frequent or not.

We have showed data integrative learning based on frequent pattern mining, or more generally frequent pair mining, where each pair is an itemset and a graph. This data and method can be more generalized in the following two points:

Different data types In this case, the pair is an itemset and a subgraph, and the algorithm repeats adding one item or one edge and checking if the new pair is frequent. Note that this can be extended into any pair of data types, such as a sequence and a graph for one instance, where adding one letter or one edge would be repeated. That is, we can keep the same algorithm and this algorithm can be used for pairs of various data types.

Combinations of more than two data types As well as other data types, not only pairs but also more than two combinations can be also another target to which the presented algorithm can be extended. For example, each instance has three data types, i.e. itemsets, sequences and graphs, where either of one item, one letter or one edge is repeatedly added and only frequent patterns are saved. Simply we can extend our current way to more then two combinations, just by changing the current pair to the number of combinations to be targeted.

9.3 Integrative Learning for Vectors and Nodes in a Graph

One instance is a vector, and at the same time, the same instance is given as a node in a graph (or multiple graphs sharing the same node set). For this situation, there are a lot of applications. Below we introduce two typical examples of the applications:

Demographic data and social networks
As shown in the beginning of this chapter, each individual is provided by demographic data to be used by various services in the society. Also the usage of each service has been also recorded for each individual. Thus the demographic data and usage record are saved in various places. Importantly they can be given as a vector from the information source. On the other hand, recently social network services (SNS) provide human connections as a large graph with individuals as nodes. Thus individuals can be characterized by vectors and at the same time nodes in one or more huge graphs.

Gene expression and gene interactions
Properties of genes have been investigated since the dawn of molecular biology. In particular, modern biology allows to examine expression of a large number of genes in a cell thoroughly and simultaneously. Thus we can repeat measuring gene expression under different experimental conditions by such an equipment, making us have a vector of each gene, where one element of a vector is the expression value measured under some condition. On the other hand, recent life sciences provide a lot of information on interactions between or among genes, such as gene regulation, signal transduction, etc. Interactions between or among proteins (gene products) are also extensively and thoroughly measured by recent high-throughput experiments. Then these results are summarized into graphs, where each node is a gene and each edge indicates an interaction of the corresponding genes.

Thus overall each gene has a vector and at the same time a node in graphs.

This input, i.e. each instance being a vector and at the same time, a node in graphs, is strongly connected to *semi-supervised learning*, which we will explain below.

Machine learning approaches can be classified by two concepts: i.e. unsupervised learning and supervised learning. Once again unsupervised learning has input with only features, while supervised learning has input with not only features but labels. On the other hand, there is a concept, called semi-supervised learning, which is between supervised learning and unsupervised learning. The problem setting for semi-supervised setting is classified into the following two types:

1. **Semi-supervised classification**
   This is supervised learning-based semi-supervised learning, where in supervised learning, labels are not assigned to all instances. That is, in general, in supervised learning, all instances have labels, while only part of instances (of training data) have labels in semi-supervised learning. This is the problem setting of *semi-supervised classification.*

2. **Semi-supervised clustering**
   This is unsupervised learning-based semi-supervised learning, particularly being based on clustering. In fact the reason why this is called semi-supervised learning is that the given constraints over instances can help clustering instances.

   The constraints over instances can be divided into two types [8]:

   - **must link:** Two instances must be in the same cluster.
   - **cannot link:** Two instances cannot be in the same cluster.

   These constraints are not labels but equivalent to teaching signals in the sense that they can fix cluster assignment, and then this type of learning is called *semi-supervised clustering.*

Below, we explain more on these two types of semi-supervised learning.

9.3.1 Semi-supervised Classification

Instances: Vectors Only

First we start with instances having vectors only (this is not integrative learning). Again supervised learning assumes that all instances have labels, while in semi-supervised classification, labels are given to only part of instances. In supervised learning, instances are given as follows:

$$(\boldsymbol{x}_1, y_1), \ldots, (\boldsymbol{x}_N, y_N). \tag{9.59}$$

The procedure of supervised learning is to estimate hypothesis h which can explain y by using $\boldsymbol{x}$:

$$y = h(\boldsymbol{x}). \tag{9.60}$$

However, in semi-supervised classification, instances with labels are limited, and so what we can do is to estimate hypothesis h just by using instances with labels y.

If h by semi-supervised classification is precisely enough, there will be no difference between semi-supervised classification and supervised learning. However, usually labels cannot be obtained so easily in real data, and semi-supervised classification means that only a limited number of labels are obtained and used. Thus when labels are limited, the performance of the obtained hypothesis h (by semi-supervised classification) would not be good enough.

Thus we need to consider a different learning approach under the setting of semi-supervised classification. In practice, we need to consider that we can use instances with no labels, and then information on instances with no labels are only features. Thus what we can do is to compare features between instances, and then we can place an assumption that if instances are similar in features, these instances are likely to have the same label.

Then a possible approach under this assumption has the following two steps:

Two steps of general semi-supervised classification

1. **Computing the similarity in features between instances**

   Again instances with no labels have features only, and then we can compute similarity between two instances by comparing their features. This means we compute the similarity of all possible instance pairs, resulting in a (instances $\times$ instances) similarity matrix. This matrix is equivalent to an adjacency matrix, which is a graph with instances as nodes. In fact what we will use in this two-step procedure is graph Laplacian $\boldsymbol{L}$ of this adjacency matrix.

2. **Learning labels by using similarity**

   As mentioned in the above, the reasonable assumption we placed was that similar instances are likely to have the same label. Thus using similarity between instances and also instances with labels, we predict the label of the instances with no labels. More in detail, we can take the best balance between the known labels and the labels predicted by similarity. Again the similarity matrix is a graph. Thus this means that we predict the labels of instances with no labels, by using edge connectivity, keeping the consistency with the known labels.

   In fact this is equivalent to label propagation we explained in Section 8.3.

   Thus, as we saw already, letting $\boldsymbol{y}$ be a vector of labels over the entire instances, $\boldsymbol{f} = (f_1, \ldots, f_N)^\top$ $(f_i \in \{-1, +1\})$ be a label assignment function and $\boldsymbol{L}$ be the graph Laplacian which was computed in the above step 1, we estimate f_i $(i = 1, \ldots, N)$:

   $$y_i \leftarrow f_i \ (i = L + 1, \ldots, N). \tag{9.61}$$

   For example, in the basic (inflexible) label propagation algorithm, shown in Section 8.3, the optimization problem can be written as follows:

   $$\arg\min_{\boldsymbol{f}} (\boldsymbol{f} - \boldsymbol{y})^\top (\boldsymbol{f} - \boldsymbol{y}) - \lambda \boldsymbol{f}^\top \boldsymbol{L} \boldsymbol{f}. \tag{9.62}$$

   We solve this problem, and $\boldsymbol{f}$ is given as follows:

   $$\hat{\boldsymbol{f}} = (\boldsymbol{I} + \lambda \boldsymbol{L})^{-1} \boldsymbol{y}. \tag{9.63}$$

   We can assign labels by using the estimated $\hat{\boldsymbol{f}}$.

Here let us think more about the above step 1 which generates a graph from given features (vectors).

As a method in which a graph is generated from features, we already explained Laplacian eigenmaps in Section 3.3.3. In Laplacian eigenmaps, we computed the distance between features. Then we generate edges (of a graph) by the criterion of either 1) the distances of two nodes which are smaller than some prespecified cut-off, or 2) the top K smallest distances for each node. Then edges are unweighted or weighted by the following heat kernel:

$$\boldsymbol{W}_{ij} = \exp(-\frac{||x_i - x_j||^2}{\tau}), \tag{9.64}$$

where parameter τ of the heat kernel is, by regarding this kernel as a normal distribution, equivalent to the variance, showing the broadness (or the skewness) of the distribution. In Laplacian eigenmaps in Section 3.3.3, τ was explained as a hyperparameter, while this parameter can be reasonably trained from given data (not like by cross-validation) [50, 51]. We explain this point below:

First a given vector has multiple features, and the distributions of feature values over instances must be different for each feature. This means that the value of τ can be decided differently for each feature. We then write edge weight $\boldsymbol{W}_{ij}$ in (9.64), following the normal distribution, as follows:

$$\boldsymbol{W}_{ij} = \exp(-\sum_{d=1}^{M} \frac{(x_{i,d} - x_{j,d})^2}{\sigma_d^2}), \tag{9.65}$$

where i and j are instances (nodes) which are connected by an edge, and $x_{i,d}$ is the d-th feature value of the i-th instance (node).

We can say that this edge should attain the following two conditions:

Condition 1: the generated graph (edges) should keep the *manifold structure* which is embedded in the original given data.

Condition 2: the weight should reflect the similarity between nodes.

That is, from the manifold structure condition, one instance should be represented by other instances (nodes) and the weights to those nodes (Condition 1) and at the same time, it is expected that each of these weights should reflect the similarity between the corresponding instances (nodes) (Condition 2). From these observations, the weight can be decided by solving the following minimization problem:

$$\min_{\{\sigma_d\}_{d=1}^{M}} \sum_{i=1}^{N} ||\boldsymbol{x}_i - \frac{1}{\boldsymbol{D}_{ii}} \sum_{j \sim i} \boldsymbol{w}_{ij} \boldsymbol{x}_j||_2^2, \tag{9.66}$$

where $i \sim j$ shows nodes i and j connected in the graph, and $\boldsymbol{D}_{ii} = \sum_j \boldsymbol{W}_{ij}$.

The gradient of the objective function of the above minimization problem (9.66) is given as follows:

$$\sum_{i=1}^{N} \frac{1}{\boldsymbol{D}_{ii}} (\boldsymbol{x}_i - \hat{\boldsymbol{x}}_i)^{\mathsf{T}} (\sum_j \frac{\partial \boldsymbol{W}_{ij}}{\partial \sigma_d} \boldsymbol{x}_j - \frac{\partial \boldsymbol{D}_{ii}}{\partial \sigma_d} \hat{\boldsymbol{x}}_i), \tag{9.67}$$

Algorithm 9.7: Semi-supervised classification (instance is a vector).

Function `Semi-supervised_classification(`$\boldsymbol{X}$, $\boldsymbol{y}$`)`

  Data: training data (Vectors): $\boldsymbol{X}$, labels partially assigned: $\boldsymbol{y}$
  Result: labels for unlabeled in $\boldsymbol{X}$

  1. Generate graph G $(= \boldsymbol{W})$, with selecting K-closest instances for each instance, and compute graph weights by a gradient-based method using the gradient of (9.67).;

  2. Run some label propagation algorithm over G, such as `Basic_label_propagation` $(\boldsymbol{W}, \boldsymbol{y})$, to output labels of unlabeled instances.;

---

where

$$\hat{\boldsymbol{x}}_i = \frac{1}{\boldsymbol{D}_{ii}} \sum_j \boldsymbol{W}_{ii} \boldsymbol{x}_j. \quad (9.68)$$

$$\frac{\partial \boldsymbol{W}_{ij}}{\partial \sigma_d} = 2\boldsymbol{W}_{ij} (\boldsymbol{x}_{id} - \boldsymbol{x}_{jd})^2 \sigma_d^{-3}. \quad (9.69)$$

$$\frac{\partial \boldsymbol{D}_{ii}}{\partial \sigma_d} = \sum_j \frac{\partial W_{ij}}{\partial \sigma_d}. \quad (9.70)$$

We can estimate $\boldsymbol{W}$ by using some optimization method, such as gradient descent, with the above gradient. Thus now we can see that the weight of the generated graph can be estimated from the given data, meaning that in the above step 1, the graph itself can be totally estimated from the given data (without hyperparameters), under the normal distribution assumption for values of each feature.

We summarize the above procedure of semi-supervised classification for instances (with vectors only). **Algorithm 9.7** shows a pseudocode of this two-step procedure. Again this algorithm for instances with vectors only, has two steps: 1) we generate a graph by using all examples with vectors only, and 2) we run label propagation over one graph for the graph generated in the first step, to have labels of unknown data.

Instances: vectors and nodes in one or more graphs

The point of the learning method for instances with vectors (only) was to generate a graph, by which the problem becomes label propagation, in which the labels are estimated by taking the balance between the known labels and edge connectivity (weights).

Then, when given not only vectors but also nodes in a graph, the labels should be estimated by taking the balance between the known labels and not only the edge connectivity of the graph generated from the given vectors but also that of the given graph as well. Thus following this observation, we can formulate the problem as follows: Letting $\boldsymbol{L}_{vec}$ be graph Laplacian of the graph generated from

Algorithm 9.8: Semi-supervised classification (instance is a vector and a node of a graph).

```
1 Function
  Semi-supervised_classification_by_vector_and_graph(X, y, G)
    Data: training data (vectors): X, labels: y, graph:G
    Result: labels for unlabeled in X
2   1. Generate graph G_vec = W from X using Step 1 of
    Semi-supervised_classification(X, y).;
3   2. Compute the combined graph Laplacian (9.72) using graph G_vec and
    given graph G = G_gra.;
4   3. Run some label propagation algorithm over the newly generated
    graph Laplacian to obtain (9.74) and output labels of unlabeled
    instances.;
```

vectors and $\boldsymbol{L}_{gra}$ be graph Laplacian of the given graph, as shown by elastic net in Section 3.3.4, we can generate a regularization term by weighing the terms derived from these two graph Laplacians, as follows:

$$\alpha \boldsymbol{f}^\top \boldsymbol{L}_{vec}\boldsymbol{f} + (1-\alpha)\boldsymbol{f}^\top \boldsymbol{L}_{gra}\boldsymbol{f} = \boldsymbol{f}^\top(\alpha \boldsymbol{L}_{vec} + (1-\alpha)\boldsymbol{L}_{gra})\boldsymbol{f}, \tag{9.71}$$

where $0 \leq \alpha \leq 1$. If $\alpha = 1$, only the graph Laplacian derived from vectors is used and the input graph is not used, while if $\alpha = 0$, only the input graph is used, and the graph generated from vectors is not used.

That is, we can just use the following graph Laplacian:

$$\boldsymbol{L}_{new} = \alpha \boldsymbol{L}_{vec} + (1-\alpha)\boldsymbol{L}_{gra}. \tag{9.72}$$

By using $\boldsymbol{L}_{new}$, the optimization problem can be formulated as follows:

$$\arg\min_{\boldsymbol{f}} (\boldsymbol{f}-\boldsymbol{y})^\top(\boldsymbol{f}-\boldsymbol{y}) - \lambda(\boldsymbol{f}^\top \boldsymbol{L}_{new}\boldsymbol{f}). \tag{9.73}$$

Bu solving this, we can estimate labels from the following function:

$$\hat{\boldsymbol{f}} = (\boldsymbol{I} + \lambda(\alpha \boldsymbol{L}_{vec} + (1-\alpha)\boldsymbol{L}_{gra}))^{-1}\boldsymbol{y}. \tag{9.74}$$

Algorithm 9.8 shows a pseudocode of the above algorithm. As you can see from this algorithm, adding nodes in a graph to the input is easily solved and we can estimate labels easily, because semi-supervised classification is an optimization problem over the graph already.

One item of note is that the above formulation (of the optimization problem), for example, (9.73) is just an example of representing each of the three terms: 1) consistency with labels, 2) consistency with the edge connectivity of the given graph and 3) consistency with the edge connectivity of the graph generated from vectors. That is, as we have shown in the framework of more flexible label propagation in Section 8.3, such as (8.64), any term can be used for each of the above three terms as long as each term takes the form of Rayleigh quotient (see Section A.8.7).

9.3.2 Semi-supervised Clustering

We here repeat the definition of semi-supervised clustering. Semi-supervised clustering is clustering with the following two types of constraints:

– **must-link:** Two instances (nodes) must be in the same cluster.

– **cannot-link:** Two instances (nodes) cannot be in the same cluster.

One important point is that for two instances (nodes), these two conditions cover all possibilities already. In fact, the above **must-link** indicates the instances (nodes) are in the same cluster. Then if two instances are not in the same cluster, this means that these two instances are in different clusters, which is equivalent to **cannot-link**. Thus, any given a pair of two inputs have to fall into these two possibilities.

In reality these two constraints would not be given to all node pairs, but if either of these two constraints are given to each of all pairs, which results in a graph. This is obvious, because must-link gives the edge connecting two nodes, and cannot-link gives no edges for two nodes. Then again in reality, it is impossible to give either of these constraints to each pair, but we interpret that these constraints provide a graph, assuming that must-link constraints must have edges and other cases (including cannot-link) have no edges. That is, the problem setting of semi-supervised clustering can be interpreted as that a graph is given as constraints.

Now the input has two: vectors and nodes in a graph, which was generated from given **must-link** constraints. Then the problem is clustering instances (nodes) keeping the balance between the consistency with the similarity (between instances) given by vectors and also the consistency with the edge connectivity of the graph given as constraints. A typical method of clustering vectors is, for example, K-means. For example, constrained K-means given in Section 3.1.2 was formulated as follows:

$$f(\boldsymbol{Z}) = \mathrm{trace}(\boldsymbol{Z}^\top \boldsymbol{E}\boldsymbol{Z}) - \lambda\, \mathrm{trace}(\boldsymbol{Z}^\top \boldsymbol{Z} - \boldsymbol{D}), \tag{9.75}$$

where $\boldsymbol{Z}$ is a cluster assignment matrix, an example having been shown in (3.2), each element of $\boldsymbol{E}$ is given by (3.11), and $\boldsymbol{D}$ is a diagonal matrix with some prespecified cluster size (the number of instances in each cluster) as the corresponding diagonal element.

On the other hand, a typical method for clustering nodes in a graph is spectral clustering. As we explained spectral clustering in Section 8.1, for example, ratio cut was formulated as follows:

$$f(\boldsymbol{Z}) = \mathrm{trace}(\boldsymbol{Z}^\top \boldsymbol{L}\boldsymbol{Z}) - \lambda\, \mathrm{trace}(\boldsymbol{Z}^\top \boldsymbol{Z} - \boldsymbol{D}), \tag{9.76}$$

where $\boldsymbol{L}$ is graph Laplacian of the graph generated by the two types of constraints: **must-link** and **cannot-link**.

Then by comparing (9.75) and (9.76), it would be very easy to understand that both clustering vectors and clustering nodes in a graph have the same form of the objective function, which is Rayleigh quotient (see Section A.8.7). Thus,

Algorithm 9.9: Semi-supervised clustering.

```
1 Function Semi-supervised_clustering(X, G, K, M)
    Data: vectors (training data): X, graph: G(= W), #clusters: K,
          #eigenvectors used: M
    Result: node clusters: C_1, ..., C_K
2   Compute Laplacian L̃ from X and G (L̃ = E − λL.;
3   Run Eigen_Decomposition(L̃, M) to have first M eigenvectors to
    generate (N × M)-matrix Z, by using the M eigenvalues.;
4   Run K-means(Z, K) over rows of Z to output clusters: C_1, ..., C_K.;
```

in semi-supervised clustering, we can combine these two objective functions (also constraints). That is, in more detail, keeping the constraint (9.76), we take (9.75) as the objective function and formulate the problem as the minimization problem:

$$
\begin{aligned}
f(\boldsymbol{Z}) &= \text{trace}(\boldsymbol{Z}^\top \boldsymbol{E}\boldsymbol{Z}) - \lambda\text{trace}(\boldsymbol{Z}^\top \boldsymbol{L}\boldsymbol{Z}) - \lambda_D\text{trace}(\boldsymbol{Z}^\top \boldsymbol{Z} - \boldsymbol{D}) && (9.77)\\
&= \text{trace}(\boldsymbol{Z}^\top (\boldsymbol{E} - \lambda\boldsymbol{L})\boldsymbol{Z}) - \lambda_D\text{trace}(\boldsymbol{Z}^\top \boldsymbol{Z} - \boldsymbol{D}) && (9.78)\\
&= \text{trace}(\boldsymbol{Z}^\top \tilde{\boldsymbol{L}}\boldsymbol{Z}) - \lambda_D\text{trace}(\boldsymbol{Z}^\top \boldsymbol{Z} - \boldsymbol{D}), && (9.79)
\end{aligned}
$$

where $\tilde{L} = \boldsymbol{E} - \lambda\boldsymbol{L}$. This, as explained in the constrained K-means and spectral clustering (ratio cut), turns into an eigenvalue problem.

That is, by considering the given constraints as a graph and using the objective function of clustering vectors and also spectral clustering, we can obtain the clusters by solving the eigenvalue problem. **Algorithm 9.9** shows a pseudocode of the above procedure.

One item of note is that in the above example, we used constrained K-means for clustering vectors and spectral clustering (ratio cut) for clustering nodes in a graph. However the combination of these two objective functions is just one example. For both vectors and nodes in a graph, as long as each of these two terms takes a form of Rayleigh quotient, we can use that term in the objective functions.

Overall we can now see that semi-supervised clustering can be solved by taking the advantage of the point that both clustering vectors and clustering nodes in a graph take the form of Rayleigh quotient [83].

9.3.3 Collaborative Matrix Factorization

So far for the input of both vectors and nodes in a graph, we have explained the two types of semi-supervised learning: semi-supervised classification and semi-supervised clustering. On the other hand, as we have explained in the chapter of learning vectors, the main approaches of unsupervised learning are two: clustering and matrix factorization. Thus as well as clustering, we can perform matrix factorization over this type of input, i.e. each instance being a vector and at the same time a node in a graph.

For learning vectors, in matrix factorization, given matrix $\boldsymbol{X}$ (instances (rows) are vectors) is factorized into two or more low-rank matrices, say $\boldsymbol{U}$ and $\boldsymbol{V}$, and a key point is that these two matrices share the row of one of the two matrices and the column of the other of the two matrices. That is, the decomposition of $\boldsymbol{X}$ into $\boldsymbol{U}$ and $\boldsymbol{V}$ can be shown as follows:

$$\boldsymbol{X} \sim \boldsymbol{U}\boldsymbol{V}^{\top}. \tag{9.80}$$

Then, to satisfy this and also generalize $\boldsymbol{U}$ and $\boldsymbol{V}$ as regularization terms, the objective function (Lagrange function) to be minimized was given as follows:

$$L(\boldsymbol{U},\boldsymbol{V}) = |\boldsymbol{X} - \boldsymbol{U}\boldsymbol{V}^{\top}||_2^2 + \lambda(||\boldsymbol{U}||_2^2 + ||\boldsymbol{V}||_2^2). \tag{9.81}$$

Under semi-supervised clustering, clustering was made to be consistent with both inputs, i.e. vectors and nodes in a graph. We can consider matrix factorization in a similar manner. That is, two low-rank matrices are obviously obtained by the decomposition of the input $\boldsymbol{X}$, and at the same time, these matrices should be consistent with the given graph constraints. In other words, this decomposition can be constrained by the input graph. More in detail, if instances are the row of $\boldsymbol{U}$, then the graph over instances can be used for the regularization term on $\boldsymbol{U}$ as constraints, as follows:

$$\mathrm{trace}(\boldsymbol{U}^{\top}\boldsymbol{L}_U\boldsymbol{U}), \tag{9.82}$$

where $\boldsymbol{L}_U$ is the graph Laplacian of the graph over instances which are given as constraints. This term has the same form having been used in label propagation (see Section 8.3) and spectral clustering (see Section 8.1.1). Minimizing this term means making $\boldsymbol{U}$ consistent with edge connectivity (or making smooth over the graph).

We can regard this constraint as an equality constraint, and then we can formulate the Lagrange function as follows:

$$L(\boldsymbol{U},\boldsymbol{V}) = ||\boldsymbol{X} - \boldsymbol{U}\boldsymbol{V}^{\top}||_2^2 + \lambda_U\mathrm{trace}(\boldsymbol{U}^{\top}\boldsymbol{L}_U\boldsymbol{U}) + \lambda(||\boldsymbol{U}||_2^2 + ||\boldsymbol{V}||_2^2) \tag{9.83}$$

Then we can take the partial derivative of the Lagrange function with respect to each of $\boldsymbol{U}$ and $\boldsymbol{V}$. The partial derivative by $\boldsymbol{V}$ is equal to (3.50), and the partial derivative with respect to $\boldsymbol{U}$ is given as follows:

$$\frac{\partial L(\boldsymbol{U},\boldsymbol{V})}{\partial \boldsymbol{U}} = -2\boldsymbol{X}\boldsymbol{V} + 2\boldsymbol{U}\boldsymbol{V}^{\top}\boldsymbol{V} + 2\lambda_U\boldsymbol{L}_U\boldsymbol{U} + 2\lambda\boldsymbol{U}. \tag{9.84}$$

In any case graph Laplacian $\boldsymbol{L}$ consists of two parts: input graph (adjacency matrix) $\boldsymbol{W}$ and diagonal matrix (where each diagonal corresponds to the node degree) $\boldsymbol{D}$. Then, we separate graph Laplacian $\boldsymbol{L}_U$ into adjacency matrix $\boldsymbol{W}_U$ and diagonal matrix $\boldsymbol{D}_U$, which both can be used in the partial derivative of $L(\boldsymbol{U},\boldsymbol{V})$ as follows:

$$\frac{\partial L(\boldsymbol{U},\boldsymbol{V})}{\partial \boldsymbol{U}} = -2\boldsymbol{X}\boldsymbol{V} + 2\boldsymbol{U}\boldsymbol{V}^{\top}\boldsymbol{V} + 2\lambda_U\boldsymbol{D}_U\boldsymbol{U} - 2\lambda_U\boldsymbol{W}_U\boldsymbol{U} + 2\lambda\boldsymbol{U}. \tag{9.85}$$

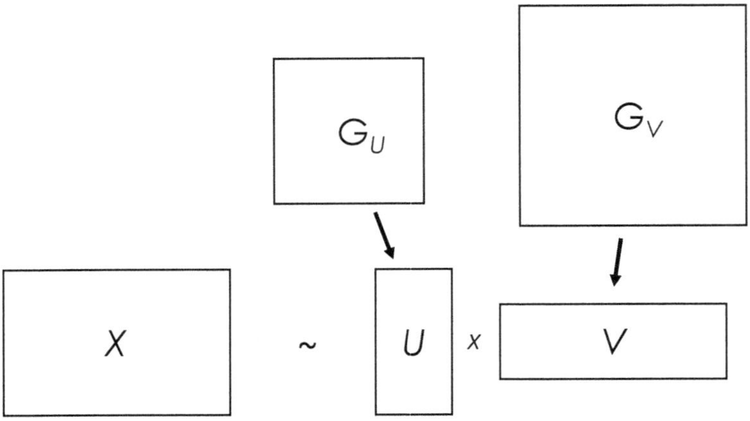

Figure 9.8: Collaborative matrix factorization: Three input matrices $\boldsymbol{X}$, $\boldsymbol{G}_U$, $\boldsymbol{G}_V$, where two graphs ($\boldsymbol{G}_U$ and $\boldsymbol{G}_V$)regulate $\boldsymbol{U}$ and $\boldsymbol{V}$.

This partial derivate gives the following multiplicative update rule as follows:

$$\boldsymbol{U}_{ik} \quad \leftarrow \quad \boldsymbol{U}_{ik} \frac{(\boldsymbol{X}\boldsymbol{V} + \lambda_U \boldsymbol{W}_U \boldsymbol{U})_{ik}}{(\boldsymbol{U}\boldsymbol{V}^\top \boldsymbol{V} + \lambda_U \boldsymbol{D}_U \boldsymbol{U} + \lambda \boldsymbol{U})_{ik}}, \tag{9.86}$$

where $\boldsymbol{U}_{ik}$ indicates the element of the i-th row and k-th column of matrix $\boldsymbol{U}$.

In practical application of machine learning and data mining, matrix factorization is used for recommendation, where $\boldsymbol{X}$ is a (users $\times$ items)-matrix. In this case, not only the user side but also the item side can have the similarity matrix, i.e. a graph. That is, not only $\boldsymbol{U}$ (users) but also $\boldsymbol{V}$ (items), can be constrained by the given graph, showing similarity of users (or items). That is, three matrices, $\boldsymbol{X}$ and two graphs can be given as input.

Fig. 9.8 shows a schematic picture of the relation of generating two low-rank matrices, $\boldsymbol{U}$ and $\boldsymbol{V}$, from three input matrices: $\boldsymbol{X}$ and two adjacency matrices, i.e. two graphs, $\boldsymbol{G}_U$ and $\boldsymbol{G}_V$. As such, given multiple matrices can be decomposed into low-rank matrices so that these low-rank matrices should be consistent with multiple input matrices. This matrix factorization is called *collaborative matrix factorization.*

In more detail, letting $\boldsymbol{L}_V$ be graph Laplacian from the graph for the $\boldsymbol{V}$ side (items side) as well as $\boldsymbol{L}_u$ be graph Laplacian for the $\boldsymbol{U}$ side (users side), we can add the following regularization term to the $\boldsymbol{V}$ side as well as the $\boldsymbol{U}$ side:

$$\text{trace}(\boldsymbol{V}^\top \boldsymbol{L}_V \boldsymbol{V}). \tag{9.87}$$

Then the Lagrange function can be as follows:

$$\begin{aligned} L(\boldsymbol{U}, \boldsymbol{V}) &= ||\boldsymbol{X} - \boldsymbol{U}\boldsymbol{V}^\top||_2^2 + \lambda_U \text{trace}(\boldsymbol{U}^\top \boldsymbol{L}_U \boldsymbol{U}) + \lambda_V \text{trace}(\boldsymbol{V}^\top \boldsymbol{L}_V \boldsymbol{V}) \\ &+ \lambda(||\boldsymbol{U}||_2^2 + ||\boldsymbol{V}||_2^2). \end{aligned} \tag{9.88}$$

Algorithm 9.10: Multiplicative update rules of collaborative matrix factorization.

Function `Collaborative_non-negative_matrix_factorization(`$\boldsymbol{X}$, G_U, G_V, K`)`

  Data: training data: $\boldsymbol{X}$, Graph A: G_U, Graph B: G_V, rank K (the size of low-rank matrices)

  Result: $\boldsymbol{U} \geq 0, \boldsymbol{V} \geq 0$

  Initialize $\boldsymbol{U}$ and $\boldsymbol{V}$.;

  Compute $\boldsymbol{L}_U$ and $\boldsymbol{L}_V$ from G_U and G_V.;

  repeat

    Update $\boldsymbol{U}$ by (9.86).;

    Update $\boldsymbol{V}$ by (9.90).;

  until *convergence*;

  Output $\boldsymbol{U} \geq 0$ and $\boldsymbol{V} \geq 0$.;

The new term on $\boldsymbol{L}_V$ has nothing to do with $\boldsymbol{U}$, and the partial derivative of $L(\boldsymbol{U}, \boldsymbol{V})$ with respect to $\boldsymbol{U}$ keeps the same as (9.85). However, the partial derivative of $L(\boldsymbol{U}, \boldsymbol{V})$ with respect to $\boldsymbol{V}$ can be given below. In particular, as well as $\boldsymbol{U}$, for $\boldsymbol{V}$, $\boldsymbol{L}_V$ can be separated into the following two costs, 1) derived from adjacency matrix (graph itself) $\boldsymbol{W}_V$ and 2) the remaining part, i.e. the diagonal $\boldsymbol{D}_V$.

$$\frac{\partial L(\boldsymbol{U}, \boldsymbol{V})}{\partial \boldsymbol{V}} = -2\boldsymbol{X}^\top \boldsymbol{U} + 2\boldsymbol{V}\boldsymbol{U}^\top \boldsymbol{U} + 2\lambda_V \boldsymbol{D}_V \boldsymbol{V} - 2\lambda_V \boldsymbol{W}_V \boldsymbol{V} + 2\lambda \boldsymbol{V}. \tag{9.89}$$

We then have the following update rule of $\boldsymbol{V}$:

$$\boldsymbol{V}_{jk} \leftarrow \boldsymbol{V}_{jk} \frac{(\boldsymbol{X}^\top \boldsymbol{U} + \lambda_V \boldsymbol{W}_V \boldsymbol{V})_{jk}}{(\boldsymbol{V}\boldsymbol{U}^\top \boldsymbol{U} + \lambda_V \boldsymbol{D}_V \boldsymbol{V} + \lambda \boldsymbol{V})_{jk}}, \tag{9.90}$$

where $\boldsymbol{V}_{jk}$ is the element of the j-th row and k-th column of matrix $\boldsymbol{V}$.

The update rule of $\boldsymbol{U}$ is the same as (9.86). **Algorithm 9.10** shows a pseudocode of collaborative matrix factorization in which the main matrix $\boldsymbol{X}$ of vectors are given as well as one similarity matrices (equivalent to adjacency matrix) is given to each of the two sides (columns and rows) of $\boldsymbol{X}$.

9.4 Integrative Learning for Nodes in Multiple Graphs

We already have explained the cases when instances are given as nodes of a graph and also multiple graphs sharing the same node set are given. In more detail, for unsupervised learning, we have presented (spectral) clustering and matrix factorization in Section 8.2 and for supervised learning (in fact semi-supervised

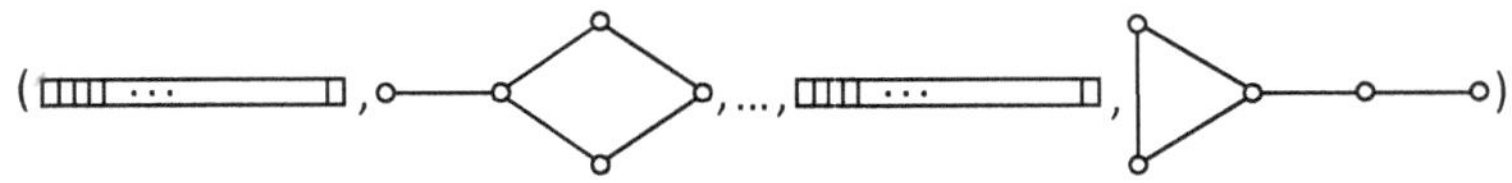

$$\Downarrow$$

$$\boldsymbol{K}^{(1)},\ \boldsymbol{K}^{(2)},\ \ldots,\ \boldsymbol{K}^{(T-1)},\ \boldsymbol{K}^{(T)}$$

$$\Downarrow\ (\boldsymbol{K} = \textstyle\sum_t^T \beta_t \boldsymbol{K}^{(t)})$$

$$\boldsymbol{K}$$

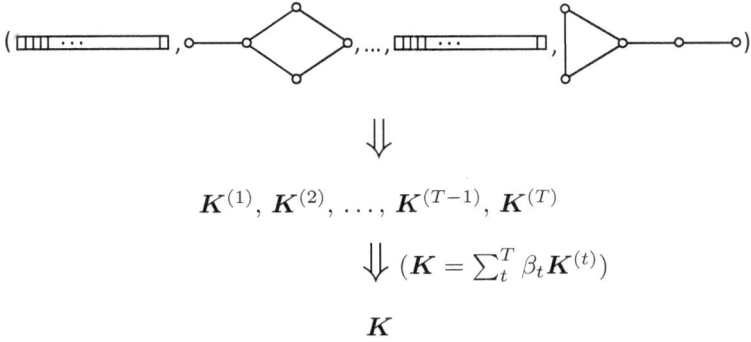

Figure 9.9: Given T datasets, T kernels: $\boldsymbol{K}^{(t)}(t = 1, \ldots, T)$ are generated and combined into one kernel being weighted by β_t, i.e. $\sum_t^T \beta_t \boldsymbol{K}^{(t)}$.

learning), we have shown label propagation in Section 8.4. Thus we can say that we have already explained integrative learning over multiple graphs, sharing a node set. One additional method, however, is that from each graph we can generate a kernel and for all graphs, multiple kernels instead of graphs are combined. We here explain this way of learning, which can be applied to any types of data. That is, we have multiple inputs, each can be converted into a kernel, and then these kernels can be used for learning. We explain this method in the next section.

9.5 Multiple Kernel Learning (MKL)

The assumption over the input here is that each instance can be given by different data types (or same types but regarded as different). We can then generate a kernel function for each data type, and combine these kernel functions into one kernel function. In fact, in this chapter already, we had a case in which one instance has two vectors, each being generated into a kernel function (Fig. 9.3) and another case in which one instance has a vector and a graph, each being generated into a kernel function (Fig. 9.6). In these two cases, we have raised *multiple kernel learning* (MKL). We here explain the procedure of MKL. Fig. 9.9 shows a schematic picture of MKL, which has two steps: 1) given T different datasets (data types), we generate T *multiple kernels*, and 2) T kernels are linearly combined with weights β_t below, where β_t $(t = 1, \ldots, T)$ are estimated so that the performance is maximized:

$$\boldsymbol{K} = \sum_t^T \beta_t \boldsymbol{K}^{(t)} \quad , \sum_t^T \beta_t = 1, \beta_t \geq 0. \tag{9.91}$$

Then the question is the method to estimate β_t $(t = 1, \ldots, T)$, i.e. MKL, and below we will explain the procedure of MKL. First we focus on the binary classification problem, for which support vector machine (SVM) (Section 3.2.11)

Algorithm 9.11: Wrapper method for Multiple Kernel Learning (learning and prediction).

Function Multiple_kernel_learning($\boldsymbol{K}^{(t)}, t = 1, \ldots, T$, $\boldsymbol{y}$, $\boldsymbol{K}_{new}$)

Data: kernels between training instances: $\boldsymbol{K}^{(t)}, t = 1, \ldots, T$, labels for training instances: $\boldsymbol{y}$, kernel between training and testing instances: $\boldsymbol{K}_{new}$

Result: estimated labels for testing instances

repeat

  Compute $\boldsymbol{\alpha}$ by solving (9.96).;

  Compute $\boldsymbol{\beta}$ by solving (9.98).;

until *convergence*;

Predict unknown labels from $\boldsymbol{\alpha}$, $\boldsymbol{y}$, $\boldsymbol{\beta}$ and $\boldsymbol{K}_{new}$ by using (9.95).;

---

is a representative method for kernel learning. Thus given test instance $\boldsymbol{x}$, SVM starts with a linear function, as follows:

$$y = \boldsymbol{w}^\top \boldsymbol{x} + b, \tag{9.92}$$

where weight $\boldsymbol{w}$ are estimated from training data by SVM to predict y. SVM however just focuses on support vectors, and then using (3.207), the above equation can be written as follows:

$$f(\boldsymbol{x}) = \sum_i \alpha_i y_i \boldsymbol{x}_i^\top \boldsymbol{x} + b. \tag{9.93}$$

In particular, we can write the inner product of the above equation by a kernel function as follows:

$$f(\boldsymbol{x}) = \sum_i \alpha_i y_i \boldsymbol{K}(\boldsymbol{x}_i, \boldsymbol{x}) + b. \tag{9.94}$$

Then we can introduce multiple kernels of (9.91) into this equation, and then we can have the following:

$$f(\boldsymbol{x}) = \sum_i \alpha_i y_i \sum_t \beta_t \boldsymbol{K}^{(t)}(\boldsymbol{x}_i, \boldsymbol{x}) + b. \tag{9.95}$$

Thus, in MKL, we need estimate the two parameters α and β of this function.

MKL has been well researched along with the boom of kernel learning and so we can say that the methods for MKL are already matured. Then the main optimization methods for MKL, called *Wrapper method*, do not solve the above both parameters $\boldsymbol{\alpha}$ and $\boldsymbol{\beta}$ at once. The point of Wrapper method is that $\boldsymbol{\alpha}$ and $\boldsymbol{\beta}$ are alternately estimated like alternating least square, i.e. repeating the following two steps: 1) $\boldsymbol{\alpha}$ is estimated, fixing $\boldsymbol{\beta}$ and 2) $\boldsymbol{\beta}$ is estimated, fixing $\boldsymbol{\alpha}$. We explain the detail of this procedure below:

First, when we fix $\boldsymbol{\beta}$, multiple kernels can be only one kernel. That is, estimating $\boldsymbol{\alpha}$ is the same as that by SVM. Then following the derivation of SVM, we can optimize the dual form given in (3.210) to (3.212) to have the optimum $\boldsymbol{\alpha}$, having the fixed $\boldsymbol{\beta}$ to generate the input kernel from multiple kernels:

$$\max_{\boldsymbol{\alpha}} \quad (\sum_i \alpha_i - \frac{1}{2}\sum_{i,j} y_i y_j \alpha_i \alpha_j (\sum_t \beta_t \boldsymbol{K}^{(t)}(\boldsymbol{x}_i, \boldsymbol{x}_j))) \tag{9.96}$$

$$\text{subject to} \quad \sum_i^N y_i \alpha_i = 0, \alpha_i \geq 0. \ (i = 1, \ldots, N) \tag{9.97}$$

Then, when we fix $\boldsymbol{\alpha}$, the objective function is the same as the above (9.96), while due to the property of $\boldsymbol{\beta}$, the problem becomes the minimization problem:

$$\min_{\boldsymbol{\beta}} \quad (\sum_i \alpha_i - \frac{1}{2}\sum_{i,j} y_i y_j \alpha_i \alpha_j (\sum_t \beta_t \boldsymbol{K}^{(t)}(\boldsymbol{x}_i, \boldsymbol{x}_j))) \tag{9.98}$$

$$\text{subject to} \quad \sum_t^T \beta_t = 1, \beta_t \geq 0. \ (t = 1, \ldots, T) \tag{9.99}$$

Thus we solve the above maximization problem (for obtaining $\boldsymbol{\alpha}$) and minimization problem (for obtaining $\boldsymbol{\beta}$) alternately, and after convergence we can predict y of test instance $\boldsymbol{x}$ by using (9.95) and obtained trained parameters.

Summarizing all above procedure, **Algorithm 9.11** shows a pseudocode of the Wrapper method of MKL. One item of note is that so far there proposed many methods already for MKL, while the difference of these methods are all in optimization methods, and the problem settings of these methods are all the same. That is, they all have solved the same problem.

9.6 Applications to Life Sciences

9.6.1 Integrative Learning of Vectors and Graphs

Frequent Pair (Sequence and Graph) Mining

In fact we already applied frequent pattern mining to biclustering in Section 3.5, resulting in Fig. 3.34. This result was obtained from interactions of drugs and targets, where targets are proteins or amino acid "sequences" and drugs are chemical compounds or molecular "graphs". That is, in this training data, each instance was a pair of a sequence and a graph. In other words, we already presented integrative learning for vectors (or sequences as a more extended data type) and graphs in Section 3.5. In this application, we used 11,219 pairs as instances, over which frequent pattern mining is run to find frequent pairs of subsequences and subgraphs. Fig. 3.34 shows the matrix of (the input 11,219 pairs × the top 10,000 frequent pairs). Also Fig. 3.36 shows the results of clustered frequent pairs of subsequences and subgraphs. From this figure, we can see that subsequences (also subgraphs) of frequent pairs are rather short (also small). Among these clusters, C8 is the most interesting cluster, in particular subsequences DRY, ISI, LGT in this

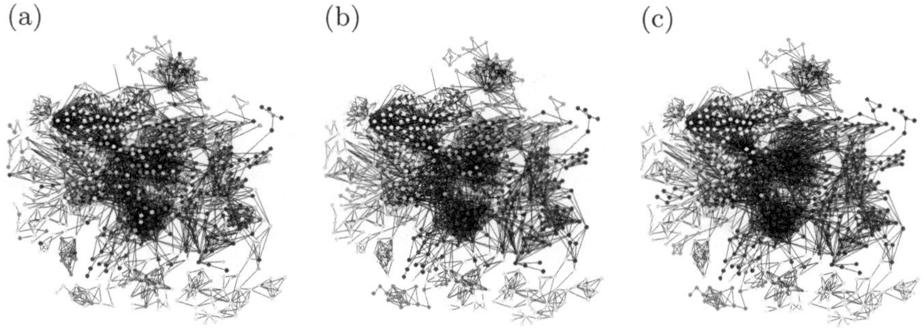

Figure 9.10: Semi-supervised clustering over real gene expression and gene network, where (a) expression only, (b) both expression and network and (c) network only are used.

cluster correspond to amino acid sequences in the location surrounding the binding site of a ligand to a typical target protein, called G-Protein Coupled Receptor (GPCR).

This result indicates that the finding of amino acid residues neighboring the ligand binding site was not obtained just by random guessing but can be automatically found/mined from data of drug-target interactions as frequent patterns without any prior knowledge. We think that the impact of the frequent pattern mining as knowledge discovery would be sizable, because this result indicates the binding site itself is predictable by using the subsequences of frequent pairs. In fact target proteins such as GPCR are membrane proteins, which are embedded in cell membranes and hard to crystallized, by which the three-dimensional (3D) structure has not been analyzed well so far. Thus, the results imply that binding sites can be estimated even without experimentally determining the 3D structures.

9.6.2 Integrative Learning of Vectors and Nodes in a Graph

Semi-supervised Clustering over Vectors with Nodes in a Graph

We present an example result of semi-supervised clustering, in which one instance has a vector and at the same time a node in a graph, where the graph is the entire dataset. Fig. 9.10 shows the result of clustering of given genes, which are represented by gene expression (vectors) and at the same time by a gene network (nodes in a graph). In this figure, genes are clustered by (a) expression only, (b) both gene expression and gene network and (c) network only.

Furthermore Fig. 9.11 shows the clustering result of the same set of genes in Fig. 9.10, by using the function category of Gene Ontoloty (GO). That is, Fig. 9.11 shows true clusters.

Gene expression are known to be very noisy data, and as shown in Fig. 9.10 (a), the results of clustering by gene expression only are not necessarily clear clusters and made us feel random division. On the other hand, Fig. 9.10 (c) looks very

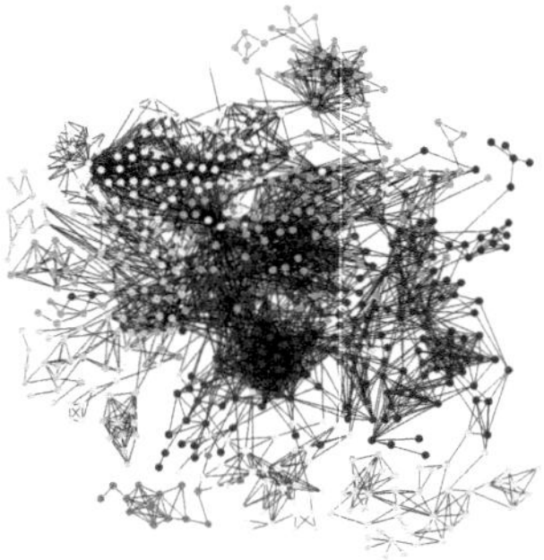

Figure 9.11: Coloring over the same genes in Fig. 9.10 by using Gene Ontology (GO).

clear clustering results, because this result is obtained by using network information only, by which clusters follow edge connectivity very well. However, as shown in Fig. 9.11, true clusters are not necessarily consistent with edge connectivity. Finally the results shown in (b) of Fig. 9.10 are more consistent with true clusters than (a) and (c), indicating that in semi-supervised learning or more generally integrative learning, using both vectors and constraints (here, graph information) is more useful than just clustering with vectors only or that with nodes in a graph only.

Chapter 10

Prediction Results Evaluation

Any machine learning method needs to be evaluated by being run over the real data and measuring its performance, even if the performance of the method can be analyzed theoretically. In this chapter, we will explain the manner for measuring the performance of machine learning methods on real data for evaluation.

Thus the evaluation manner of a machine learning method changes, depending on the problem setting to which the method is applied. Simply speaking, in unsupervised learning, training and test data are the same. That is, after training a method by using training data in an unsupervised manner, this trained model is evaluated by the same training data. On the other hand, in supervised learning, test data must be prepared being independent of training data. In this case, test data can be obtained by either of the following three methods:

1. **Different Information Sources:** This is the case we can use different data for testing from training. In other words, this is one example of test data, which can be obtained from a different information source from that of training data.

   We just raise simple examples.

   Difference in time-series: A (users×items) matrix of E-commerce is gradually filled as time goes by. That is, in the beginning, most elements are not filled, and the matrix is very sparse. However, one year or later, sparseness of this matrix might be relaxed a bit. Thus we can set a cut-off value against time to generate training and test data. That is, the elements filled before the cut-off value can be used for training data, and elements filled after the cut-off can be test data.

   Life science applications: The data from a different database from those used for training can be another type of test data, which are retrieved from a different information source. However, in this case, usually

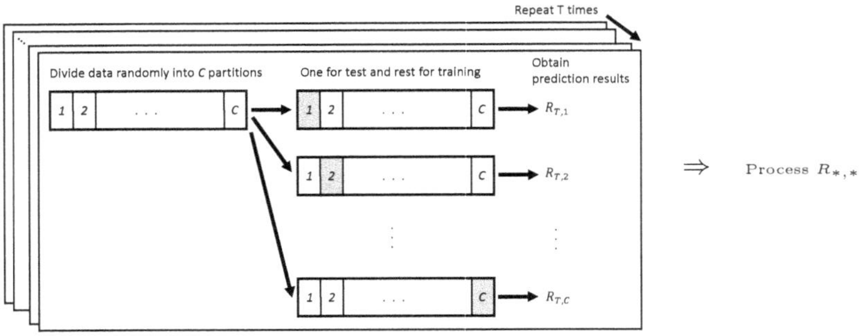

Figure 10.1: Schematic picture of $T \times C$-cross validation.

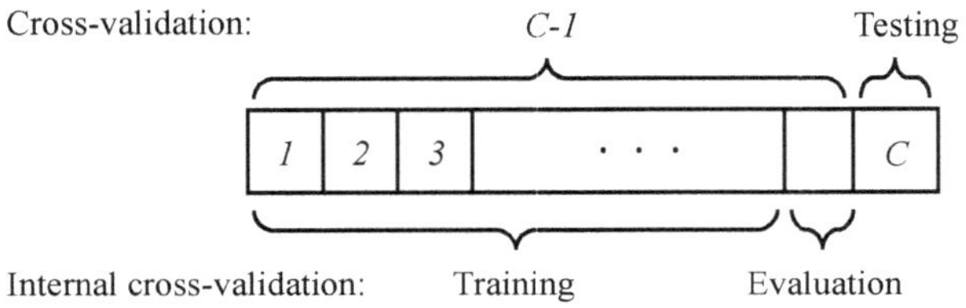

Figure 10.2: Cross-validation and internal cross-validation.

biological databases have heavy overlaps, and this type of overlaps needs to be removed carefully.

2. **Cross-validation:** The $T \times C$*-cross validation* means that the following is repeated T times: the given data is randomly divided into C folds with an equal size and *one of C folds for test and the rest folds for training* is repeated C times. That is, we can have $T \times C$ prediction results, and usually they are averaged. Usually five or ten is used for T and C. Fig. 10.1 shows a schematic procedure of $T \times C$-cross validation. **Algorithm 10.1** shows a pseudocode of the procedure of $T \times C$-cross validation.

   In addition to this procedure, in training, to estimate the optimum parameters of the model. The training data are further divided into training and evaluation data, and by using evaluation data, we can decide parameter values. Thus in this case, using $C - 1$ folds for training data, we can do cross-validation further, which is called *internal cross-validation.* Fig. 10.2 shows a schematic picture of the internal cross-validation.

3. **Bootstrapping** As shown in Fig. 3.23 already, generating data by sampling with replacement from the original given data is called *bootstrapping.* The last idea is that we generate test data by using bootstrapping.

Once test data are prepared, we can make prediction and have prediction results. We definitely need some measure for evaluating the prediction results.

Algorithm 10.1: $T \times C$ Cross-validation.

Function `T×C-cross_validation`(D, T, C)
  Data: given instances: D, #iterations: T, #partitions: C
  Result: $\{R_{1,1}, \ldots, R_{T,C}\}$
  for $t \leftarrow 1$ **to** T **do**
    Divide D into C folds: D_c $(c = 1, \ldots, T)$.;
    for $c \leftarrow 1$ **to** C **do**
      Use D_c for test, and the remaining folds for training.;
      Train model by training folds only, and test data for prediction.;
      Record prediction results in $R_{t,c}$.;
  Output $\{R_{1,1}, \ldots, R_{T,C}\}$.;

Below we explain evaluation measures thoroughly. In particular, an important point is to use the evaluation measure most appropriate for the problem setting.

Evaluation measures are explained under each of the following four problem settings:

1. **Supervised learning and binary labels** Evaluation over binary labels (binary classification) is the most standard problem setting for supervised learning. In this case, a typical example is one label for positives and the other for negatives.

   This is not necessarily standard, but even if we have multiple (more than two) labels, we can use binary label-based evaluation. For example, we can regard one of the multiple labels as positives and all other labels as negatives and repeat this for each of multiple labels. Then usually we can take the average (or the weighted average) over all results.

2. **Supervised learning and continuous label** Evaluation over the continuous label is for regression.

3. **Unsupervised learning (clustering)** When we evaluate the clustering results, we usually assume that there are true (gold-standard) clusters. We explain methods for evaluating clusters particularly, while usually we have more than two clusters and true clusters, meaning that this is equivalent to supervised learning (classification) with multiple labels.

   Thus even for clustering, we can use binary classification evaluation methods, implying that approaches for evaluating binary classification are very important in machine learning.

4. **Ranking** The ranking results are ordered instances of test data, and this order is evaluated. Even if these instances ordered by using scores predicted by some learning (or prediction) method, we do not use these scores (and also the classes which can be specified by the scores). This point is the difference from the evaluation of supervised learning. In other words, if the prediction scores are used for evaluation, for example, evaluation methods

Table 10.1: Four possible cases for two true values and two predicted values.

		True value	
		Positive	Negative
Prediction	Positive	true positive	false positive Type I error
value	Negative	false negative Type II error	true negative

for the continuous label (or discrete label, depending on the problem setting) can be used to examine the prediction results. However, only the order of the test instances are examined, by which some method special for evaluating the order is needed.

10.1 Supervised Learning and Binary Labels

We describe the methods for evaluating instances with true binary labels. Binary labels can be regarded as positives and negatives. Note that in supervised learning, even if the labels are not binary, they can be transformed into binary labels and approaches for evaluating binary labels can be used, as follows:

Multiple (more than two) labels: We explained this case briefly already. Given multiple labels, again we can focus on only one label, and this label can be regarded as positives, and the other labels can be regarded as negatives, resulting in a binary classification problem. This generates multiple binary classification problems. That is, if we have five labels, we can have five binary classification problems.

Continuous label: We can set up some reasonable cut-off value against the continuous values, and if the predicted value (score) of some test instance is larger than the cut-off value, the test instance is regarded as a positive; otherwise, the test instance can be a negative.

10.1.1 Prediction without Scores

We explain the case in which prediction results are either of positives or negatives and there are no more information than those binary results. In this case, true values are binary and also prediction values (scores) are also binary, and so as shown in Table 10.1, we have only four (two $\times$ two) cases (true positive, pseudo positive, pseudo negative and true negative) for each test instance. Then we can count the test instances, which fall into each of the four cases, and then as shown in Table 10.2, we can generate a table with numbers in the four partitions, which is called 2×2 *contingency table.*

Using the notation shown in Table 10.2, we raise measures to evaluate the 2×2 contingency table. In fact first eight measures are very basic ones, which can be used for other measures further, then we will present five measures, which are

Table 10.2: 2×2 contingency table.

		True value Positive	True value Negative	
Prediction value	Positive	N_{TP}	N_{FP}	$N_{.P}$
	Negative	N_{FN}	N_{TN}	$N_{.N}$
		N_P	N_N	N

N_{TP}: number of true positives predicted as positives.
N_{FN}: number of true positives but predicted as negatives.
N_{FP}: number of true negatives but predicted as positives.
N_{TP}: number of true negatives predicted as negatives.
$N_P := N_{TP} + N_{FN}$.
$N_N := N_{FP} + N_{TN}$.
$N_{.P} := N_{TP} + N_{FP}$.
$N_{.N} := N_{FN} + N_{TN}$.
$N := N_P + N_N = N_{.P} + N_{.N} = N_{TP} + N_{FN} + N_{FP} + N_{TN}$.

standard measures used for evaluating binary classification results. After these 13 measures, we explain five hypothesis tests over 2 × 2 contingency tables.

First we raise four measures, corresponding to the four partitions of Table 10.1. These four measures in Table 10.1 can be rewritten in the table with the same division, being summarized into Table 10.3. Again these four measures are not necessarily used by themselves but rather used to establish other measures.

True Positive Rate (TPR): $\dfrac{N_{TP}}{N_P}$.

Definition: Ratio of correctly predicted true positives to all true positives.

Other names: Recall, Sensitivity, Probability of detection.

False Positive Rate (FPR): $\dfrac{N_{FP}}{N_N}$.

Definition: Ratio of wrongly predicted true negatives to all true negatives.

Other names: Fallout.

Notes: FPR + TPR = 1.

False Negative Rate (FNR): $\dfrac{N_{FN}}{N_P}$.

Definition: Ratio of wrongly predicted true positives to all true positives.

True Negative Rate (TNR): $\dfrac{N_{TN}}{N_N}$.

Definition: Ratio of correctly predicted true negatives to all true negatives.

Other names: Specificity.

Table 10.3: The same table as Table 10.1, but with different names in cells.

		True value	
		Positive	Negative
Prediction value	Positive	TPR recall, sensitivity	FPR fallout
	Negative	FNR	TNR specificity

Notes: FNR + TNR = 1.

We then raise other four measures, again corresponding to the four partitions of Table 10.1. These four measures in Table 10.1 can be also rewritten in the table with the same partitions, being summarized into Table 10.4. Again these measures are not necessarily used by themselves but rather used to establish other measures.

Positive Predictive Value (PPV): $\frac{N_{TP}}{N_{\cdot P}}$.

Definition: Number of correctly predicted positives to predicted positives.

Other names: Precision.

False Discovery Rate (FDR): $\frac{N_{FP}}{N_{\cdot P}}$.

Definition: Number of wrongly predicted positives to predicted positives.

Notes: PPV + FDR = 1.

False Omission Rate (FOR): $\frac{N_{FN}}{N_{\cdot N}}$.

Definition: Number of wrongly predicted negatives to predicted negatives.

Negative Predictive Value (NPV): $\frac{N_{TN}}{N_{\cdot N}}$

Definition: Number of correctly predicted negatives to predicted negatives.

Notes: FOR + NPV = 1.

We further raise five other measures, which can be also computed from the notation shown in Table 10.2. Different from the previous eight measures, however, each of these five measures can be used to evaluate binary classification results by itself.

Accuracy: $\frac{N_{TP}+N_{TN}}{N}$.

Table 10.4: The same table as Table 10.1, but with other different names in cells.

		True value	
		Positive	Negative
Prediction	Positive	PPV precision	FDR
value	Negative	FOR	NPV

Definition: Ratio of correctly predicted test instances to all test instances.

Other names: Correct Classification Rate (CCR).

Note 1: Amount: Higher or larger accuracy is better.

Note 2: Affected by data imbalanceness: The accuracy is not robust against the imbalanceness between positives and negatives. This is clear by thinking about a simple, extreme example, say that the ratio of N_P to N_N in test instances is 99 to 1. In this case, even the accuracy of 99% can be achieved by predicting all test instances as positives. Although the accuracy is well used in many applications, the accuracy should be avoided to use, particularly for the case that the binary labels are very imbalanced.

Error rate: $\frac{N_{FP}+N_{FN}}{N}$.

Definition: Ratio of wrongly predicted test instances to all test instances.

Note 1: Amount: Smaller value is better.

Note 2: Accuracy + Error rate = 1:

Note 3: Affected by data imbalanceness: Error rate is 1 - Accuracy, and so the error rate has the same problem as the accuracy in terms of not being robust against test data imbalanceness.

Matthew's Correlation Coefficient (MCC): $\frac{N_{TP} \cdot N_{TN} - N_{FP} \cdot N_{FN}}{\sqrt{N_{.P} N_{.N} N_P N_N}}$.

Idea: The first idea is to multiply the correctly predicted positives by the correctly predicted negatives. Then this value is (in some sense) normalized, by being subtracted by the number obtained by multiplying wrongly predicted positives by wrongly predicted negatives.

Both these two terms are affected by the number of the entire instances. So the above two values are further normalized by the number obtained by multiplying four: the number of true positives (N_P), the number of true negatives (N_N), the number of predicted positives ($N_{\cdot P}$) and the number of predicted negatives ($N_{\cdot N}$).

Extreme cases and range: If prediction results are totally correct, $N_{TP} = N_{\cdot P} = N_P$, $N_{TN} = N_{\cdot N} = N_N$ and $N_{FP} = N_{FN} = 0$. That is, both the denominator and numerator are $N_{TP}N_{TN}$, meaning that MCC is +1. Also if predictions are totally wrong, $N_{FP} = N_{\cdot P} = N_N$, $N_{FN} = N_{\cdot N} = N_P$ and $N_{TP} = N_{TN} = 0$. That is, the numerator is $-N_{FP}N_{FN}$ and the denominator is $N_{FP}N_{FN}$, meaning that MCC is -1. So in summary, any prediction result can be scored from -1 (worst) to +1 (best).

Applications: MCC has been often used in bioinformatics.

Note: Indeed the range is clearly defined from -1 to +1, while the behavior between this range is unclear. In this sense, MCC also might not be an ideal measure to understand the prediction results.

F_1 score (F-score, F-measure): $\frac{2}{\frac{1}{precision} + \frac{1}{recall}}$.

Definition: Harmonic mean of precision and recall.

Idea: To have a higher precision value, the recall should be smaller. Also having high precision values for high recall values is very hard. Thus the idea behind F_1 score is to check if both values can be kept high, and this score can compared with other methods.

Applications: The F_1 score has been well used in natural language processing, particularly information retrieval.

Note: As mentioned above, the idea is to regard the point which makes both precision and recall large as the point to be compared with other methods. This idea is understandable, while there are no clear reasons behind why the harmonic mean should be taken. In this sense, this is an *ad hoc* measure.

Relative risk: $\frac{\frac{N_{TP}}{N_{\cdot P}}}{\frac{N_{FN}}{N_{\cdot N}}}$.

Definition: Ratio of the number of correctly predicted positives to the number of predicted positives is divided by the ratio of the number of wrongly predicted negatives to the number of predicted negatives.

Idea: Both the numerator and denominator are ratios. Then the numerators of the both ratios are on the number of positives, meaning that this measure compares the number of positives in the predicted positives with that in the predicted negatives. You can see the note below, to understand more what the relative risk examines.

Other names: Risk ratio (RR).

Note: In reality, this measure is not necessarily used for binary classification only, in which there are only two labels, such as positives and negatives. Instead, a more possible case for relative risk is that true labels are different from predicted labels. For example, true labels are patients and non-patients (healthy people), and predicted labels are smokers and non-smokers. In this case, relative ratio examines the ratio of patients/smokers to patients/non-smokers. That is, this is the ratio of the number of patients among smokers to the number of patients among non-smokers. Thus if this is far larger than 1, the disease would be related with smoking, and if this is close to 1, the disease has nothing to do with smoking. Also if this is far smaller then 1, the disease would be rather related with non-smoking.

Jaccard index (JI) [47]: $\frac{N_{TP}}{N_{TP}+N_{FN}+N_{FP}}$.

Definition: Ratio of correctly predicted positives to all instances (except correctly predicted negatives)

Other names: Tanimoto similarity index [73].

Sometimes Tanimoto similarity index is given as the negative logarithm of Jaccard Index.

Note: Usage This is often used for evaluating clustering results of unsupervised learning more than classification. This is possible, because as mentioned earlier, we assume true clusters when we evaluate clustering results, and regard two test instances in the same true cluster as a positive; otherwise a negative. This becomes a binary classification problem, and from true clusters and predicted clusters, we can generate a 2×2-contingency table, for which any evaluation method of Supervised Learning and Binary Labels (without scores) can be run. Then Jaccard Index is often used for clustering in the above manner (See Section 10.3.1).

We further explain five hypothesis testing over the 2×2 contingency matrix. In particular, standard hypothesis testing examines the independence between rows or between columns.

Fisher's exact test:

$$\frac{\begin{pmatrix} N_{\cdot P} \\ N_{TP} \end{pmatrix}\begin{pmatrix} N_{\cdot N} \\ N_{FN} \end{pmatrix}}{\begin{pmatrix} N \\ N_P \end{pmatrix}} = \frac{N_{\cdot P}!N_{\cdot N}!N_P!N_N!}{N_{TP}!N_{FP}!N_{FN}!N_{TN}!N!}. \tag{10.1}$$

Property: This test exactly examines the independence between rows or between columns.

χ^2 **test:**

$$N\frac{(N_{TP}N_{TN} - N_{FP}N_{FN})^2}{N_{\cdot P}N_{\cdot N}N_PN_N}.$$

Property: Due to the high computational cost of Fisher's exact test, χ^2 test has been used as an approximation of Fisher's exact test for a long time.

Distribution: χ^2 test follows the χ^2 distribution with the degree of 1, approximately.

χ^2 **test with Yate's correction for continuity:**

$$N\frac{(|N_{TP}N_{TN} - N_{FP}N_{FN}| - \frac{N}{2})^2}{N_{\cdot P}N_{\cdot N}N_PN_N}.$$

Property: As well as χ^2 test, due to the high computational load of Fisher's exact test, this test is also an approximation of Fisher's exact test.

Distribution: This test follows the χ^2 distribution with the degree of one approximately but more precisely than the regular χ^2 test.

Z-test: $\frac{|TPR-FPR|}{\sqrt{p(1-p)(\frac{1}{N_P}+\frac{1}{N_N})}}$, where $p = \frac{N_{TP}+N_{FP}}{N}$.

Property: This is a test on the difference between TPR and FPR.

Distribution: This test follows the standard normal distribution, approximately.

Z-test with correction for continuity:

$$\frac{|TPR - FPR| - \frac{(\frac{1}{N_P}+\frac{1}{N_N})}{2}}{\sqrt{p(1-p)(\frac{1}{N_P}+\frac{1}{N_N})}}$$, where p is the same as that already defined.

Property: This is also a test on the difference between TPR and FPR.

Distribution: This test follows the standard normal distribution, approximately, more precisely than Z-test static.

10.1.2 Prediction with Scores

As prediction results, a machine learning method often provides some scores for test instances, where instances are predicted as positives more with larger scores and reversely as negatives more with smaller scores. By using these scores for instances, we can evaluate the prediction results more precisely. That is, we can evaluate the hypotheses and models more strictly.

Measure 1: Receiver Operator Characteristic (ROC) Curve

Receiver Operator Characteristic (ROC) curve [38] is the most well-used and standard evaluation method for binary classification. The most important point of the ROC curve is that this measure is invariant against the imbalanceness in the number of positives and negatives in test data. Thus we do not have to care the ratio of positives to negatives (or its reverse).

Below we can explain the way to write the ROC curve in three manners, starting with a simpler one to a more careful one:

1. **Simplest explanation, probably hard to understand:** The simplest explanation is that we sort the test instances in the descending order of predicted scores, and show FPR for the x-axis and TPR for the y-axis.

2. **A bit more understandable explanation:** However, the above explanation would not be so easily understood. A more concrete explanation would be below: We set a cut-off value at the top k instances of the sorted test instances, meaning that instances with larger scores than the cut-off value are regarded as positives and those with smaller scores as negatives. Then, suppose that the numbers of positives and negatives in the instances with larger scores are k_P and k_N, FPR is $\frac{k_N}{N_N}$ and TPR is $\frac{k_P}{N_P}$, and so $(\frac{k_N}{N_N}, \frac{k_P}{N_P})$ is plotted.

   We examine the above cut-off value from the top to the bottom of the sorted test instances and plot the above (FPT, TPR) for each cut-off.

3. **Careful explanation:** The above two explanations are not wrong, but still not easily understandable. Below is a more intuitive, and in practice easily executable explanation: We first sort test instances in the descending order

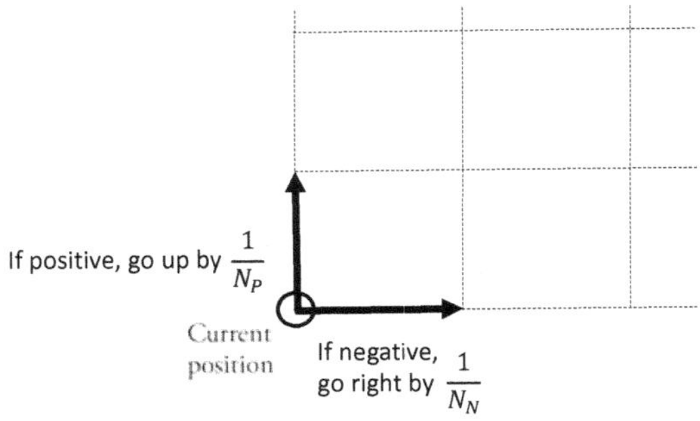

Figure 10.3: Procedure for drawing an ROC curve.

Algorithm 10.2: ROC curve.

Function `ROC curve`(*scores*, N_P, N_N)
  Data: scores for instances, #true positives: N_P, #true negatives: N_N
  Result: ROC curve
  Initialization: sort instances according to scores, resulting in $\boldsymbol{x}_1, \ldots, \boldsymbol{x}_N$.;
  $k_P \leftarrow 0$.;
  $k_N \leftarrow 0$.;
  for $k \leftarrow 1$ **to** N **do**
    if $\boldsymbol{x}_k$ *is predicted as a positive* **then**
      $k_P \leftarrow k_P + 1$;
    else
      $k_N \leftarrow k_N + 1$;
    Plot $(\frac{k_N}{N_N}, \frac{k_P}{N_P})$.;
    Draw the line from the last point to the new point.;

---

of the scores. We then check each test instance from the top to the bottom, repeating the following for each instance: if the instance is a positive, we draw the line in the direction of the y-axis by $\frac{1}{N_P}$; otherwise we draw the line in the direction of x-axis by $\frac{1}{N_N}$. Fig. 10.3 shows this procedure (step) for each instance, where at each current position, we move either of the two: if positive, go up by $\frac{1}{N_P}$; otherwise go right by $\frac{1}{N_N}$. **Algorithm 10.2** shows a pseudocode of this procedure of drawing the ROC curve for given prediction results.

We can see that the ROC curve simply examines the appearance of positives (y-axis) and also that of negatives (x-axis). An important point is that the ROC curve is a monotone increasing curve, which never decreases. If your ROC curve decreases, something is wrong with your writing of the ROC curve.

To understand the ROC curve more, we now think about two extreme cases

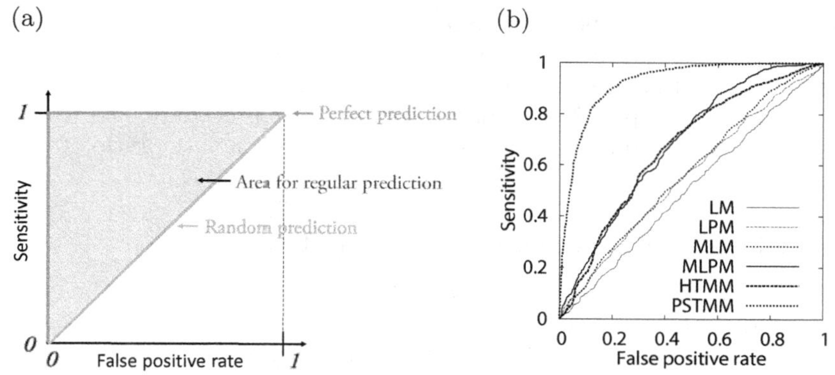

Figure 10.4: ROC curves: (a) extreme examples (b) real example.

first and then any other case between these two.

Random prediction (the most useless predictor):
If prediction does not work at all, even if the instances are sorted, positives and negatives will appear randomly, according to their ratio, i.e. $N_P : N_N$. This means that we will draw lines randomly with the same length parallelly to the x-axis as that to the y-axis. Thus over the two-dimensional space of the x-axis and y-axis, the ROC curve will be the line between the x-axis and y-axis, i.e. the diagonal line. Fig. 10.4 (a) shows an example of the diagonal line, which corresponds to random guessing (random prediction).

Perfectly correct prediction (the most useful predictor):
On the other hand, if predictions are complete (and if instances are sorted in the descending order of the scores), N_P positives appear first and then N_N negatives appear. Thus, the ROC curve must be drawn on the y-axis ($x = 0$) first, and then the curve will be further drawn from $y = 0$ to 1 (on $y = 1$). Fig. 10.4 (a) shows this perfect prediction case as one example.

Other prediction (intermediate between the above two extremes):
Any other prediction result can be between the random prediction and perfectly correct prediction. Thus the ROC curve of another method is located between the diagonal and the perfect line. Fig. 10.4 (a) shows the region in which another prediction can be taken. In practice, another prediction result neither goes upper nor the more left than the perfect prediction, while it can be lower or more right than random guessing. Fig. 10.4 (b) shows one real example of ROC curves. Simply speaking, a better performance method is more distant from the diagonal. In this figure, the performance of PSTMM is best, while LM, LPM and MLM are close to the diagonal, being equivalent in performance to random guessing. As such, one important characteristic of the ROC curve is that multiple methods can be visually compared in performance with each other over the two dimensional space.

As explained in the above, the ROC curve is drawn by the size of $\frac{1}{N_P}$ or $\frac{1}{N_N}$ for each instance. This means the y- and x- axes are for positives and negatives, respectively, and the size of each axis is normalized to be 1 by the number of entire instances (i.e. positives or negatives). By this procedure, the ROC curve is robust against the imbalanceness of the data labels, i.e. the ratio of positives to negatives.

Measure 2: Area Under the ROC curve (AUC)

The ROC curve is very useful in the sense that the performance can be visualized and also the performance difference can be visually captured. However, there would be a case, where for two methods, even visually the performance advantage/disadvantage is not so clear. For example, in Fig. 10.4, although PSTMM is clearly advantageous in performance, it is unclear which of MLPM and HTMM is better than the other.

Area under the ROC curve (AUC) is a measure, which summarizes the performance shown by the ROC curve into only one value. Again the ROC curve becomes the diagonal for random guessing and any method is above the diagonal (while the two lines of $x = 0$ and $y = 1$ shows the perfect prediction, and any method is below this). Then, one idea is to check how much the prediction result (ROC curve) is above the diagonal. That is, we can check the space of the area between the ROC curve by some prediction results and the diagonal. This space becomes zero by random guessing, and will be $\frac{1}{2}$ if the prediction is perfect. AUC exactly follows this idea, and plus $\frac{1}{2}$ to the above space. Thus AUC takes a value within the range from 0.5 to 1 (if AUC is significantly less than 0.5, the prediction should be reverse totally).

We can compute AUC after we draw the ROC curve, but AUC can be computed more easily just by using the numbers of positives and negatives in test data (i.e. N_P と N_N), and the rank sum of negative instances in test data. Thus, letting RSum_N be the rank sum of negative instances in test data, AUC can be computed as follows [37]:

$$\mathrm{AUC} = \frac{\mathrm{RSum}_N - N_N(\frac{N_N+1}{2})}{N_P N_N}, \tag{10.2}$$

where N_P and N_N are obtained from test data, regardless of prediction results. Thus, once if we have prediction results, we can just compute the sum of the ranks of negative instances and compute AUC by using the rank sum of negative instances. Below, we raise some points related with AUC.

Most standard measure: Since the ROC curve is invariant to the ratio of labels in test data, AUC also has the same property. AUC is the most standard measure for binary classification, and so AUC is a measure to be used primarily for evaluating binary classification prediction results.

Gini index (Gini coefficient) Gini index is a measure originally proposed in economics. Also Gini index can be used as a criterion for data partitioning in decision tree (See 3.2.3). In fact one of the early days decision tree methods, called CART (Classification And Regression Tree) [15], uses the Gini index. Gini index and AUC are equivalent, holding the following equation.

$$\text{Gini index} + 1 \quad = \quad 2 \times \text{AUC}. \tag{10.3}$$

That is,

$$\text{AUC} \quad = \quad \frac{1}{2} + \frac{\text{Gini index}}{2}. \tag{10.4}$$

As mentioned in the above, AUC is obtained by adding $\frac{1}{2}$ to the area between the prediction results and the diagonal. Thus Gini index is just a half of the area of the above space.

Ranked sum tests: AUC is equivalent to Mann-Whitney U test (which is called by other names, such as Mann-Whitney-Wilcoxon (MWW), Wilcoxon ranked some test or Wilson-Mann-Whitney test). This testing examines the difference in real values of two given groups, where the null hypothesis is that the given two groups have no difference.

Measure 3: Precision-Recall (PR) Curve

When we draw the ROC curve and examine the sorted test instances, we go up by $\frac{1}{N_P}$ if an instance is a positive. This amount, $\frac{1}{N_P}$, is always the same for any point of the x-axis. However maybe this amount, $\frac{1}{N_P}$, should not be the same and can be changed. In fact in the two-dimensional space of the ROC curve, instances close to 0 in the x-axis have higher scores and those close to 1 have lower scores, while in the ROC curve and AUC, these scores are not considered and the amount, $\frac{1}{N_P}$, is equal over all range of the x-axis.

One idea for this issue is that instance with higher scores should be evaluated more. For example, if a lot of costs are needed to have labels, we will be able to check labels only for test instances with high predicted scores. In this case, we are interested in only data with high predicted scores, and more emphasis should be placed on the prediction with higher scores.

The *precision-recall (PR) curve* realizes this idea in an *ad hoc* manner. The precision-recall curve is drawn focusing on only one label, usually only positives, rather than both positives and negatives. That is, instead of negative instances for the x-th axis in the ROC curve, both the x-th and y-th axes are measures for positives in the precision-recall curve. In more detail, checking the sorted instances from the top, for each positive instance, recall (or TPR) and precision (or PPV) are plotted in the x-axis and y-axis, respectively. **Algorithm 10.3** shows a pseudocode of this procedure. After finishing the entire test instances, the x-axis is $x = 1$ and the precision is equal to just the ratio of positive examples to the entire examples, i.e. $\frac{N_P}{N}$. This means that for any test dataset, the last point is $(1, \frac{N_P}{N})$.

To understand the precision-recall curve more, we consider two extreme cases and another prediction result which falls between the two extreme cases.

Algorithm 10.3: Precision-recall curve.

Function `Precision-recall curve`(*scores*, N_P, N_N)

  Data: Instances:$D_i (i = 1, \dots, N)$ and their predicted scores, #positive instance: N_P、#negative instances: N_N

  Result: precision-recall curve

  Initialization: sort instances in the descending order of their scores, resulting in $\boldsymbol{x}_1, \dots, \boldsymbol{x}_N$.;

  $k_P \leftarrow 0$, $k_N \leftarrow 0$;

  for $k \leftarrow 1$ **to** N **do**

    if $\boldsymbol{x}_k$ *is predicted as a positive* **then**

      $k_P \leftarrow k_P + 1$;

      Plot $(\frac{k_P}{N_P}, \frac{k_P}{k_P + k_N})$.;

      Draw the line from the last point to the new point.;

    else

      $k_N \leftarrow k_N + 1$;

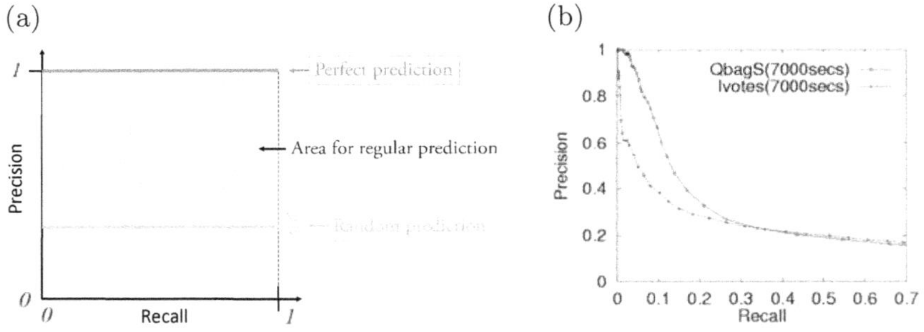

Figure 10.5: (a) Extreme cases of precision-recall curves, and (b) real curves.

Random prediction (the most useless predictor): Random prediction just shows the distribution (of positives and negatives) of data, meaning that precision is a constant. Thus for all recall values from 0 to 1, the same precision value (simply equal to the ratio of positives to all instances) is taken. This is shown in Fig. 10.5 (a), where $y = \frac{N_P}{N}$ for all 0–1 of x.

Perfectly correct prediction (the most useful predictor): When we sort instances predicted perfectly, positives appear first and then negatives appear. In this case, while the first positives are appearing, precision is always 1. Then negative instances are not considered, since the precision-recall curve just focuses on only one label, usually positives only. This means that y is always 1 for the range from 0 to 1 of x. In Fig. 10.5 (a), this line is shown in the top of this figure.

Other prediction (intermediate between the above two extremes):

Any other prediction has to be drawn below the perfect prediction and above the random prediction. This region is shown in Fig. 10.5 (a). In general top prediction (with higher scores) should be predicted well, meaning higher precision. This means that precision is higher for x with closer to 0, and precision is gradually lower with increasing x, finally turning into $y = \frac{N_P}{N}$ at $x = 1$. The ROC curve was a monotone increasing curve, while the precision-recall curve is not necessarily a monotone curve. That is, entirely the precision-recall curve should be decreasing, while it may increase. Also it's y value can be smaller than the final precision, i.e. $y = \frac{N_P}{N}$.

Fig. 10.5 (b) is a real example of two precision-recall curves. These curves were obtained by applying two different machine learning methods to the same synthetic data. Since the data are synthetic, the curves are rather smooth. From this figure, we can see that the two compared methods have a significant difference of precision in the region of smaller recall values. This indicates that these two methods have decisive differences in predicting instances with higher scores. As such, to emphasize the difference in the predictive performance for the instances mostly important, the precision-recall curves are very useful.

Measure 4: Area Under the Precision-Recall curve (AUPR)

As well as AUC for the ROC curve, we can compute the size of the area between the precision-recall curve and $y = 0$ for the region from 0 to 1 of the x-axis. This size is called *Area Under the Precision-Recall curve (AUPR)*. In detail, after sorting test instances according to the prediction scores, we can check the test instances from the top to the bottom, and we can just sum up all precisions we can obtain every time a positive test instance appears. Unfortunately there are no methods proposed for computing AUPR easily. Again the precision-recall curve realizes the idea of weighing instances with higher scores more in an *ad hoc* manner, and so there are no connections to other known measures.

Measure 5: Precision at N (P@N)

The motivation of the precision-recall curve was it better to weigh instances with higher scores more. To realize this motivation more directly, we can compare the precision of instances with high scores directly. For example, we can compare the precision of the top 10% of positive instances. This idea is in general called *P@N (Precision at N)*, where N is a cut-off value, like the above 10%, and P@N is the precision value specified by the cut-off percentage.

Measure 6: F_1 score (F-score, F-measure)

The F_1 score was already defined as the harmonic mean of precision and recall, when predicted instances have no scores (meaning that predicted scores are just binary). When we have scores, we can sort instances due to scores and check the instances from the top (like the cut-off value), computing the F_1 scores for each cut-off value. Then we can take the maximum of the computed F_1 scores for the F_1 value to evaluate the given prediction results to some set of test instances by some machine learning method.

A clear problem of this method is "taking the maximum", which does not necessarily consider the performance over all test instances. This point true of P@N. Again AUC and AUPR are the sum (or average) of sensitivity and precision, respectively, over all test instances and all positive test instances, respectively, while P@N and F_1 score check only the values at one cut-off value. This difference might be understood more by using (car, horse or whatever) racing: The average speed comparison from the start to the end is, technically, the same as AUC and AUPR (this is an usual manner of evaluating car racing results), while P@N and F_1 score are like comparison by the highest speed or the fastest lap. Thus it would be understood that P@N and F_1 score are more tricky, unstable and biased measures than AUC and AUPR.

10.2 Supervised Learning and Continuous Labels

Let $y_1^*, \ldots, y_N^*$ be true labels and $\hat{y}_1, \ldots, \hat{y}_N$ be predicted labels.

10.2.1 Root Mean Square Deviation

The standard and most used measure is *Root Mean Square Deviation (RMSD).*

$$\text{RMSD} = \sqrt{\frac{1}{N}\sum_{i=1}^{N}|y_i^* - \hat{y}_i|^2}. \tag{10.5}$$

RMSD is the measure, which computes the distance between true labels and predicted labels by using the square of the quadratic distance. A smaller value is better, but the value totally depends upon the given data and is neither an absolute value nor some absolute range. Also instead of RMSD we can use *Mean Square Distance (MSD)*, which is without taking the square, but the distance comparison is kept in MSD.

10.2.2 Transforming into Binary Labels

We can set up a cut-off value for numerical (continuous) values, to transform continuous labels into binary labels. By doing this, we can use any measure for supervised learning with binary labels, particularly those for prediction with scores, such as AUC.

10.3 Unsupervised Learning (Clustering)

As mentioned earlier, the assumption behind the evaluation on clustering is that true (gold-standard) clusters are given. Let $\mathcal{C}^* = \{C_1^*, \ldots, C_{K^*}^*\}(K^* = |\mathcal{C}^*|)$ be a

Table 10.5: $K^* \times \hat{K}$ contingency table for clustering.

		True clustering			
		C_1^*	$\cdots$	$C_{K^*}^*$	
Predicted clustering	$\hat{C}_1$	$N_{1,1}$	$\cdots$	N_{1,K^*}	$\lvert\hat{C}_1\rvert$
	$\vdots$	$\vdots$	$\ddots$	$\vdots$	$\vdots$
	$\hat{C}_{\hat{K}}$	$N_{\hat{K},1}$	$\cdots$	$N_{\hat{K},K^*}$	$\lvert\hat{C}_{\hat{K}}\rvert$
		$\lvert C_1^*\rvert$	$\cdots$	$\lvert C_{K^*}^*\rvert$	N

set of true clusters and $\hat{\mathcal{C}} = \{\hat{C}_1, \ldots, \hat{C}_{\hat{K}}\}(\hat{K} = |\hat{\mathcal{C}}|)$ be a set of predicted clusters. In the combinations of two cluster sets, using the number of test instances which fall into one cluster of the true set and also one cluster of the predicted set, we can generate a $K^* \times \hat{K}$-*contingency table*. Table 10.5 shows an example of this contingency table.

There are two approaches for evaluating clustering results.

1. **With $K^* \times \hat{K}$-contingency table,** $K^* \times \hat{K}$-contingency table itself is evaluated.

2. **Without $K^* \times \hat{K}$ contingency table,** evaluation can be done in another manner.

Below we raise three measures: Rand Index (RI), Adjusted Rand Index (ARI) and Normalized Mutual Information (NMI). ARI and NMI evaluate the contingency table itself, and so these measures are categorized into the number one of the above. On the other hand, RI does not use the contingency table, and so this measure is categorized into the number two of the above.

10.3.1 RI (Rand Index)

Rand Index (RI) [70] considers all possible pairs of test instances, and then regards one pair as one instance. As we already know true clusters, if the pair is in the same true cluster, it is a positive; otherwise a negative. Given the number of instances, N, the number of all pairs (i.e. instances for RI), $N^{(p)}$, can be given as follows:

$$N^{(p)} = \binom{N}{2} = \frac{N(N-1)}{2}. \tag{10.6}$$

Depending on whether each pair is in the same cluster or not, we can write a 2×2-contingency table, as shown in Table 10.6. Over this 2×2 contingency table, we can use any of evaluation methods (with no scores) we showed for supervised learning and binary labels in Section 10.1.1. For example, Jaccard Index (JI) can be used for clustering rather than classification, over the 2×2-contingency table.

Table 10.6: 2×2 contingency table for clustering.

		The same cluster in true clustering		
		Positive	Negative	
The same cluster	Positive	$N^{(p)}_{TP}$	$N^{(p)}_{FP}$	$N^{(p)}_{\cdot P}$
by prediction	Negative	$N^{(p)}_{FN}$	$N^{(p)}_{TN}$	$N^{(p)}_{\cdot N}$
		$N^{(p)}_{P}$	$N^{(p)}_{N}$	$N^{(p)}$

In fact Rand Index (RI) is equivalent to applying the accuracy over the 2×2-contingency table. By using Table 10.6, we can compute Rand Index as follows:

$$\text{Rand Index} = \frac{N^{(p)}_{TP} + N^{(p)}_{TN}}{N^{(p)}}. \tag{10.7}$$

RI takes 1 if predicted clusters are totally consistent with true clusters, while RI takes 0 if predicted clusters are totally inconsistent with true clusters. Practically, the value of RI can be too high and cannot be necessarily easy to compare with each other. This is the same problem that by the accuracy for evaluating binary classification results. Adjusted Rand Index (ARI) in the next section improves this problem of Rand Index.

10.3.2 Adjusted Rand Index (ARI)

Adjusted Rand Index (ARI) [46] is not an extension of Rand Index but a hypothesis test, rather just borrowing the idea of *Random* (Rand) Index.

ARI can be computed in the following: Let $|C^*_1|, \ldots, |C^*_{K^*}|$ and $|\hat{C}_1|, \ldots, |\hat{C}_{\hat{K}}|$ be the size of true and predicted clusters, respectively. ARI assumes hypergeometric distribution to generate clusters randomly and examines the practical number of instances in clusters: $N_{1,1}, \ldots, N_{\hat{K},K^*}$, against the null hypothesis (which generates clusters randomly):

$$\text{ARI} = \frac{\sum_{i=1}^{\hat{K}} \sum_{j=1}^{K^*} \binom{N_{i,j}}{2} - \text{ExpI}(\mathcal{C}^*, \hat{\mathcal{C}})}{\text{MaxI}(\mathcal{C}^*, \hat{\mathcal{C}}) - \text{ExpI}(\mathcal{C}^*, \hat{\mathcal{C}})}, \tag{10.8}$$

where

$$\text{MaxI}(\mathcal{C}^*, \hat{\mathcal{C}}) = \frac{1}{2}(\text{Comb}(\mathcal{C}^*) + \text{Comb}(\hat{\mathcal{C}})), \tag{10.9}$$

$$\text{Comb}(\mathcal{C}^*) = \sum_{i=1}^{K^*} \binom{|\mathcal{C}^*|}{2}, \tag{10.10}$$

$$\text{Comb}(\hat{\mathcal{C}}) = \sum_{j=1}^{\hat{K}} \binom{|\hat{\mathcal{C}}|}{2}, \text{ and} \tag{10.11}$$

$$\text{ExpI}(\mathcal{C}^*, \hat{\mathcal{C}}) = \frac{\text{Comb}(\mathcal{C}^*)\text{Comb}(\hat{\mathcal{C}})}{\frac{N(N-1)}{2}} \tag{10.12}$$

ARI becomes 1 when predicted clusters are totally consistent with true clusters, but the range of ARI is not necessarily from zero to one.

As shown in Rand Index, we can focus on instance pairs, which leads to 2×2-contingency table, shown in Table 10.6. For 2×2-contingency table, ARI is defined as follows [92]:

$$\mathrm{ARI} = \frac{2(N_{TP}^{(p)} N_{TN}^{(p)} - N_{FN}^{(p)} N_{FP}^{(p)})}{(N_{FP}^{(p)} + N_{TN}^{(p)})(N_{FP}^{(p)} + N_{TP}^{(p)}) + (N_{TN}^{(p)} + N_{FN}^{(p)})(N_{FN}^{(p)} + N_{TP}^{(p)})} \tag{10.13}$$

$$= \frac{2(N_{TP}^{(p)} N_{TN}^{(p)} - N_{FN}^{(p)} N_{FP}^{(p)})}{N_{N}^{(p)} N_{\cdot P}^{(p)} + N_{\cdot N}^{(p)} N_{P}^{(p)}}. \tag{10.14}$$

The numerator of the above definition of ARI is the same as Matthew's Correlation Coefficient, while the denominator is different. However the same four terms $(N_N, N_{\cdot P}, N_{\cdot N}, N_P)$ in the denominator of Matthew's Correlation Coefficient are used in ARI as well.

10.3.3 Normalized Mutual Information (NMI)

Given predicted clustering results, we can count the number of predicted instances in cluster $\hat{C}_i$ to obtain the probability distribution $P(\hat{C}_i)(i = 1, \ldots, K)$ over predicted clusters:

$$P(\hat{C}_i) = \frac{|\hat{C}_i|}{N}. \tag{10.15}$$

Using them, we can compute entropy $H(\hat{\mathcal{C}})$, which can examine the bias of the probability distribution, i.e. predicted clusters (see Section 3.2.2 for decision stump which uses entropy and Section A.2.2 for entropy):

$$H(\hat{\mathcal{C}}) := -\sum_{i=1}^{\hat{K}} P(\hat{C}_i) \log_2(P(\hat{C}_i)). \tag{10.16}$$

In addition, we can consider not only predicted clusters but also true clusters, and allow to consider the difference of the predicted clusters from the true clusters. This is possible just by simply considering the distributions of instances into the two sets of clusters, i.e. $\hat{\mathcal{C}}, \mathcal{C}^*$ and compute the entropy over them:

$$H(\hat{\mathcal{C}}, \mathcal{C}^*) := -\sum_{i=1}^{\hat{K}} \sum_{j=1}^{K^*} P(i,j) \log_2 P(i,j), \tag{10.17}$$

where, by using Table 10.5, $P(i,j)$ can be computed as follows:

$$P(i,j) = \frac{N_{i,j}}{N}. \tag{10.18}$$

This entropy is a straight-forward extension of the entropy with just one variable, and shows the uncertainty of the two variables.

Similarly we can define conditional entropy of predicted clusters $\hat{\mathcal{C}}$, given true clusters $\mathcal{C}^*$, as follows:

$$H(\hat{\mathcal{C}}|\mathcal{C}^*) \quad := \quad -\sum_{i=1}^{\hat{K}}\sum_{j=1}^{K^*} P(i,j)\log_2 P(i|j), \tag{10.19}$$

where $P(i|j)(=\frac{P(i,j)}{P(j)})$ can be computed from (10.15) and (10.18).

We now define *mutual information*, which examines the dependency of two variables as information, which is also the bias of two variables by using two sets of clusters, i.e. predicted clusters $\hat{\mathcal{C}}$ and true clusters $\mathcal{C}^*$, as follows (the definition below include the above entropy and conditional entropy to show the relationship between them and mutual information):

$$\begin{aligned}
\mathrm{MI}(\hat{\mathcal{C}};\mathcal{C}^*) \quad &:= \quad \sum_{i=1}^{|\hat{\mathcal{C}}|}\sum_{j=1}^{|\mathcal{C}^*|} P(i,j)\log_2 \frac{P(i,j)}{P(i)P(j)} &(10.20)\\
&= \quad H(\hat{\mathcal{C}}) - H(\hat{\mathcal{C}}|\mathcal{C}^*) &(10.21)\\
&= \quad H(\mathcal{C}^*) - H(\mathcal{C}^*|\hat{\mathcal{C}}) &(10.22)\\
&= \quad H(\hat{\mathcal{C}}) + H(\mathcal{C}^*) - H(\hat{\mathcal{C}}|\mathcal{C}^*). &(10.23)
\end{aligned}$$

Also see Section A.3.1 for more detail of mutual information.

The maximum value of mutual information is either the entropy of true clusters or that of predicted clusters:

$$\mathrm{MI}(\hat{\mathcal{C}};\mathcal{C}^*) \leq \min\{H(\hat{\mathcal{C}}), H(\mathcal{C}^*)\}. \tag{10.24}$$

This means that the range of MI depends on given data, and we cannot evaluate the prediction results by some absolute value. *Normalized Mutual Information* (NMI) improves this range problem of mutual information and normalizes the range into between zero to one. In fact the manner of normalization is not necessarily only one but more than one. We here raise two typical ones (in general the former is considered as normalized mutual information).

$\mathbf{NMI_{SG}}$ by geometric mean [85]

$$\begin{aligned}
\mathrm{NMI_{SG}}(\hat{\mathcal{C}};\mathcal{C}^*) \quad &:= \quad \frac{\mathrm{MI}(\hat{\mathcal{C}};\mathcal{C}^*)}{\sqrt{H(\hat{\mathcal{C}})H(\mathcal{C}^*)}} &(10.25)\\
&= \quad \frac{\mathrm{MI}(\hat{\mathcal{C}};\mathcal{C}^*)}{\mathrm{MI}(\hat{\mathcal{C}};\hat{\mathcal{C}})\mathrm{MI}(\mathcal{C}^*;\mathcal{C}^*)}. &(10.26)
\end{aligned}$$

NMI$_{\mathbf{AJ}}$ by simple mean [5, 55]

$$\mathrm{NMI}_{\mathrm{AJ}}(\hat{\mathcal{C}};\mathcal{C}^*) := \frac{\mathrm{MI}(\hat{\mathcal{C}};\mathcal{C}^*)}{H(\hat{\mathcal{C}})+H(\mathcal{C}^*)} \tag{10.27}$$

$$= \frac{\mathrm{MI}(\hat{\mathcal{C}};\mathcal{C}^*)}{\mathrm{MI}(\hat{\mathcal{C}};\hat{\mathcal{C}})+\mathrm{MI}(\mathcal{C}^*;\mathcal{C}^*)}. \tag{10.28}$$

Note that both these two manners of NMI satisfy the following range:

$$0 \le \mathrm{NMI}(\hat{\mathcal{C}};\mathcal{C}^*) \le 1. \tag{10.29}$$

Furthermore these two NMI have the same conditions for the upper and lower limitation of the range.

$$\mathrm{NMI}(\hat{\mathcal{C}};\mathcal{C}^*) = 1 \quad \longleftrightarrow \quad \hat{\mathcal{C}} = \mathcal{C}^*. \tag{10.30}$$

$$\mathrm{NMI}(\hat{\mathcal{C}};\mathcal{C}^*) = 0 \quad \longleftrightarrow \quad P(i,j)=0 \ \text{ or } \ P(i,j)=P(i)\cdot P(j) \ \text{ for all } \ (i,j). \tag{10.31}$$

Also we briefly introduce three other manners of normalization for NMI. All these three manners take the range between zero and 1.

1. Joint entropy [101].

$$\mathrm{NMI}_{\mathrm{Joint}}(\hat{\mathcal{C}};\mathcal{C}^*) := \frac{\mathrm{MI}(\hat{\mathcal{C}};\mathcal{C}^*)}{H(\hat{\mathcal{C}},\mathcal{C}^*)}. \tag{10.32}$$

2. Maximum entropy [55].

$$\mathrm{NMI}_{\mathrm{Max}}(\hat{\mathcal{C}};\mathcal{C}^*) := \frac{\mathrm{MI}(\hat{\mathcal{C}};\mathcal{C}^*)}{\max\{H(\hat{\mathcal{C}}),H(\mathcal{C}^*)\}} \tag{10.33}$$

$$= \frac{\mathrm{MI}(\hat{\mathcal{C}};\mathcal{C}^*)}{\max\{\mathrm{MI}(\hat{\mathcal{C}};\hat{\mathcal{C}}),\mathrm{MI}(\mathcal{C}^*;\mathcal{C}^*)\}}. \tag{10.34}$$

3. Minimum entropy [55].

$$\mathrm{NMI}_{\mathrm{Max}}(\hat{\mathcal{C}};\mathcal{C}^*) := \frac{\mathrm{MI}(\hat{\mathcal{C}};\mathcal{C}^*)}{\min\{H(\hat{\mathcal{C}}),H(\mathcal{C}^*)\}} \tag{10.35}$$

$$= \frac{\mathrm{MI}(\hat{\mathcal{C}};\mathcal{C}^*)}{\min\{\mathrm{MI}(\hat{\mathcal{C}};\hat{\mathcal{C}}),\mathrm{MI}(\mathcal{C}^*;\mathcal{C}^*)\}}. \tag{10.36}$$

10.3.4 Other Measures for Clustering

Numerous measures have been proposed for clustering, and we introduced the most typical three measures. We raise two reviews, one summarizing historical methods well [96] and the other thoroughly investigating information content-based approaches such as NMI [92]. In particular the latter article introduces a method combining NMI and ARI, called *Adjusted-for-Chance MI.* In practice when we evaluate the clustering results with many clusters, the value of NMI is likely to be very small (for example, around 0.1 even though the range is between 0 and 1).

Table 10.7: Example of binary relevance $\mathrm{Rel}_q(k)$ for k-th rank and some query q.

rank (k)	1	2	3	4	5	...
$\mathrm{Rel}_q(k)$	0	1	0	0	1	...
$\mathrm{Prec}_q(k)$	0	1/2	1/3	1/4	2/5	...

10.4 Ranking

The problem setting of *ranking* comes from the following two practical situations:

Information retrieval (or search) in World Wide Web (WWW): We usually search the most relevant homepages or documents in World Wide Web (WWW) to the casted query, and receive the results (of found homepages or documents) through the browser. This is generally the main problem of information retrieval.

Recommendation in E-commerce: Similarly we search the most relevant goods or items in the sites of E-commerce to the casted query and the relevant goods or items are presented in the web site.

As such, we receive multiple recommended homepages, documents, items or goods (hereafter we call them instances), depending on each situation. However they cannot be presented at the same time and so have to be *ranked* to be presented. Then we have evaluation measures for such ranking results. Below is assumptions we make to evaluate the ranking results.

Predicted scores: Evaluation is done for ranking only. The predictor may provide scores to have ranked results, while we do not consider any predicted scores. We just focus on the ranking results only.

True (gold-standard) results: It is very hard to have true (gold-standard) ranking results. Thus, instead we have labels (relevance), showing how much each ranked instance is relevant to the casted query.

There are two cases on relevance: binary (showing if the instance is relevant to the query or not) and multiple levels (i.e. scores showing the degree of how much each instance is relevant to the query).

Under the above assumptions, we explain the measures to evaluate ranking results. We divide the measures into two cases: 1) binary labels (relevance) and 2) multiple labels, where we explain two measures for each of these two cases.

Let $\mathcal{D}$ be the set of ranked instances and $d \in \mathcal{D}$ be each instance in the set.

10.4.1 Binary Labels Relevant to the Query

True values are binary, showing if each instance is relevant to the query or not, i.e. 1 if relevant to the query; otherwise 0. Let $\mathrm{Rel}_q(k) \in \{0, 1\}$ be the true label for query q and for the instance of rank k.

The problem setting is that we consider not only one but multiple queries. That is, we cast several queries and want to rank instances in the order of relevance to the entire multiple queries. Thus let Q be a set of queries, and $q \in Q$ be a query in the set.

Mean Reciprocal Rank (MRR)

Reciprocal rank means the inverse of the rank. *Mean Reciprocal Rank* uses only the highest rank K_q, which is relevant to query q, where K_q can be defined as follows:

$$K_q : \mathrm{Rel}_q(K_q) = 1 \text{ where } \mathrm{Rel}_q(k) = 0 \text{ for all } k < K_q. \tag{10.37}$$

This means that K_q becomes smaller for an instance relevant to q at a higher rank.

MRR is given as the sum of the inverse of K_q over all queries in Q, as follows:

$$\mathrm{MRR} := \frac{1}{|Q|} \sum_q \frac{1}{K_q}. \tag{10.38}$$

That is, MRR is larger, as instances are ranked higher on average for a larger number of queries in Q. Table 10.7 shows an example of ranked instances, where $k = 2$ when $\mathrm{Rel}_q(k)$ is 1 at the highest rank, and then $K_q = 2$. Thus, if we use this query only, MRR is simply computed as follows:

$$\mathrm{MRR} = \frac{1}{2}. \tag{10.39}$$

MRR uses the highest rank only, which would be a problem, because even by some accident, MRR can be larger if an instance is ranked high. Also this does not matter if the number of instances ranked high. This point is improved by Mean Average Precision of the subsequent section.

Mean average precision (MAP)

Let $\mathrm{Prec}_q(k)$ be precision for query q and the instance of rank k. Again precision is the positive predictive value used for evaluating supervised learning and binary labels. That is, $\mathrm{Prec}_q(k)$ is the ratio of relevant instances in the top k instances (which are all regarded as relevant) to k, and can be computed as follows:

$$\mathrm{Prec}_q(k) = \frac{\sum_i^k \mathrm{Rel}_q(i)}{k}. \tag{10.40}$$

Average precision of query q is given as the sum of precision for query q and the instance of rank k over k, where precision of only relevant instances are used:

$$\text{Average precision} := \sum_k \mathrm{Rel}_q(k)\mathrm{Prec}_q(k). \tag{10.41}$$

Table 10.8: Example of binary relevance $\mathrm{Rel}_q(k)$ for k-th rank and discount $(\frac{1}{\log_2(k+1)})$ and also final nDCG.

rank (k)	1	2	3	4	5	...
$\mathrm{Rel}_q(k)$	4	1	2	0	3	...
Log= $1/(\log_2(k+1))$	1	0.631	0.5	0.431	0.387	...
$\mathrm{Rel}_q(k)\cdot$ Log	4	0.631	1	0	1.161	...

This means, when the number of relevant instances is fixed, if these relevant instances are ranked higher, average precision becomes larger. However, if the number of relevant pinstances is changeable, average precision can be larger as the number of relevant instances is larger, even if they are not ranked high enough.

Thus a possible idea against this issue is to divide the average precision by the number of relevant instances, which results in *Mean Average Precision* (MAP), as follows:

$$\mathrm{MAP} \quad := \quad \frac{1}{|\mathcal{R}_q|} \sum_k \mathrm{Rel}_q(k)\mathrm{Prec}_q(k), \tag{10.42}$$

where $\mathcal{R}_q$ is the set of relevant instances.

In fact we think about binary relevance, which means that $\mathrm{Rel}_q(k)$ takes a binary, i.e. 0 or 1, and only the cases of 1 are considered. For example, in Table 10.7, $\mathrm{Rel}_q(k)$ is 1, when $k = 2$ and $k = 5$. Thus, MAP can be computed for Table 10.7, as follows:

$$\mathrm{MAP} \quad = \quad \frac{1}{2}(1\dot{1}/2 + 1\dot{2}/5) = \frac{1}{2}\frac{9}{10} = \frac{9}{20}. \tag{10.43}$$

If we consider multiple queries, MAP can be given as follows:

$$\mathrm{MAP} \quad := \quad \sum_q \frac{1}{|\mathcal{R}_q|} \sum_k \mathrm{Rel}_q(k)\mathrm{Prec}_q(k). \tag{10.44}$$

10.4.2 Multiple Labels Relevant to Query

True values showing the relevance (score) to a query can be multiple labels or even continuous values. This is an extension of binary relevance, and so the methods to be introduced here can be used for binary relevance. Let $\mathrm{Rel}_q(k)(\in \mathbb{R})$ be (true) relevance to query q for the instance of rank k.

Normalized Discounted Cumulative Gain (nDCG)

Cumulative gain (CG) is simply the sum of the relevance of the top K instances.

$$\mathrm{CG}_{q,K} \quad := \quad \sum_k^K \mathrm{Rel}_q(k). \tag{10.45}$$

Rankings with the same sum of relevances have the same cumulative gain, even if the rankings are different.

To solve this problem, *Discounted Cumulative Gain* (DCG) uses a heuristic, which places a weight on the relevance at each rank, so that relevance at higher ranks should be more weighed. As the weight, DCG uses the inverse of the logarithm of the rank: As shown in Table 10.8, the inverse of the logarithm of the rank is 1 for rank 1 and getting smaller as the rank is going down. An interesting property of the inverse of the logarithm of the rank is that for higher ranks, this value is getting smaller largely when the rank is going down, while for lower ranks, this value becomes smaller very slowly as the rank is going down. This means that if the top, particularly the number one, is a relevant instance, DCG becomes larger. DCG can be given as follows:

$$\mathrm{DCG}_{q,K} \quad := \quad \sum_{k}^{K} \frac{\mathrm{Rel}_q(k)}{\log_2(k+1)}. \tag{10.46}$$

The example of Table 10.8 shows that $\mathrm{DCG}_{q,5} = 6.792$.

Furthermore, to compare with other queries, the range of the value must be normalized into between 0 and 1. In fact *Normalized Discounted Cumulative Gain* (nDCG) [16] divides the obtained DCG by the maximum value of DCG (among possible rankings), as follows:

$$\mathrm{nDCG}_{q,K} \quad := \quad \frac{\mathrm{DCG}_{q,K}}{\mathrm{IDCG}_{q,K}}, \tag{10.47}$$

where $\mathrm{IDCG}_{q,K}$ is the maximum value of the possible ranking.

For the example in Table 10.8, the maximum value can be obtained by that $\mathrm{Rel}_q(k)$ is 4, 3, 2, 1 and 0 for $k = 1, 2, 3, 4$ and 5, resulting in that $\mathrm{IDCG}_q, 5$ is 7.324. Finally,

$$\mathrm{nDCG}_{q,5} \quad = \quad \frac{\mathrm{DCG}_{q,5}}{\mathrm{IDCG}_{q,5}} = \frac{6.792}{7.324} = 0.927. \tag{10.48}$$

This result indicates that nDCG is rather high and the ranking in Table 10.8 is not so bad.

Expected Reciprocal Rank (ERR)

In nDCG, relevances are weighed by the ranks: if ranks are higher, weights are larger. In this case, the value to be multiplied is fixed due to the rank, i.e. the inverse of the logarithm of the rank. Then the relevance is discounted in the same way for any query q. However the discounting way can be changed due to the relevance of instances to the query q, even when the rank is the same. This motivation generates another criterion, called *Expected Reciprocal Rank* (ERR) [19].

First, we modify the relevance as follows:

$$R(g) := \frac{2^g - 1}{2^{g_{\max}}}, \tag{10.49}$$

where g is the original relevance. That is, for example, if the relevance has five stages: 0 to 4, $R(0) = 0$, $R(1) = \frac{2^1-1}{2^4} = \frac{1}{16}$, $R(2) = \frac{2^2-1}{2^4} = \frac{3}{16}$,... This means, bigger relevances are more weighted.

Using the modified relevances, expected reciprocal rank is defined as follows:

$$\mathrm{ERR} := \sum_{k=1}^{K} \frac{1}{k} (\prod_{i=1}^{k-1} (1 - R_i)) R_k. \tag{10.50}$$

We now consider each of the three parts inside of $\sum$, as follows:

$$\frac{1}{k}, \quad \prod_{i=1}^{k-1} (1 - R_i), \quad R_k. \tag{10.51}$$

The third is the relevance, meaning that cumulative gain is equal to sum up the third term, R_k. Then if you want this term discounted, you can add the first term to the third term as follows:

$$\frac{1}{k} R_k, \tag{10.52}$$

which is more drastically discounted than $\frac{1}{\log_2(i+1)} R_k$, i.e. nDCG.

ERR uses the second term to discount this more. The point of the second term is to use the relevance of the instances, which are ranked higher than rank k. That is, if the relevances of the higher ranked instances are weighed, the current relevance (of rank k) is discounted more. This would be consistent with the general behavior of users who lose the interest of lower ranked instances if they can see the highly relevant instances at higher ranks [19].

10.5 Summary

We have explained the measures for evaluating four machine learning categories: 1) supervised learning and binary labels, 2) supervised learning and continuous labels, 3) unsupervised learning (clustering) and 4) ranking.

Over all we can summarize the most standard or recommended measures to be used in each of these categories, as follows:

Supervised learning and binary labels: ROC curve, AUC, Precision-recall curve, AUPR

Supervised learning and continous labels: RMSD

Unsupervised learning (clustering): NMI

Ranking: nDCG

Appendix A

Basics and Derivations in the Main Text

A.1 Sampling

There are two sampling methods: sampling with replacement and without replacement.

Given a set of elements, when you repeatedly sample an element out of this set, if you sample one element out of the same original set every time, this sampling is called *sampling with replacement*. Sampling with replacement means that each sampling is independent of another sampling. On the other hand, under the same situation as the above, if you remove the instance you sampled every time, this sampling is called *sampling without replacement*. Sampling without replacement means that each sampling is dependent on former sampling.

A.2 Statistics: One Variable (Univariate)

First, let $\boldsymbol{X}$ be a random variable, and $\boldsymbol{x} = (x_1, \cdots, x_N)^\top$ be observed values.

Probability distribution can be defined in the following two ways, depending on discrete or continuous random variable:

Discrete random variable: Probability mass function (PMF) p for each observed value. See Section A.2.5 for detailed distributions.

Continuous random variable: Probability density function (PDF) f for each range of observed values. See Section A.2.6 for detailed distributions.

We first explain basic statistics and then describe PMF and PDF, for which we show those relevant to machine learning, particularly those in this book. A detailed reference on PMF and PDF is, for example, [58].

A.2.1 Basic Statistics

Mean

Mean is the expectation value over observed values. For discrete values, the mean μ is computed as follows:

$$\mu = \sum_{n=1}^{N} p(x_i)x_i, \tag{A.1}$$

where p is a probability mass function.

For continuous values, the mean μ is given as follows:

$$\mu = \int f(x)x dx, \tag{A.2}$$

where f is a probability density function.

Below we assume a discrete random variable. Also we assume that a probability mass function is always a uniform distribution, meaning that this function does not affect the statistics we introduce at all.

Thus again,

$$\mu = \frac{\sum_{n=1}^{N} x_n}{N}. \tag{A.3}$$

Normalization by Mean

$$\bar{x}_n = x_n - \mu \quad \text{for} \quad n = 1, \ldots, N. \tag{A.4}$$

We write the normalized vector as follows:

$$\bar{\boldsymbol{x}} = (\bar{x}_1, \ldots, \bar{x}_N)^\mathsf{T}. \tag{A.5}$$

Variance

The variance is defined as the expectation value of the square of the normalized (by the mean), observed value. Thus for discrete value, the variance can be computed as follows:

$$s = \sum_{i=1}^{N} p(x_n)(x_i - \mu)^2 = \frac{1}{N}\sum_{i=1}^{N}(x_i - \mu)^2 = \frac{1}{N}\sum_{i=1}^{N} ||x_i - \mu||_2^2. \tag{A.6}$$

This is equivalent to the L^2 norm (see Section A.6). The root of the variance is called standard deviation, written as σ. That is, $s = \sigma^2$.

We can expand (A.6) further as follows:

$$
\begin{aligned}
s &= \frac{1}{N}\sum_{i=1}^{N}(x_i-\mu)^2 = \frac{1}{N}\sum_{i=1}^{N}(x_i^2-2x_i\mu+\mu^2) && \text{(A.7)}\\
&= \frac{1}{N}\sum_{i=1}^{N}x_i^2 - 2\mu\frac{\sum_{i=1}^{N}x_i}{N} + \frac{1}{N}\mu^2\sum_{i=1}^{N}1 && \text{(A.8)}\\
&= \frac{1}{N}\sum_{i=1}^{N}x_i^2 - 2\mu^2+\mu^2 = \frac{1}{N}\sum_{i=1}^{N}x_i^2-\mu^2. && \text{(A.9)}
\end{aligned}
$$

Thus we can see that the variance is the squared observed value minus the squared mean. We can modify this by using $\bar{\boldsymbol{x}}$, which is obtained by normalizing $\boldsymbol{x}$ by the mean, as follows:

$$
s = \frac{1}{N}\sum_{i=1}^{N}(x_i-\mu)^2 = \frac{1}{N}\sum_{i=1}^{N}\bar{x}_i^2 = \frac{1}{N}\bar{\boldsymbol{x}}^{\mathsf{T}}\bar{\boldsymbol{x}}. \tag{A.10}
$$

This means that if we normalize the observed values already, the variance is equivalent to the inner product.

Also (A.9) can be further developed into the following:

$$
s = \frac{1}{2N^2}\sum_{i=1}^{N}\sum_{j=1}^{N}(x_i-x_j)^2. \tag{A.11}
$$

This means that the variance is the squared sum over the difference of the observed values. The derivation from (A.9) to (A.11) is shown in Section A.9.8.

A.2.2 Entropy

Entropy is a metric on bias of a probability distribution of a discrete random variable. As shown in Fig. 3.13, if entropy is larger, the bias (uncertainty and information content) is also larger.

$$
H(\boldsymbol{X}) := -\sum_{i\in\mathcal{S}} p_i\log_2(p_i), \tag{A.12}
$$

where let $\mathcal{S}$ be a set of discrete values to be taken by a discrete variable.

A.2.3 Sigmoid function

Sigmoid function is called *logistic function*. When we transform the input real value v into a value in the range of $[0,1]$, the simplest method is to place a cut-off value c against v and uses the step function, which outputs 1 if v is larger than c; otherwise 0 (See Fig. 3.18 (a)):

$$
\begin{aligned}
&\text{if} \quad v>c \quad \text{then} \quad 1, && \text{(A.13)}\\
&\text{otherwise} \quad 0. && \text{(A.14)}
\end{aligned}
$$

However, this change of $0 \to 1$ by the step function is too drastic and needs a smooth change between 0 and 1, which can be performed by the sigmoid function (See Fig. 3.18 (b)):

$$\sigma(x) = \frac{1}{1+\exp(-a \cdot x)}, \tag{A.15}$$

where a is a constant and the sigmoid function can be converged into the step function for $a \to \infty$. Since $a = 1$ is normally used, so below we use $a = 1$. Below are several properties:

Derivative:

$$\frac{d\sigma(x)}{dx} = \frac{\exp(-x)}{(1+\exp(-x))^2} = \sigma(x)(1-\sigma(x)). \tag{A.16}$$

Transformation into a probability: The sigmoid function can be interpreted as the probability that the output is one (or zero), given x:

$$p(y=1|x) = \sigma(x) = \frac{1}{1+\exp(-a \cdot x)}. \tag{A.17}$$

$$p(y=0|x) = 1-\sigma(x) = \frac{\exp(-a \cdot x)}{1+\exp(-a \cdot x)}. \tag{A.18}$$

Using these two equations, we can compute the odds between $y = 1$ and $y = 0$:

$$\frac{p(y=1|x)}{p(y=0|x)} = \exp(-a \cdot x). \tag{A.19}$$

A.2.4 Probability Distributions

Probability distributions can be devided into *discrete probability distributions* and *continuous probability distributions*. The formula to represnet a discrete probability distribution is called a *probability mass function*. Similarly the continous probability distribution is represented by a corresponding *probability density function*.

A.2.5 Discrete Probability Distribution (Probability Mass Function (PMF))

Binomial Distribution

We consider the trial which has two outcomes: a success or a failure. In particular we consider a probabilistic trial, called *Bernoulli trial*, with two outcomes: a success with the probability of θ ($0 \leq \theta \leq 1$) or a failure with the probability of $1-\theta$.

Binomial distribution provides the probability of k successes out of n Bernoulli trials:

$$p(n,k,\theta) = \binom{n}{k}\theta^k(1-\theta)^{n-k} = C(n,k)\theta^k(1-\theta)^{n-k} \quad (A.20)$$

$$= \frac{n!}{k!(n-k)!}\theta^k(1-\theta)^{n-k} \quad (K=0,\ldots,N), \quad (A.21)$$

where $C(n,k)$ is called *binomial coefficient.*

Also we call the part corresponding to the Bernoulli trial the *Bernoulli distribution:*

$$p(k,\theta) = \theta^k(1-\theta)^{1-k} \quad (k = 0 \text{ or } 1). \quad (A.22)$$

Multinomial Distribution

Multinomial distribution is an extension of binomial distribution for k outcomes, instead of binary outcomes. We consider a probabilistic trial with k outcomes: for the i-th outcome, the success with the probability of θ_i ($i = 1,\ldots,k$).

Multinomial distribution is the probability of n_i ($i = 1,\ldots,k$) successes out of n trials ($\sum_i n_i = n$, $\sum_i \theta_i = 1$):

$$p(n_1,\ldots,n_k,\theta_1,\ldots,\theta_k) = \frac{n!}{\prod_{i=1}^k n_i!}\prod_{i=1} k\theta_i^{n_i} \quad (A.23)$$

$$= \frac{\Gamma(\sum_i n_i + 1)}{\prod_i \Gamma(n_i+1)}\prod_{i=1}^k \theta_i^{n_i}, \quad (A.24)$$

where $\Gamma(\cdot)$, given a positive integer as the input, is as follows:

$$\Gamma(n) = (n-1)!. \quad (A.25)$$

Again multinomial distribution is an extension of binomial distribution, and so the following two special cases exist:

$n = 1$, $k = 2$: Bernoulli distribution.

$n > 1$, $k = 2$: Binomial distribution.

A.2.6 Continous Probability Distribution (Probability Density Function (PDF))

Normal Distribution

The *normal (Gaussian or Gauss) distribution* is the most common continous probability distribution, which is represented as follows:

$$f(\theta|\mu,\sigma^2) = \frac{1}{\sqrt{2\pi\sigma^2}}e^{-\frac{(\theta-\mu)^2}{2\sigma^2}}, \quad (A.26)$$

where μ and σ^2 are the mean and variance, respectively, over the given data.

The shoulder of the exponential in the right-hand side of (A.26) is the square loss (L^2 norm), which is also equivalent to the variance. Thus, given data, when we take the least squares method (squared loss) in the optimization, that means that we assume that the given data follows the normal distribution.

When $\mu = 0$ and $\sigma^2 = 1$, the following distribution is called the *standard normal distribution*:

$$f(\theta) = \frac{1}{\sqrt{2\pi}} e^{-\frac{\theta^2}{2}} \quad \text{(A.27)}$$

Beta Distribution

The beta distribution corresponds to the binomial distribution of probability mass function. The beta distribution is given as

$$f(\theta|\alpha, \beta) = \frac{1}{B(\alpha, \beta)} \theta^{\alpha-1}(1-\theta)^{\beta-1}, \quad \text{(A.28)}$$

where beta function $B(\alpha, \beta)$ corresponds to the binomial coefficient for continuous values:

$$B(\alpha, \beta) = \frac{\Gamma(\alpha)\Gamma(\beta)}{\Gamma(\alpha+\beta)} = \frac{(\alpha-1)!(\beta-1)!}{(\alpha+\beta-1)!} \simeq \frac{1}{\binom{\alpha+\beta-2)!}{(\alpha-1)!}}. \quad \text{(A.29)}$$

Dirichlet Distribution

The Dirichlet distribution corresponds to the multinomial distributions of probability mass function. The Dirichlet distribution is given as follows:

$$f(\theta_1, \ldots, \theta_K | \alpha_1, \ldots, \alpha_K) = \frac{1}{Z(\boldsymbol{\alpha})} \prod_i \theta_i^{\alpha_i - 1}, \quad \text{(A.30)}$$

where $\sum_i \theta_i = 1$, $\theta_i \geq 0$, and

$$Z(\boldsymbol{\alpha}) = \frac{\prod_i \Gamma(\alpha_i)}{\Gamma(\sum_i \alpha_i)}. \quad \text{(A.31)}$$

The followings are important properties of the Dirichlet distribution:

Expectation value:

$$E(\theta_i) = \int \theta_i f(\boldsymbol{\theta}|\boldsymbol{\alpha}) = \frac{1}{Z(\boldsymbol{\alpha})} \int \theta_i \prod_j \theta_j^{\alpha_j - 1} \quad \text{(A.32)}$$

$$= \frac{\alpha_i}{\sum_j \alpha_j}. \quad \text{(A.33)}$$

Generalization of the beta distribution: Similar to the relation between the binomial and multinomial distributions, the Dirichlet distribution is a generalization of the beta distribution, and is equal to the beta distribution for $K = 2$.

A.3 Statistics: Two Variables

Let $\boldsymbol{X}$ and $\boldsymbol{Y}$ be two random variables, with $\boldsymbol{x} = (x_1, \cdots, x_N)^\mathsf{T}$ and $\boldsymbol{y} = (y_1, \cdots, y_N)^\mathsf{T}$ be two sets of observed values.

A.3.1 Basic Statistics

Inner Product

The inner product of two vectors can be given as follows: $\boldsymbol{x}^\mathsf{T}\boldsymbol{y} = \sum_{i=1}^{N} x_i y_i$.

Orthogonal vectors, $\boldsymbol{x}$ and $\boldsymbol{y}$: $\boldsymbol{x}^\mathsf{T}\boldsymbol{y} = \sum_{i=1}^{N} x_i y_i = 0$.

Orthonormal vectors, $\boldsymbol{x}$ and $\boldsymbol{y}$: Both orthogonal vectors and $\boldsymbol{x}^\mathsf{T}\boldsymbol{x} = 1$ and $\boldsymbol{y}^\mathsf{T}\boldsymbol{y} = 1$.

Covariance

Covariance shows the strength of the correlation between two random variables $\boldsymbol{X}$ and $\boldsymbol{Y}$. Formerly covariance is defined to first normalize each variable by subtracting the expectation value from the original value and then take the expectation value of that obtained by multiplying the two variables:

$$\mathrm{cov}(\boldsymbol{X}, \boldsymbol{Y}) := E((\boldsymbol{X} - E(\boldsymbol{X}))(\boldsymbol{Y} - E(\boldsymbol{Y}))), \tag{A.34}$$

where $E(\boldsymbol{X})$ is the expectation value of $\boldsymbol{X}$.

More concretely,

$$\mathrm{cov}(\boldsymbol{X}, \boldsymbol{Y}) = \sum_i \frac{(x_i - \mu_X)(y_i - \mu_Y)}{N}, \tag{A.35}$$

where $\mu_X = \frac{\sum_i x_i}{N}$ and $\mu_Y = \frac{\sum_i y_i}{N}$, i.e. the means. That is, as shown in Section A.2.1, in the numerator the value in each variable is normalized to make the mean zero. Letting $\bar{x}_i = x_i - \mu_X, \bar{y}_i = y_i - \mu_Y (i = 1, \ldots, N)$ be the normalized values with the means of zero, the covariance can be written as follows:

$$\mathrm{cov}(\boldsymbol{X}, \boldsymbol{Y}) = \sum_i \frac{\bar{x}_i \bar{y}_i}{N}, \tag{A.36}$$

which is equivalent to the inner product of the normalized values.

We can consider a special case of the covariance with $\boldsymbol{Y} = \boldsymbol{X}$, as follows:

$$\mathrm{cov}(\boldsymbol{X}, \boldsymbol{X}) = \sum_i \frac{(x_i - \mu_X)(x_i - \mu_X)}{N} \tag{A.37}$$

$$= \frac{\sum_i (x_i - \mu_X)^2}{N}. \tag{A.38}$$

In this case, covariance becomes variance, and the derivation of (A.11) can be applied to the above variance (the same as (A.9)). The variance can be written as σ_X^2, where σ is the standard deviation. Thus, covariance can be written in σ_{XY}^2.

Also by modifying (A.35), we can write the covariance as follows:

$$\begin{aligned} \mathrm{cov}(\boldsymbol{X}, \boldsymbol{Y}) &= \frac{1}{2N^2} \sum_i \sum_j (x_i - x_j)(y_i - y_j) && \text{(A.39)} \\ &= \frac{1}{N^2} \sum_i \sum_{j>i} (x_i - x_j)(y_i - y_j). && \text{(A.40)} \end{aligned}$$

See Section A.9.9 for the derivation from (A.35) to (A.40).

Correlation Coefficient

As well as covariance, *correlation coefficient* (generally called *Pearson correlation coefficient*) shows the strength of correlation between two random variables $\boldsymbol{X}$ and $\boldsymbol{Y}$. The correlation coefficient, $\mathrm{cor}(\boldsymbol{X}, \boldsymbol{Y})$, can be defined as follows:

$$\begin{aligned} \mathrm{cor}(\boldsymbol{X}, \boldsymbol{Y}) &:= \frac{\mathrm{cov}(\boldsymbol{X}, \boldsymbol{Y})}{\sigma_{\boldsymbol{X}} \sigma_{\boldsymbol{Y}}} && \text{(A.41)} \\ &= \frac{\sigma^2_{\boldsymbol{XY}}}{\sigma_{\boldsymbol{X}} \sigma_{\boldsymbol{Y}}} && \text{(A.42)} \\ &= \frac{\sigma^2_{\boldsymbol{XY}}}{\sqrt{\sigma^2_{\boldsymbol{X}} \sigma^2_{\boldsymbol{Y}}}}. && \text{(A.43)} \end{aligned}$$

Mutual Information

We can start with the basics of information theory. In Section A.2.2, we already define the entropy for one variable (univariate), and now define the entropy for two variables, assuming discrete random variable (continuous variable takes the integral instead of the summation):

$$H(\boldsymbol{X}, \boldsymbol{Y}) := -\sum_{i,j} p(x_i, y_j) \log p(x_i, y_j). \tag{A.44}$$

As well as that for one variable, this entropy is a measure for uncertainty.

In fact we can define multiple, different types of entropy for two variables. The above entropy is called the joint entropy. Given random variable $\boldsymbol{Y}$, the conditional entropy of $\boldsymbol{X}$ can be given as follows:

$$H(\boldsymbol{X}|\boldsymbol{Y}) := \sum_{i,j} p(x_i, y_j) \log p(x_i|y_j). \tag{A.45}$$

Conditional entropy shows the uncertainty of random variable $\boldsymbol{X}$, given random variable $\boldsymbol{Y}$.

Mutual information shows the dependency between two random variables. Letting both $\boldsymbol{X}$ and $\boldsymbol{Y}$ be discrete random variables, the mutual information can be defined as follows:

$$I(\boldsymbol{X}; \boldsymbol{Y}) := \sum_i \sum_j p(x_i, y_j) \log(\frac{p(x_i, y_j)}{p(x_i)p(y_j)}). \tag{A.46}$$

Intuitively mutual information examines the overlap between two random variables. Below are the properties of mutual information:

Non negativity: $p(x_i, y_j) \geq p(x_i)p(y_j)$ leads to that the inside of the logarithm is no less than one. So mutual information is non-negative:

$$I(\boldsymbol{X}; \boldsymbol{Y}) \geq 0. \tag{A.47}$$

Connection to Kullback-Leibler (KL) divergence: Mutual information examines the distance between $p(x_i, y_j)$ and $p(x_i)p(y_j)$. In fact regarding the distance between two probability distributions (for example, $p(x)$ and $q(x)$), *Kullback-Leibler* (KL) divergence [24] is a well-known measure, which is defined as follows:

$$D_{KL}(p||q) := \sum_x p(x) \log \frac{p(x)}{q(x)}. \tag{A.48}$$

KL-divergence is non-symmetric and different from a standard distance, which must be symmetric.

However mutual information can be written by using the KL-divergence, as follows:

$$I(\boldsymbol{X}; \boldsymbol{Y}) = D_{KL}(p(x_i, y_j)||p(x_i)p(y_j)). \tag{A.49}$$

Note that mutual information is symmetric:

$$I(\boldsymbol{X}; \boldsymbol{Y}) = I(\boldsymbol{Y}; \boldsymbol{X}). \tag{A.50}$$

Connection to Entropy: Mutual information can be written by entropy as follows:

$$I(\boldsymbol{X}; \boldsymbol{Y}) = H(\boldsymbol{X}) - H(\boldsymbol{X}|\boldsymbol{Y}) \tag{A.51}$$
$$= H(\boldsymbol{Y}) - H(\boldsymbol{Y}|\boldsymbol{X}) \tag{A.52}$$
$$= H(\boldsymbol{X}) + H(\boldsymbol{Y}) - H(\boldsymbol{X}, \boldsymbol{Y}). \tag{A.53}$$

A.4 Matrix

Let $\boldsymbol{X}$ be a matrix with N rows and M columns. This matrix is equivalent to N instances (vectors) with M features, i.e. vectors, in machine learning.

Below we explain the details. In fact some part might be used already before we explain its detail.

A.4.1 Basics

Transpose: $\boldsymbol{X}^\top$: The transpose of $\boldsymbol{X}$ is the $(M \times N)$-matrix obtained by flipping $\boldsymbol{X}$ over the diagonal. If $\boldsymbol{X}^\top = \boldsymbol{X}$, $\boldsymbol{X}$ is symmetric.

Square matrix: If $N = M$, $\boldsymbol{X}$ is called a *square matrix.*

Diagonal matrix: If $\boldsymbol{X}$ is a square matrix and all elements are zero except diagonals, $\boldsymbol{X}$ is called a *diagonal matrix.* A diagonal matrix is often written as $\boldsymbol{D}$.

Lower triangle (upper triangle) matrix: *Lower triangle (upper triangle) matrix* is a matrix, in which elements below (above) the diagonals are all zero.

Identity matrix: If $\boldsymbol{X}$ is a diagonal matrix and all diagonal elements are 1, $\boldsymbol{X}$ is called an *identity matrix.* An identity matrix is often written as $\boldsymbol{I}$, particularly $\boldsymbol{I}_K$ with the size of rows (and columns) of K.

Inverse matrix: $\boldsymbol{X}^{-1}$: The matrix $\boldsymbol{X}$, which satisfies the following, is called an *inverse matrix*:

$\boldsymbol{X}\boldsymbol{X}^{-1} = \boldsymbol{I}$ and $\boldsymbol{X}^{-1}\boldsymbol{X} = \boldsymbol{I}$.

Orthogonal matrix: The matrix $\boldsymbol{X}$, which satisfies the following, is called an *orthogonal matrix*: $\boldsymbol{X}^\top = \boldsymbol{X}^{-1}$.

Thus $\boldsymbol{X}^\top\boldsymbol{X} = \boldsymbol{I}$ and $\boldsymbol{X}\boldsymbol{X}^\top = \boldsymbol{I}$.

In the orthogonal matrix, column vectors are called *orthonormal vectors / orthogonal unit vectors.*

Rank: rank($\boldsymbol{X}$) is the number of column vectors (or low vectors), which are linearly independent in $\boldsymbol{X}$.

The matrix $\boldsymbol{X}$ is called *full rank*, if rank($\boldsymbol{X}$) $= \min(M, N)$.

Trace: The sum of the diagonal elements of a matrix is called a *trace.* The trace is important, because a variety of matrix computation can be done more easily by trace. For example, given matrix $\boldsymbol{X}$, the square sum of all elements of $\boldsymbol{X}$, i.e. a matrix norm, can be computed by the trace of $\boldsymbol{X}\boldsymbol{X}^\top$ (or $\boldsymbol{X}^\top\boldsymbol{X}$).

A.4.2 Singular Matrix

All matrices can be classified into the following two types:

Singular matrix: We call a matrix, which does not have any inverse matrices, a *singular matrix.*

For singular matrix $\boldsymbol{X}$, there exists $\boldsymbol{x}$ ($\neq \boldsymbol{0}$) which satisfies $\boldsymbol{X}\boldsymbol{x} = \boldsymbol{0}$.

Nonsingular matrix: We call a matrix, which has an inverse matrix, a *nonsingular matrix.*

Also we call this matrix in another way, such as a regular matrix and an invertible matrix.

Nonsingular matrix $\boldsymbol{X}$ satisfies the following properties:

Rank: A nonsingular matrix is a full rank matrix.

Eigenvalues: In fact eigenvalues are treated in the subsequent section, while the following proposition holds for eigenvalues of a nonsingular matrix:

Proposition A.1 (Invertible matrix theorem). ***X*** *is a nonsingular matrix* $\Leftrightarrow$ *Any singular value (eigenvalue) of* ***X*** *cannot be zero.*

Proof.

$\Rightarrow$

Eigenvalue decomposition (not defined yet though) is given in (A.55). If $\boldsymbol{X}$ is a nonsingular matrix, there exists an inverse matrix of $\boldsymbol{X}$. We can multiply the inverse matrix of $\boldsymbol{X}$ from the left-hand side of (A.55), then the left-hand side of (A.55) is $\boldsymbol{a}$. On the other hand, if an eigenvalue (of the right-hand side) can be zero, the right-hand side becomes zero, which contradicts the left-hand side, which is $\boldsymbol{a}$. Thus an eigenvalue of a non-singular matrix cannot be zero.

$\Leftarrow$

If an eigenvalue becomes zero, from (A.55),

$$\boldsymbol{X}\boldsymbol{a} = \lambda\boldsymbol{a} = \boldsymbol{0} \tag{A.54}$$

This means that $\boldsymbol{X}$ cannot be a non-singular matrix and $\boldsymbol{X}$ must be a singular matrix. That is, if an eigenvalue cannot be zero, $\boldsymbol{X}$ can be a nonsingular matrix. □

Zero vector: $\boldsymbol{X}\boldsymbol{x} = \boldsymbol{0}$ holds only if $\boldsymbol{x} = \boldsymbol{0}$.

A.4.3 Eigenvalue Problem

Suppose that $\boldsymbol{X}$ is a square matrix ($N = M$), λ and $\boldsymbol{a}$, which satisfy

$$\boldsymbol{X}\boldsymbol{a} = \lambda\boldsymbol{a}. \tag{A.55}$$

are called an eigenvalue and an eigenvector of $\boldsymbol{X}$.

For multiple eigenvalues and eigenvectors,

$$\boldsymbol{X}\begin{pmatrix}\boldsymbol{a}_1 \cdots \boldsymbol{a}_N\end{pmatrix} = \begin{pmatrix}\boldsymbol{a}_1 \cdots \boldsymbol{a}_N\end{pmatrix}\begin{pmatrix}\lambda_1 & 0 & 0\\ 0 & \ddots & 0\\ 0 & 0 & \lambda_N\end{pmatrix}. \tag{A.56}$$

$$\boldsymbol{X}\boldsymbol{A} = \boldsymbol{A}\boldsymbol{\Lambda}, \tag{A.57}$$

where $\boldsymbol{\Lambda}$ is a diagonal matrix, where the diagonal elements of $\boldsymbol{\Lambda}$ are eigenvalues. That is, given square matrix $\boldsymbol{X}$, we call the problem of obtaining matrix Λ and eigenvector matrix $\boldsymbol{A}$, which satisfy (A.57), an *eigenvalue problem.*

Eigenvector matrix $\boldsymbol{A}$ must be an orthogonal matrix. That is, multiple eigenvectors $\boldsymbol{a}_1, \ldots, \boldsymbol{a}_N$ are orthogonal each other.

If eigenvector matrix $\boldsymbol{A}$ is a non-singular matrix, $\boldsymbol{X}$ can be written as follows:

$$\boldsymbol{X} = \boldsymbol{A}\boldsymbol{\Lambda}\boldsymbol{A}^{-1}. \tag{A.58}$$

This is called *eigenvalue decomposition.*

A.4.4 Generalized Eigenvalue Problem

Given two square matrices $\boldsymbol{X}$ and $\boldsymbol{Y}$, we consider scalar λ and vector $\boldsymbol{a}$, which satisfy the following:

$$\boldsymbol{X}\boldsymbol{a} = \lambda \boldsymbol{Y}\boldsymbol{a}. \tag{A.59}$$

Then we think about multiple pairs of λ and $\boldsymbol{a}$ and obtain the following:

$$\boldsymbol{X}\boldsymbol{A} = \boldsymbol{Y}\boldsymbol{A}\boldsymbol{\Lambda}. \tag{A.60}$$

We call the problem of obtaining eigenvector matrix $\boldsymbol{A}$ and diagonal matrix $\boldsymbol{\Lambda}$ with eigenvalues for diagonals in (A.60), a *generalized eigenvalue problem.*

This problem can be reduced to a standard eigenvalue problem, if $\boldsymbol{Y}$ is a non-singular matrix, by multiplying the inverse matrix of $\boldsymbol{Y}$ from the left-hand side of (A.60), as follows:

$$(\boldsymbol{Y}^{-1}\boldsymbol{X})\boldsymbol{A} = \boldsymbol{A}\boldsymbol{\Lambda}. \tag{A.61}$$

Also if $\boldsymbol{Y}$ is a positive semidefinite matrix (See Section A.4.6), similarly the generalized eigenvalue problem can be solved by a standard eigenvalue problem, shown in (A.63). That is, if $\boldsymbol{Y}$ is a positive semidefinite matrix, by using Cholesky decomposition (See **Proposition A.5**):

$$\boldsymbol{Y} = \boldsymbol{L}\boldsymbol{L}^{\top}. \tag{A.62}$$

(A.60) can be the following eigenvalue problem.

$$(\boldsymbol{L}^{-1}\boldsymbol{X}\boldsymbol{L}^{-\top})\boldsymbol{A} = \boldsymbol{A}\boldsymbol{\Lambda}. \tag{A.63}$$

A.4.5 Singular Value Decomposition (SVD)

Singular value decomposition is a generalization of eigenvalue decomposition. In the above generalized eigenvalue decomposition, $\boldsymbol{A}$ must be non-singular matrix, and $\boldsymbol{X}$ must be a square matrix. On the other hand, given $\boldsymbol{X}$ with arbitrary numbers of rows (and columns), singular value decomposition is given as follows:

$$\boldsymbol{X} = \boldsymbol{U}\boldsymbol{\Sigma}\boldsymbol{V}^{\top}, \tag{A.64}$$

where $\boldsymbol{U}$ and $\boldsymbol{V}$ are square matrices, i.e. $(N \times N)$- and $(M \times M)$-matrices, respectively. Thus Σ is a $(N \times M)$- orthogonal matrix. Also Σ is a diagonal matrix.

We can obtain $\boldsymbol{U}$ and $\boldsymbol{V}$ by thinking about the following two derivations for $\boldsymbol{X}\boldsymbol{X}^\top$ and $\boldsymbol{X}^\top\boldsymbol{X}$, respectively:

$$\begin{aligned} \boldsymbol{X}\boldsymbol{X}^\top &= \boldsymbol{U}\boldsymbol{\Sigma}\boldsymbol{V}^\top(\boldsymbol{U}\boldsymbol{\Sigma}\boldsymbol{V}^\top)^\top = \boldsymbol{U}\boldsymbol{\Sigma}\boldsymbol{V}^\top\boldsymbol{V}\boldsymbol{\Sigma}^\top\boldsymbol{U}^\top && \text{(A.65)} \\ &= \boldsymbol{U}\boldsymbol{\Sigma}\boldsymbol{\Sigma}^\top\boldsymbol{U}^\top = \boldsymbol{U}\boldsymbol{\Sigma}^2\boldsymbol{U}^\top. && \text{(A.66)} \end{aligned}$$

$$\begin{aligned} \boldsymbol{X}^\top\boldsymbol{X} &= (\boldsymbol{U}\boldsymbol{\Sigma}\boldsymbol{V}^\top)^\top\boldsymbol{U}\boldsymbol{\Sigma}\boldsymbol{V}^\top = \boldsymbol{V}\boldsymbol{\Sigma}^\top\boldsymbol{U}^\top\boldsymbol{U}\boldsymbol{\Sigma}\boldsymbol{V}^\top && \text{(A.67)} \\ &= \boldsymbol{V}\boldsymbol{\Sigma}^\top\boldsymbol{\Sigma}\boldsymbol{V}^\top = \boldsymbol{V}\boldsymbol{\Sigma}^2\boldsymbol{V}^\top. && \text{(A.68)} \end{aligned}$$

From the above derivations, $\boldsymbol{U}$ and $\boldsymbol{V}$ are orthogonal matrices, which are eigenvector matrices obtained by eigenvalue decomposition of $\boldsymbol{X}\boldsymbol{X}^\top$ and $\boldsymbol{X}^\top\boldsymbol{X}$, respectively. $\boldsymbol{\Sigma}$ are eigenvalues obtained and shared by their eigenvalue decomposition.

A.4.6 Positive Semidefinite Matrix

Given arbitrary non-zero vector $\boldsymbol{a}$, if $\boldsymbol{a}^\top\boldsymbol{X}\boldsymbol{a} > 0$, we call $\boldsymbol{X}$ a *positive definite matrix*. Also if $\boldsymbol{a}^\top\boldsymbol{X}\boldsymbol{a} \geq 0$, we call $\boldsymbol{X}$ a *positive semidefinite matrix.*

Proposition A.2. *$\boldsymbol{X}$ is a positive semidefinite matrix $\Leftrightarrow$ eigenvalues of $\boldsymbol{X}$ are nonnegatives, where $\boldsymbol{X}$ is a square matrix ($N = M$).*

Proof.
$\Rightarrow$

$\boldsymbol{X}$ is a positive semidefinite matrix, and so for arbitrary vector $\boldsymbol{a}$, $\boldsymbol{a}^\top\boldsymbol{X}\boldsymbol{a} \geq 0$. Also from the eigenvalue decomposition of $\boldsymbol{X}$, we obtain $\boldsymbol{X} = \boldsymbol{V}\boldsymbol{\Lambda}\boldsymbol{V}^\top$.

Thus $\boldsymbol{a}^\top\boldsymbol{V}\boldsymbol{\Lambda}\boldsymbol{V}^\top\boldsymbol{a} \geq 0$ must hold. Letting $\boldsymbol{b} = \boldsymbol{V}^\top\boldsymbol{a}$, $\boldsymbol{b} = (b_1, \ldots, b_N)^\top$ and $\mathrm{diag}(\boldsymbol{\Lambda}) = (\lambda_1, \ldots, \lambda_N)^\top$, we can have the following derivation:

$$\boldsymbol{a}^\top\boldsymbol{V}\boldsymbol{\Lambda}\boldsymbol{V}^\top\boldsymbol{a} = (\boldsymbol{V}^\top\boldsymbol{a})^\top\boldsymbol{\Lambda}\boldsymbol{V}^\top\boldsymbol{a} = \boldsymbol{b}^\top\boldsymbol{\Lambda}\boldsymbol{b} = \sum_{i=1}^{N}\lambda_i b_i^2. \qquad \text{(A.69)}$$

Thus (A.69) needs be nonnegative for arbitrary b_i ($i = 1, \ldots, N$), and to satisfy this, all λ_i ($i = 1, \ldots, N$) needs to be nonnegative.
$\Leftarrow$

From eigenvalue decomposition, $\boldsymbol{X} = \boldsymbol{V}\boldsymbol{\Lambda}\boldsymbol{V}^\top$. Also since the eigenvalue is nonnegative, we can have the following derivation:

$$\begin{aligned} \boldsymbol{a}^\top\boldsymbol{X}\boldsymbol{a} &= \boldsymbol{a}^\top\boldsymbol{V}\boldsymbol{\Lambda}\boldsymbol{V}^\top\boldsymbol{a} = (\boldsymbol{V}^\top\boldsymbol{a})^\top\sqrt{\boldsymbol{\Lambda}}^\top\sqrt{\boldsymbol{\Lambda}}\boldsymbol{V}^\top\boldsymbol{a} && \text{(A.70)} \\ &= ||\sqrt{\boldsymbol{\Lambda}}\boldsymbol{V}^\top\boldsymbol{a}||^2 \geq 0. && \text{(A.71)} \end{aligned}$$

Thus $\boldsymbol{X}$ is a positive semidefinite matrix. □

Based on this proposition, we can have more properties on positive semidefinite matrices.

Proposition A.3. *Using* **Proposition A.1** *also, positive semidefinite matrix is always a nonsingular matrix.*

Proposition A.4. *The inverse matrix of a positive semidefinite matrix is also a positive semidefinite matrix.*

Proof.

$$\boldsymbol{X}\boldsymbol{a} = \lambda \boldsymbol{a} \quad \rightarrow \quad \boldsymbol{X}^{-1}\boldsymbol{a} = \frac{1}{\lambda}\boldsymbol{a}. \tag{A.72}$$

Since $\lambda > 0$, the eigenvalue is nonzero, and the inverse matrix is also a positive semidefinite matrix. □

Proposition A.5. *$\boldsymbol{X}$ is a positive semidefinite matrix $\Leftrightarrow \boldsymbol{X} = \boldsymbol{A}^\top \boldsymbol{A}$.*

Proof.
$\Rightarrow$

Any eigenvalue of a positive semidefinite matrix is nonnegative, and so for the eigenvalue decomposition of $\boldsymbol{X}$, $\boldsymbol{X} = \boldsymbol{V}\boldsymbol{\Lambda}\boldsymbol{V}^\top$, we can have $\sqrt{\boldsymbol{\Lambda}}$ as a matrix with numerical values. Letting $\boldsymbol{A} = \sqrt{\boldsymbol{\Lambda}}\boldsymbol{V}^\top$,

$$\boldsymbol{A}^\top \boldsymbol{A} = \boldsymbol{V}\sqrt{\boldsymbol{\Lambda}}\sqrt{\boldsymbol{\Lambda}}\boldsymbol{V}^\top = \boldsymbol{V}\boldsymbol{\Lambda}\boldsymbol{V}^\top = \boldsymbol{X}. \tag{A.73}$$

$\Leftarrow$

For an arbitrary vector $\boldsymbol{a}$,

$$\boldsymbol{a}^\top \boldsymbol{X}\boldsymbol{a} = \boldsymbol{a}^\top \boldsymbol{A}^\top \boldsymbol{A}\boldsymbol{a} = (\boldsymbol{A}\boldsymbol{a})^\top \boldsymbol{A}\boldsymbol{a} = ||\boldsymbol{A}\boldsymbol{a}||^2 \geq 0. \tag{A.74}$$

This decomposition is called *Cholesky decomposition.* □

A.5 Kernel Function

A.5.1 Inner Product Space

Inner product space is defined as the space, which satisfies the following three properties:

1. **Symmetry**
2. **Linearity**
3. **Positive semidefiniteness**

Note that these three properties are also the conditions for kernel functions in the *reproducing kernel Hilbert space* (RKHS), and eventually a kernel matrix.

A.5.2 Definition of Kernel Functions

For two input numerical vectors $\boldsymbol{x}_1, \boldsymbol{x}_2 \in \mathcal{X}^N$, kernel function $\kappa(\boldsymbol{x}_1, \boldsymbol{x}_2)$ can be defined as function $\kappa : \mathcal{X}^N \times \mathcal{X}^N \to R$, as follows[1]:

$$\kappa(\boldsymbol{x}_1, \boldsymbol{x}_2) = \langle \phi(\boldsymbol{x}_1), \phi(\boldsymbol{x}_2) \rangle_{\mathcal{F}}, \tag{A.75}$$

where $\phi(\boldsymbol{x})$ projects $\boldsymbol{x}$ with the size of N into inner product space $\mathcal{F}$. Note that $\boldsymbol{x}_1$ and $\boldsymbol{x}_2$ are replacable each other, and so the kernel function has the symmetry property. Also the projection into the inner product space allows the kernel function to have the two other properties of the inner product space: linearity and positive semidefiniteness. These two points are shown later and also used for proving other properties of kernel functions (and further a kernel matrix).

Notes on Kernel Functions

Implicit project functions: The key advantage of a kernel function is that, in (A.75), we do not have to have project function $\phi(\boldsymbol{x})$ explicitly. Instead what we need is to define kernel function $\kappa(\boldsymbol{x}_1, \boldsymbol{x}_2)$ explicitly for any arbitrary two instances $\boldsymbol{x}_1$ and $\boldsymbol{x}_2$.

Positive semidefiniteness: A kernel function (and also a kernel matrix) must be positive semidefinite, which makes the kernel function having the same property as the inner product, and also brings a lot of benefits, especially in optimization. For example, we already have seen that a generalized eigenvalue problem can be reduced to a regular eigenvalue problem (as shown in (A.63)), by using positive semidefiniteness. As such, the positive semidefiniteness of a kernel function is useful for solving the optimization problem.

Kernel Matrix

Both rows and columns of a kernel matrix correspond to instances, and elements are the outputs of the kernel functions corresponding to the instances of rows and columns. Kernel functions are symmetric, and so a kernel matrix is symmetric.

Positive Semidefinite Property of Kernel Matrix

Kernel functions are defined as follows:

$$\kappa(\boldsymbol{x}_i, \boldsymbol{x}_j) = \langle \phi(\boldsymbol{x}_i), \phi(\boldsymbol{x}_j) \rangle_{\mathcal{F}}. \tag{A.76}$$

Thus, for kernel matrix $\boldsymbol{K}$,

$$\begin{aligned} \boldsymbol{a}^\mathsf{T} \boldsymbol{K} \boldsymbol{a} &= \sum_{i,j} a_i K_{i,j} a_j = \sum_{i,j} a_i \langle \phi(\boldsymbol{x}_i), \phi(\boldsymbol{x}_j) \rangle_{\mathcal{F}} a_j & \text{(A.77)} \\ &= \langle \sum_i a_i \phi(\boldsymbol{x}_i), \sum_j a_j \phi(\boldsymbol{x}_j) \rangle_{\mathcal{F}} = || \sum_i a_i \phi(\boldsymbol{x}_i) ||_{\mathcal{F}}^2 \geq 0. & \text{(A.78)} \end{aligned}$$

[1]Strictly speaking kernel functions can be defined in reproducing kernel Hilbert space (RKHS), which has more properties than the inner product space, while for simplicity, we define kernel functions with the three properties in the inner product space.

That is, kernel matrices are always positive semidefinite matrices. This means that to use a matrix as a kernel matrix, the matrix has to satisfy the positive semidefinite property.

A.5.3 Properties of Kernel Functions

We raise useful properties of kernel functions to generate new kernel functions from existing kernel functions.

Addition

Adding a kernel function to another kernel function results in a kernel function:

$$K(\boldsymbol{x}, \boldsymbol{x}') = K_1(\boldsymbol{x}, \boldsymbol{x}') + K_2(\boldsymbol{x}, \boldsymbol{x}'), \tag{A.79}$$

where K_1 and K_2 keep the positive semidefinite properties, and so the followings are satisfied: $\boldsymbol{u}^\top K_1 \boldsymbol{u} \geq 0$, $\boldsymbol{u}^\top K_2 \boldsymbol{u} \geq 0$.

That is, $\boldsymbol{u}^\top K \boldsymbol{u} = \boldsymbol{u}^\top (K_1 + K_2)\boldsymbol{u} \geq 0$.

Multiplication (by a number)

A function, obtained by multiplying a kernel function by some positive, numerical value, is also a kernel function.

$$K(\boldsymbol{x}, \boldsymbol{x}') = aK_1(\boldsymbol{x}, \boldsymbol{x}'), a \in \mathcal{R}^+. \tag{A.80}$$

$\boldsymbol{u}^\top K_1 \boldsymbol{u} \geq 0$ and $a\boldsymbol{u}^\top K_1 \boldsymbol{u} \geq 0$.

Also from the above two properties, we can see that kernel functions have the linearity property.

Multiplication (Product of Two Kernels)

The product of two kernel functions is also a kernel function.

$$K(\boldsymbol{x}, \boldsymbol{x}') = K_1(\boldsymbol{x}, \boldsymbol{x}')K_2(\boldsymbol{x}, \boldsymbol{x}'). \tag{A.81}$$

Product of Two Functions

The product of two functions below is a kernel function.

$$K(\boldsymbol{x}, \boldsymbol{x}') = f(\boldsymbol{x})f(\boldsymbol{x}'). \tag{A.82}$$

More generally, letting $\psi : \boldsymbol{x} \to f(\boldsymbol{x})$, $K(\boldsymbol{x}, \boldsymbol{x}') = \langle \psi(\boldsymbol{x}), \psi(\boldsymbol{x}') \rangle$.

Kernel with Interaction Term

For symmetric, positive semidefinite matrix $\boldsymbol{A}$, the following function can be a kernel function.

$$K(\boldsymbol{x}, \boldsymbol{x}') = \boldsymbol{x}^\top \boldsymbol{A} \boldsymbol{x}'. \tag{A.83}$$

$\boldsymbol{A}$ is a positive semidefinite kernel, and then $\boldsymbol{A} = \boldsymbol{B}^\top \boldsymbol{B}$. $\boldsymbol{x}^\top \boldsymbol{A} \boldsymbol{x} = \boldsymbol{x}^\top \boldsymbol{B}^\top \boldsymbol{B} \boldsymbol{x} = (\boldsymbol{B}\boldsymbol{x})^2 \geq 0$. Thus, $K(\boldsymbol{x}, \boldsymbol{x}')$ is a positive semidefinite matrix and kernel function.

A.5.4 Examples of Kernel Functions

As explained in the main part of this book, building proper kernel functions depending on applications is important in kernel learning. However, for cases without any good knowledge on applications, we use standard kernel functions, which will be shown below:

Linear Kernel

Linear kernel is equal to the inner product:

$$\kappa(\boldsymbol{x}_1, \boldsymbol{x}_2) = \boldsymbol{x}_1^\top \boldsymbol{x}_2. \tag{A.84}$$

Polynomial Kernel

Polynomial kernel is defined below:

$$\kappa(\boldsymbol{x}_1, \boldsymbol{x}_2) = (\boldsymbol{x}_1^\top \boldsymbol{x}_2 + c)^d, \tag{A.85}$$

where c and d are hyperparameters. As d increases, the kernel space increases and also the computational cost increase.

Gaussian Kernel

Gaussian kernel (or called radial basis function kernel as well) is defined as follows:

$$\kappa(\boldsymbol{x}_1, \boldsymbol{x}_2) = \exp\left(-\frac{||\boldsymbol{x}_1 - \boldsymbol{x}_2||^2}{2\sigma^2}\right), \tag{A.86}$$

where σ is a hyperparameter.

Sigmoid Kernel

Sigmoid kernel can be defined as follows:

$$\kappa(\boldsymbol{x}_1, \boldsymbol{x}_2) = \tanh(c\boldsymbol{x}_1^\top \boldsymbol{x}_2 + \theta), \tag{A.87}$$

where c and θ are hyperparameters.

A.6 Norm

Norm is a function, which gives nonnegative values to vectors (not only vectors and also matrices, etc.) in a vector space. These values represent the size, length or distance of a vector in the space, and in this sense, norm defines the vector space. Thus for the vector space V^M (size: M), norm f is defined as $f : V^M \to \mathbb{R}$. In machine learning, the vector space defined by norm can be used as a space of parameter values which should be optimized in training, and so norm defines the possible space or range, which parameter values can take.

Norm $\boldsymbol{u} \in V^M$ and $\boldsymbol{v} \in V^M$ need satisfy the following three conditions:

Zero vector property: If $f(\boldsymbol{u}) = 0$, $\boldsymbol{u}$ must be a zero vector.

Absolutely homogeneous (scalable) property: $f(a\boldsymbol{u}) = |a| f(\boldsymbol{u})$.

Triangle equality or subadditive property: $f(\boldsymbol{u} + \boldsymbol{v}) \leq f(\boldsymbol{u}) + f(\boldsymbol{v})$.

A.6.1 Vectors

We raise the following L^p norm for the vector space.

L^P norm

L^p norm can be divided into three cases: $1 \leq p$, $0 < p < 1$ and $p = 0$, according to the value of p.

$1 \leq p$: For vector space V^M, we can consider vector $\boldsymbol{x} = \{x_1, x_2, ..., x_M\}$ and $\boldsymbol{x} \in V^M$, and L^p norm of vector $\boldsymbol{x}$ can be defined as follows:

$$||\boldsymbol{x}||_p := (|x_1|^p + |x_2|^p + \cdots + |x_M|^p)^{\frac{1}{p}}. \tag{A.88}$$

We raise special cases below:

$p = \infty$: By taking $p \to \infty$ of (A.88),

$$||\boldsymbol{x}||_\infty := \max\{|x_1|, |x_2|, \ldots, |x_M|\}. \tag{A.89}$$

This is called *infinity norm* or *maximum norm.*

$p = 2$:

$$||\boldsymbol{x}||_2 := (|x_1|^2 + |x_2|^2 + \cdots + |x_M|^2)^{\frac{1}{2}}. \tag{A.90}$$

This is called *Euclidean norm* or L^2 *norm.* The defined space is the Euclidean space.

$p = 1$:

$$||\boldsymbol{x}||_1 := |x_1| + |x_2| + \cdots + |x_M| = \sum_{i=1}^{M} |x_i|. \tag{A.91}$$

This is called *Manhattan norm* or L^1 *norm.*

$0 < p < 1$: In this range of p, (A.88) cannot satisfy the triangle inequality. So the following is used as L^p norm:

$$||\boldsymbol{x}||_p := |x_1|^p + |x_2|^p + \cdots + |x_M|^p. \tag{A.92}$$

This satisfies the triangle inequality, while the absolutely homogeneous property is not satisfied. Thus strictly speaking, this cannot be a norm, but for convenience, the above is used as a norm.

p = 0: In (A.92), by letting $p \to 0$, we can obtain the norm below:

$$||\boldsymbol{x}||_0 := |x_1|^0 + |x_2|^0 + \cdots + |x_M|^0, \tag{A.93}$$

where in machine learning, we define that $|x_i|^0$ is 1 if $|x_i|$ is nonzero; otherwise zero.

This definition means that $||\boldsymbol{x}||_0$ is the number of nonzero elements. We then call this norm L^0 *norm*, while again the absolutely homogeneous property is not satisfied, meaning that this is not a norm strictly but used in machine learning to capture the number of nonzero elements in a vector for convenience.

A.6.2 Matrix

For matrices, L^p norm (for vectors) cannot be defined in only one way. Below we introduce three rather reasonable extensions from L^p norm for vectors to the norms for matrices. Let $V^{N\times M}$ be a vector space for matrices, and x_{ij} be the (i,j)-element of $\boldsymbol{X} \in V^{N\times M}$.

Element-wise Matrix Norm

Element-wise matrix norm is defined as follows:

$$||\boldsymbol{X}||_p := (\sum_{i=1}^{N}\sum_{j=1}^{M} |x_{ij}|^p)^{\frac{1}{p}}. \tag{A.94}$$

We can examine several special cases below:

$p \to \infty$ Taking $p \to \infty$ in (A.94), we can have the following:

$$||\boldsymbol{X}||_\infty := \max_{i,j} |x_{ij}|. \tag{A.95}$$

This is called *max norm*.

$p = 2$

$$||\boldsymbol{X}||_2 := (\sum_{i=1}^{N}\sum_{j=1}^{M} |x_{ij}|^2)^{\frac{1}{2}}. \tag{A.96}$$

This is called *Frobenius norm* or *Hilbert-Schmidt norm*. This is the most standard norm used in machine learning, and so often written as Frobenius norm in the following way:

$$||\boldsymbol{X}||_F := (\sum_{i=1}^{N}\sum_{j=1}^{M} |x_{ij}|^2)^{\frac{1}{2}}. \tag{A.97}$$

We are raising three different types of matrix norms, while these norms are all Frobenius norm for $p = 2$.

Also when we just say "a matrix norm", which usually indicate the square of the Frobenius norm, as follows:

$$||\boldsymbol{X}||_2^2 := \sum_{i=1}^{N}\sum_{j=1}^{M}|x_{ij}|^2. \tag{A.98}$$

That is, this norm is equal to the squared sum of all elements of a given matrix.

$\underline{p=1}$

$$||\boldsymbol{X}||_1 := (\sum_{i=1}^{N}\sum_{j=1}^{M}|x_{ij}|). \tag{A.99}$$

Schatten p-norm

The Schatten p-norm is defined as follows:

$$||X||_p \quad = \quad (\sum_{i}^{\min\{N,M\}} \boldsymbol{s}_i(\boldsymbol{X})^p)^{\frac{1}{p}}, \tag{A.100}$$

where $\boldsymbol{s}(\boldsymbol{X})_i$ is a singular value of $\boldsymbol{X}$, and an eigenvalue of $|\boldsymbol{X}| := \sqrt{\boldsymbol{X}^\top \boldsymbol{X}}$.

Again we can examine several special cases, below:

$\underline{p=2}$**:** This is equal to the Frobenius norm.

$\underline{p=1}$**:**

$$||X||_p \quad = \quad \sum_{i}^{\min\{N,M\}} \boldsymbol{s}_i(\boldsymbol{X}). \tag{A.101}$$

This is equal to the nuclear norm and the trace norm.

Vector Induced Matrix Norm

We can just examine the special cases only:

$\underline{p \to \infty}$

$$||\boldsymbol{X}||_\infty = \max_{1\le i\le N} \sum_{i=1}^{M}|x_{ij}|. \tag{A.102}$$

$\underline{p=2}$ This is equal to the Frobenius norm.

$\underline{p=1}$

$$||\boldsymbol{X}||_1 = \max_{1\le j\le N} \sum_{i=1}^{N}|x_{ij}|. \tag{A.103}$$

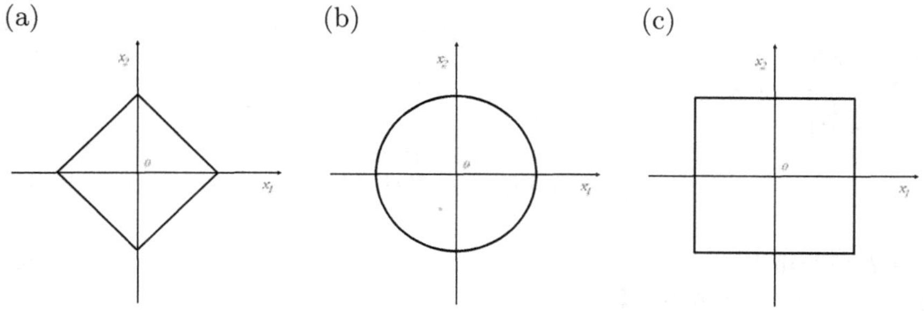

Figure A.1: L^p norm: (a) $p = 1$, (b) $p = 2$ and (c) $p \to \infty$.

The norm of $p \to \infty$ and that of $p = 1$ have very similar forms. The difference between them is that rows are used for $p \to \infty$ while columns are used for $p = 1$.

Also again, all three norms have the Frobenius norm for $p = 2$.

A.6.3 Computation of Matrix Norm

Computing norms for vectors is rather easy. We here raise several points on computing norms for matrices.

As we mentioned in the earlier section, the standard norm in machine learning can be given as follows:

$$||\boldsymbol{X}||_2^2 := \sum_{i=1}^{N}\sum_{j=1}^{M} |x_{ij}|^2. \tag{A.104}$$

By using the trace, this can be written as follows:

$$||\boldsymbol{X}||_2^2 := \mathrm{trace}(\boldsymbol{X}\boldsymbol{X}^\top) = \mathrm{trace}(\boldsymbol{X}^\top\boldsymbol{X}). \tag{A.105}$$

This means that the partial derivative of the matrix norm can be written as follows:

$$\frac{\partial ||\boldsymbol{X}||_2^2}{\partial \boldsymbol{X}} = \boldsymbol{X}^\top. \tag{A.106}$$

or

$$\frac{\partial ||\boldsymbol{X}||_2^2}{\partial \boldsymbol{X}} = \boldsymbol{X}. \tag{A.107}$$

A.6.4 Regularization – Constraints by Norm

As shown in (3.96), given matrix $\boldsymbol{X}$ and label vector $\boldsymbol{y}$, linear regression can be formulated to optimize regression coefficient $\boldsymbol{\beta}$, as follows:

$$\begin{aligned} \boldsymbol{y} &= h(\boldsymbol{X}) && \text{(A.108)} \\ &= \boldsymbol{X}\boldsymbol{\beta} + \epsilon. && \text{(A.109)} \end{aligned}$$

In this formulation, $\boldsymbol{\beta}$ can be any numerical vector, with high flexibility, and $\boldsymbol{\beta}$ might overfit given $\boldsymbol{X}$ and $\boldsymbol{y}$. Thus we need use a constraint for $\boldsymbol{\beta}$ to reduce the high flexibility of $\boldsymbol{\beta}$. In fact, as shown in (3.100), we usually introduce some norm as the constraint term for $\boldsymbol{\beta}$, as follows:

$$\hat{\boldsymbol{\beta}} = \arg\min_{\boldsymbol{\beta}}(||\boldsymbol{y} - \boldsymbol{X}\boldsymbol{\beta}||^2 + \lambda||\boldsymbol{\beta}||_2^2), \tag{A.110}$$

where L^2 norm is used, and $\boldsymbol{\beta}$ is optimized to minimize the objective function, keeping the constraint by the norm.

We call this type of introduction of the constraint term over the parameters to be optimized *regularization*. Also the constraint term is called a *regularization term* or *regularizer*. This parameter regularization is widely used in machine learning to avoid overfitting of parameters to be estimated. Another advantage of the regularization term is that this term can relax the regular matrix assumption in the derivation process of optimization (See Section 3.2.5).

Suppose that we have only two features (x_1,x_2), Fig. A.1 shows possible values taken by the two features, due to the structure of the norm. As shown in (b), L^2 ($p = 2$ for L^p norm) shows a circle for possible values on the two-dimensional space, where the solution is the contact between the linear regression and this circle. The point is in the circle, the contact point has no bias, and any place in the circle can be the contact point equally. On the other hand, as shown in (a), L^1 norm is a linear line connecting the axis of x_1 and the axis of x_2. The contact of this line and the linear regression is only on the axis of x_1 (or x_2), meaning that either of x_1 or x_2 is zero. This implies L^1 norm is likely to take a sparse solution always.

A.7 Nodes in a Graph

Graphs are explained in Section 2.3.5, and nodes in a graph are explained in Section 2.3.6 well. Again when we think about nodes in a graph as instances, nodes are all unique. Thus a graph is always equivalent to an adjacency matrix. An adjacency matrix is a symmetric matrix in which nodes are for both rows and columns. Elements in the adjacency matrix have two cases: 1) binary values showing the existence of edges and 2) weights being placed on edges, showing the significance of the edges.

A.7.1 Graph Laplacian

Definition

Let $\boldsymbol{W}_{ij}$ be the (i, j) element of an adjacency matrix, and $d_i = \sum_j \boldsymbol{W}_{ij}$ be the number of edges (degree) of node i. Let $\boldsymbol{D} = \text{diag}(d_1, \ldots, d_N)$ be a diagonal matrix, where d_i is the degree of node i.

Graph Laplacian can be given as follows:

$$\boldsymbol{L} = \boldsymbol{D} - \boldsymbol{W}. \tag{A.111}$$

Thus, if $\boldsymbol{W}$ is a binary matrix, the diagonal of $\boldsymbol{L}$ is the degree of the corresponding node, and nondiagonal of $\boldsymbol{L}$ is -1 if there is an edge in the two corresponding nodes; otherwise zero.

Normalized graph Laplacian is obtained by normalizing the graph Laplacian by its diagonal matrix, as follows:

$$\bar{\boldsymbol{L}} = \boldsymbol{D}^{-\frac{1}{2}}\boldsymbol{L}\boldsymbol{D}^{-\frac{1}{2}}. \tag{A.112}$$

More in detail, the element of $\bar{\boldsymbol{L}}$ is given as follows (assuming $\boldsymbol{W}$ is a binary adjacency matrix).

$$\bar{\boldsymbol{L}}_{ij} = \begin{cases} 1 & \text{if} \quad i = j, \\ -\frac{1}{\sqrt{d_i d_j}} & \text{if} \quad i \quad \text{and} \quad j \quad \text{are adjacent}, \\ 0 & \text{otherwise}. \end{cases} \tag{A.113}$$

Graph Laplacian is used to measure the similarity between each node and neighboring nodes.

Clustering into Two Clusters

We now consider the problem of clustering given nodes in a graph into two clusters. Let $\boldsymbol{z} = (z_1, \ldots, z_N)^\top$ be a binary vector ($z_i \in \{-1, 1\}$ for all i), indicating the cluster (two clusters are indicated by -1 or 1), in which each node i is.

We place the objective function F to be minimized for this problem, as follows:

$$F = \sum_{i,j} \boldsymbol{W}_{ij}||z_i - z_j||^2. \tag{A.114}$$

We then consider to minimize F, where z_i and z_j should have the same value if node i and node j are connected, i.e. $\boldsymbol{W}_{ij} = 1$. This means that the value of $\boldsymbol{z}$ should be consistent with the edge connectivity of the given graph, i.e. adjacency matrix $\boldsymbol{W}$.

We can modify (A.114) into the following:

$$F = \boldsymbol{z}^\top(\boldsymbol{D} - \boldsymbol{W})\boldsymbol{z} = \boldsymbol{z}^\top\boldsymbol{L}\boldsymbol{z}. \tag{A.115}$$

As explained in the above, minimizing F is equivalent to obtaining binary vector $\boldsymbol{z}$ consistent with adjacency matrix $\boldsymbol{W}$. In other words, this corresponds to labeling binary values optimally consistent with edges of a given graph. Furthermore (A.115) indicates that graph Laplacian measures the consistency to adjacency matrix, or smoothness of the given graph.

Also from (A.114) and (A.115), for any arbitrary vector $\boldsymbol{z} = (z_1, \ldots, z_N)^\top$, the following holds:

$$\boldsymbol{z}^\top\boldsymbol{L}\boldsymbol{z} = \boldsymbol{z}^\top(\boldsymbol{D} - \boldsymbol{W})\boldsymbol{z} = \sum_{i,j} \boldsymbol{W}_{ij}||z_i - z_j||^2 \geq 0. \tag{A.116}$$

Thus graph Laplacian has the positive semidefinite property.

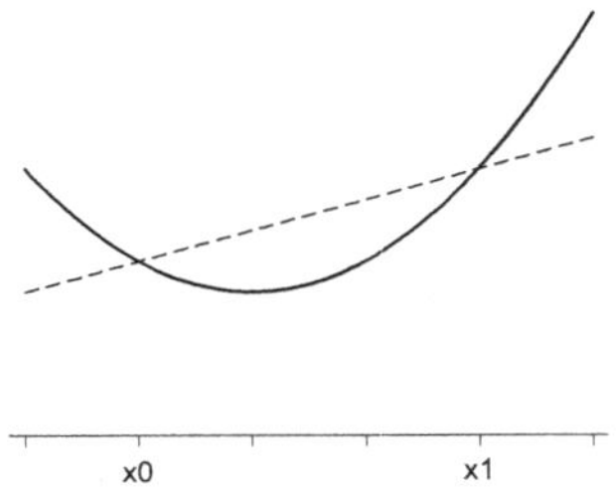

Figure A.2: A typical example of convex functions.

Real Applications

In real applications, a graph is *sparse*. That is, for a graph with N nodes, the possible number of edges reaches $\frac{N(N-1)}{2}$, while in reality the number of edges from one node is a very small number on average. For example, in a so-called citation graph of scientific papers, each node is an article and an edge indicates either of the connected articles is cited by the other, the number of citations per article, i.e. the degree of one node, is normally around ten to twenty.

Thus, a graph is very sparse, and so an adjacency matrix corresponding to a graph is also sparse, meaning that only a very small number of elements are nonzeros (and most of elements are zeros). At the same time, as you can see from (A.113), the number of nonzeros in graph Laplacian is totally the same as that of an adjacency matrix, meaning that graph Laplacian is also sparse. Furthermore, as shown in (A.72), the inverse matrix of graph Laplacian shares the same eigenvectors with graph Laplacian, meaning that the inverse matrix is also sparse. In addition to them, graph Laplacian and its inverse matrix are all positive semidefinite matrices and so rather easy to use in machine learning. For example, the inverse matrix of graph Laplacian is a positive semidefinite matrix, showing similarity between nodes, and can be used as a kernel matrix.

A.8 Optimization

A.8.1 Convex Optimization

Convex Function

A convex function $f(x)$ satisfies the following property in the range from x_0 to x_1 ($x_0 < x_1$).

$$f(tx_0 + (1-t)x_1) \leq tf(x_0) + (1-t)f(x_1). \tag{A.117}$$

where $t \in [0, 1]$.

Fig. A.2 is one example of a convex function, satisfying (A.117). In this figure, the dotted line (linear function) corresponds to the right-hand side of (A.117), while the left-hand side is the solid line, implying function f is always lower than the linear function.

Also from the figure, we can see that this property is always kept even if x_0 and x_1 are shifted. Overall the point of convex functions is that the local optimal point is consistent with the global optimum.

Convex Optimization

Convex optimization is to formulate the optimization problem as a convex function and take the advantage of the property that locally optimal solution becomes the globally optimum solution if the problem is convex. Thus in convex optimization, we obtain the global optimum by solving the optimization problem using a local optimization method.

A.8.2 Least Squares

The *least squares method* is, given data $\boldsymbol{X} = \{(\boldsymbol{x}_1, y_1), \ldots, (\boldsymbol{x}_N, y_N)\}$, to estimate function f to minimize the square sum of residual r_i over i. Taking the square sum means that the residual is assumed to follow the normal distribution.

$$\arg\min_f \sum_i^N r_i^2 = \arg\min_f \sum_i^N (y_i - f(\boldsymbol{x}_i)). \tag{A.118}$$

If function f is a linear function, this is the solution of linear regression. In statistical science, this is called *ordinary least squares* (OLS).

Also we can assume that the residual does not follow the normal distribution. Under this assumption, the linear model is called *generalized linear model.*

Least squares is the criterion of the optimization, and does not specify the optimization method.

A.8.3 Alternating Least Squares (ALS)

For some problem with two different types of parameters, when we fix one parameter, estimating the other of the two can be a convex problem. In this case, this problem is called a *biconvex* problem. One solution of this problem is to repeat estimating one type of parameter keeping the other parameters fixed (in other words, regarding them as a constant) alternately. In particular, when the problem is to minimize the least squares loss, this solution is called *alternating least squares* (ALS).

ALS is a high-level procedure for the biconvex problem. Each of the two alternate steps can have any proper optimization method.

A.8.4 Steepest Descent

Steepest descent, a typical optimization method, is a repetitive method to obtain a local optimal solution. Each step of the repetition takes the direction of the gradient for maximization (or the negative of the gradient for minimization). More mathematically, steepest descent obtains the local optimal solution of function $f(\boldsymbol{x}) = f(x_1, \ldots, x_M)$ with M variables $\boldsymbol{x} = (x_1, \ldots, x_M)^\mathsf{T}$. The direction of the gradient is given by $\nabla f(\boldsymbol{x})$. Thus a repetitive formula of the steepest descent updates the value of variable $\boldsymbol{x}$ (where $f(\boldsymbol{x}^{(0)})$ is the initial value, $t > 0$ and solving the minimization problem) as follows:

$$\boldsymbol{x}^{(t)} \leftarrow \boldsymbol{x}^{(t-1)} - \gamma_{t-1} \nabla f(\boldsymbol{x}^{(t-1)}). \tag{A.119}$$

That is,

$$\begin{pmatrix} x_1^{(t)} \\ \vdots \\ x_M^{(t)} \end{pmatrix} \leftarrow \begin{pmatrix} x_1^{(t-1)} \\ \vdots \\ x_M^{(t-1)} \end{pmatrix} - \gamma_{t-1} \begin{pmatrix} \frac{\partial f(\boldsymbol{x}^{(t-1)})}{\partial x_1^{(t-1)}} \\ \vdots \\ \frac{\partial f(\boldsymbol{x}^{(t-1)})}{\partial x_M^{(t-1)}} \end{pmatrix}. \tag{A.120}$$

This update achieves the following inequality:

$$f(\boldsymbol{x}^{(0)}) \geq f(\boldsymbol{x}^{(1)}) \geq f(\boldsymbol{x}^{(2)}) \ldots \tag{A.121}$$

That is, the convergence to a local minimum is guaranteed.

A.8.5 Method of Lagrange Multipliers

Method of Lagrange multipliers is a method to obtain a local optimal under the equality constraints. Mathematically, for M variables $x_1, \ldots, x_M$, method of Lagrange multipliers obtains a local optimal of $f(x_1, \ldots, x_M)$ under the constraints of $g(x_1, \ldots, x_M) = 0$. More generally for g functions, Lagrange function $L(x_1, \ldots, x_n)$ can be defined as follows:

$$L(x_1, \ldots, x_M) \quad = \quad f(x_1, \ldots, x_M) - \sum_{k=1}^{K} \lambda_k g_k(x_1, \ldots, x_M), \tag{A.122}$$

where λ_k is called a *Lagrange multiplier.*

When f takes a local optimal, the following equations should be satisfied:

$$\frac{\partial L(x_1, \ldots, x_M)}{\partial x_1} = \cdots = \frac{\partial L(x_1, \ldots, x_M)}{\partial x_M} = 0. \tag{A.123}$$

That is, the following equations should be satisfied, when f can be the local optimal.

$$\frac{\partial f(x_1, \ldots, x_M)}{\partial x_1} - \sum_{k=1}^{K} \lambda_k \frac{\partial g_k(x_1, \ldots, x_M)}{\partial x_1} \quad = \quad 0. \tag{A.124}$$

$$\vdots$$

$$\frac{\partial f(x_1, \ldots, x_M)}{\partial x_M} - \sum_{k=1}^{K} \lambda_k \frac{\partial g_k(x_1, \ldots, x_M)}{\partial x_M} \quad = \quad 0. \tag{A.125}$$

Letting $\boldsymbol{x} = (x_1, \dots, x_M)^\top$, the above equations can be summarized into the following:

$$\nabla_{\boldsymbol{x}} L(\boldsymbol{x}) = \nabla_{\boldsymbol{x}} f(\boldsymbol{x}) - \sum_{k=1}^{K} \lambda_k \nabla_{\boldsymbol{x}} g_k(\boldsymbol{x}) = 0. \tag{A.126}$$

Thus, we can obtain optimal values (of variables) which satisfy the above equation.

A.8.6 Karush-Kuhn-Tucker (KKT) conditions

Under both the equality and inequality constraints, the Karush-Kuhn-Tucker (KKT) conditions obtain an optimal value of function f. Note that the method of Lagrange multipliers can be used under the equality conditions only, while KKT conditions can be used under the both equality and inequality constraints.

Mathematically, for M variables, $x_1, \dots, x_M$, the KKT conditions obtain the local optimal of $f(x_1, \dots, x_M)$, under the constraints of $g_k(x_1, \dots, x_M) \leq 0, k = 1, \dots, K$ and $h_l(x_1, \dots, x_M) = 0, l = 1, \dots L$. That is, the optimization problem is as follows:

$$\min_{\boldsymbol{x}} f(\boldsymbol{x}) \tag{A.127}$$

$$\text{subject to} \quad g_k(\boldsymbol{x}) \leq 0 \ (k = 1, \dots, K) \tag{A.128}$$

$$\text{and} \quad h_l(\boldsymbol{x}) = 0 \ (l = 1, \dots, L). \tag{A.129}$$

This problem can be formulated as follows:

$$\min_{\boldsymbol{x}} f(\boldsymbol{x}) + \sum_{k}^{K} \alpha_k g_k(\boldsymbol{x}) + \sum_{l}^{L} \beta_l h_l(\boldsymbol{x}). \tag{A.130}$$

For solving this problem, the following four KKT conditions are given:

Stationary condition:

$$\nabla_{\boldsymbol{x}} f(\boldsymbol{x}) + \sum_{k}^{K} \alpha_k \nabla_{\boldsymbol{x}} g_k(\boldsymbol{x}) + \sum_{l}^{L} \beta_l \nabla_{\boldsymbol{x}} h_l(\boldsymbol{x}) = 0. \tag{A.131}$$

(for the maximization problem, the condition is below:)

$$\nabla_{\boldsymbol{x}} f(\boldsymbol{x}) - \sum_{k}^{K} \alpha_k \nabla_{\boldsymbol{x}} g_k(\boldsymbol{x}) + \sum_{l}^{L} \beta_l \nabla_{\boldsymbol{x}} h_l(\boldsymbol{x}) = 0. \tag{A.132}$$

Complementary slackness:

$$\alpha_k \cdot g_k(\boldsymbol{x}) = 0 \text{ for all } k. \tag{A.133}$$

Primal feasibility:

$$g_k(\boldsymbol{x}) \leq 0, \qquad h_l(\boldsymbol{x}) = 0 \text{ for all } k \text{ and } l. \tag{A.134}$$

Dual feasibility:

$$\alpha_k \geq 0 \text{ for all } k. \tag{A.135}$$

The solution of (A.130) needs to satisfy all four KKT conditions. Among the above four conditions, the first condition is the same as the condition of the method of Lagrange multipliers. The most important point is the second condition, in which either of α_k or $g(\boldsymbol{x})$ must be zero. This condition as well as the last condition leads $\alpha_k \geq 0$ and $g(\boldsymbol{x}) = 0$, which are very useful to obtain a local optimal solution.

A.8.7 Rayleigh Quotient

Let positive semidefinite matrices $\boldsymbol{A}$ and $\boldsymbol{B}$ be input (training) data, and let nonzero vector $\boldsymbol{z}$ be a parameter vector to be estimated. The numerator is that $\boldsymbol{A}$ is sandwiched by $\boldsymbol{z}$ and its transpose $\boldsymbol{z}^\top$, as follows:

$$\boldsymbol{z}^\top \boldsymbol{A}\boldsymbol{z}. \tag{A.136}$$

Also the denominator has a similar form, which has $\boldsymbol{B}$ instead of $\boldsymbol{A}$:

$$\boldsymbol{z}^\top \boldsymbol{B}\boldsymbol{z}. \tag{A.137}$$

The following relation with these two forms is called *generalized Rayleigh quotient*:

$$R(\boldsymbol{A}, \boldsymbol{B}, \boldsymbol{z}) = \frac{\boldsymbol{z}^\top \boldsymbol{A}\boldsymbol{z}}{\boldsymbol{z}^\top \boldsymbol{B}\boldsymbol{z}}. \tag{A.138}$$

Furthermore, when the denominator of (A.138) is simpler like in (A.139), this is called *Rayleigh quotient*:

$$R(\boldsymbol{A}, \boldsymbol{z}) = \frac{\boldsymbol{z}^\top \boldsymbol{A}\boldsymbol{z}}{\boldsymbol{z}^\top \boldsymbol{z}}. \tag{A.139}$$

Note that since $\boldsymbol{B}$ is a positive semidefinite matrix, generalized Rayleigh quotient can be always reduced to Rayleigh quotient, along with the following procedure. That is, we can decompose $\boldsymbol{B}$ as $\boldsymbol{B} = \boldsymbol{C}\boldsymbol{C}^\top$, and letting $\boldsymbol{x} = \boldsymbol{C}^\top \boldsymbol{z}$ and $\boldsymbol{D} = \boldsymbol{C}^{-1}\boldsymbol{A}(\boldsymbol{C}^\top)^{-1}$,

$$\frac{\boldsymbol{z}^\top \boldsymbol{A}\boldsymbol{z}}{\boldsymbol{z}^\top \boldsymbol{B}\boldsymbol{z}} = \frac{\boldsymbol{z}^\top \boldsymbol{A}\boldsymbol{z}}{\boldsymbol{z}^\top \boldsymbol{C}\boldsymbol{C}^\top \boldsymbol{z}} = \frac{(\boldsymbol{C}^\top \boldsymbol{z})^\top \boldsymbol{C}^{-1}\boldsymbol{A}(\boldsymbol{C}^\top)^{-1}\boldsymbol{C}^\top \boldsymbol{z}}{(\boldsymbol{C}^\top \boldsymbol{z})^\top \boldsymbol{C}^\top \boldsymbol{z}} = \frac{\boldsymbol{x}^\top \boldsymbol{D}\boldsymbol{x}}{\boldsymbol{x}^\top \boldsymbol{x}} (= R(\boldsymbol{D}, \boldsymbol{x})). \tag{A.140}$$

The problem of minimizing (or maximizing) Rayleigh quotient can be equivalent to that the numerator is the objective function and the denominator is a constraint. That is, the numerator is minimized (or maximized) under the denominator. Thus if the problem is a minimization problem, letting C be a constant, the problem can be the following minimization problem:

$$\min_{\boldsymbol{z}} \quad \boldsymbol{z}^\top \boldsymbol{D}\boldsymbol{z} \tag{A.141}$$

$$\text{subject to} \quad \boldsymbol{z}^\top \boldsymbol{z} = C. \tag{A.142}$$

This problem can be solved by the following method of Lagrange multipliers. Lagrange function:

$$L(\boldsymbol{z}) = \boldsymbol{z}^\top \boldsymbol{D}\boldsymbol{z} - \lambda(\boldsymbol{z}^\top \boldsymbol{z} - C). \tag{A.143}$$

By taking the derivatives of the both sides,

$$\frac{dL(\boldsymbol{z})}{d\boldsymbol{z}} \quad \Rightarrow \quad \boldsymbol{D}\boldsymbol{z} - \lambda\boldsymbol{z} \quad \Rightarrow \quad \boldsymbol{D}\boldsymbol{z} = \lambda\boldsymbol{z}. \tag{A.144}$$

The most right-hand side of (A.144) shows that the minimization or maximization problem given by Rayleigh quotient can be an eigenvalue problem. By solving this eigenvalue problem, we can obtain eigenvectors, which are equivalent to $\boldsymbol{z}$ eventually from input $\boldsymbol{A}$ and $\boldsymbol{B}$. That is, if an optimization problem is formulated as in the form of (general) Rayleigh quotient, this optimization problem can be solved as an eigenvalue problem[2].

A lot of methods in machine learning and multi-variate analysis can be formulated into Rayleigh quotient. For example, K-means (Section 3.1.2), principal component analysis (Section 3.3.2), spectral clustering (Section 8.1.1) and canonical correlation analysis (Section 9.1.2), etc. can be formulated into the form of Rayleigh quotient [104]. Furthermore, although this method is not introduced in this book, *Fisher discriminant analysis* also has the above property. Again the objective function and constraints of these problems can be the form of generalized eigenvalue problem. Their objective functions and constraints take the form of Rayleigh quotient (or generalized Rayleigh quotient), turning into an eigenvalue problem. However solving an eigenvalue problem needs high complexity, and so the eigenvalue problem is often solved by a heuristics.

A.8.8 Maximum Likelihood Estimation

Given data, the maximum likelihood (ML) estimator can be obtained so that the likelihood of the model over the given data is maximized, as follows:

$$\hat{\Phi} = \arg\max_{\Phi} P_{\Phi}(\boldsymbol{x}), \tag{A.145}$$

where Φ is model parameters, and $P_{\Phi}(\boldsymbol{x})$ is the likelihood of given data $\boldsymbol{x}$ by the model (with parameter Φ).

[2] If the given matrix is a positive semidefinite matrix, we do not have to consider a generalized eigenvalue problem. Instead we can transform the problem into a Rayleigh quotient, and the problem can be solved by a Rayleigh quotient.

A.8.9 Expectation-Maximization (EM) Algorithm

The Expectation-Maximization (EM) algorithm is a local optimization algorithm for maximum likelihood estimation.

Notation

Observable variable	$\boldsymbol{x}$.
Latent variable	$\boldsymbol{z}$.
Parameters (Parameter set)	Φ.
Probability distribution	P.

Idea

The model we consider here has latent variables, and so the model parameters Φ cannot be estimated only by observable variables so that they can satisfy (A.145). Then instead of estimating Φ', which maximizes $P_\Phi(\boldsymbol{x})$, we estimate Φ', which maximizes the following Q function:

$$Q(\Phi;\Phi') = \sum_{\boldsymbol{z}} P_\Phi(\boldsymbol{x},\boldsymbol{z}) \log P_{\Phi'}(\boldsymbol{x},\boldsymbol{z}), \tag{A.146}$$

where Φ is the current parameters (parameter set).

Proposition A.6. *Estimating Φ', which maximizes the following Q function, is equivalent to estimating Φ', which maximizes $P_\Phi(\boldsymbol{x})$.*

Proof.

$$Q(\Phi;\Phi') - Q(\Phi;\Phi) = \sum_{\boldsymbol{z}} P_\Phi(\boldsymbol{x},\boldsymbol{z}) \log P_{\Phi'}(\boldsymbol{x},\boldsymbol{z}) - \sum_{\boldsymbol{z}} P_\Phi(\boldsymbol{x},\boldsymbol{z}) \log P_{\Phi}(\boldsymbol{x},\boldsymbol{z}) \tag{A.147}$$

$$= \sum_{\boldsymbol{z}} P_\Phi(\boldsymbol{x},\boldsymbol{z}) \log \frac{P_{\Phi'}(\boldsymbol{x},\boldsymbol{z})}{P_\Phi(\boldsymbol{x},\boldsymbol{z})} \tag{A.148}$$

$$\leq \sum_{\boldsymbol{z}} P_\Phi(\boldsymbol{x},\boldsymbol{z}) \left(\frac{P_{\Phi'}(\boldsymbol{x},\boldsymbol{z})}{P_\Phi(\boldsymbol{x},\boldsymbol{z})} - 1\right) \quad (\leftarrow \log x \leq x - 1) \tag{A.149}$$

$$= \sum_{\boldsymbol{z}} (P_{\Phi'}(\boldsymbol{x},\boldsymbol{z}) - P_\Phi(\boldsymbol{x},\boldsymbol{z})) \tag{A.150}$$

$$= P_{\Phi'}(\boldsymbol{x}) - P_\Phi(\boldsymbol{x}). \tag{A.151}$$

In summary,

$$Q(\Phi;\Phi') > Q(\Phi;\Phi) \quad \Rightarrow \quad P_{\Phi'}(\boldsymbol{x}) > P_\Phi(\boldsymbol{x}). \tag{A.152}$$

That is, estimating new Φ' to increase Q function results in estimating Φ', which makes the likelihood larger. □

Algorithm A.1: EM (Expectation Maximization) algorithm.

```
Function Expectation_Maximization_Algorithm(X, Φ)
    Data: X, Φ
    Result: Φ̂
    Initialization: Set initial parameter values randomly.;
    repeat
        E-step: Compute Q(Φ; Φ');
        M-step: Φ̂ ← arg max_Φ' Q(Φ; Φ');
    until convergence;
```

Algorithm

Algorithm A.1 shows a pseudocode of the EM (Expectation-Maximization) algorithm.

Maximum likelihood by the EM algorithm is, comparing with the Bayes estimation in Section A.8.10, called *point estimation*, because only a set of values are obtained for the parameter set. On the other hand, Bayes learning, for example, LDA (latent Dirichlet allocation) [11] in Section 4.2.1 estimate (not a point but) posterior distribution by using prior distribution. Thus Bayes learning can be more robust and flexible than point estimation, while Bayes learning needs quite a large amount of time for parameter estimation.

A.8.10 Bayes Estimation (Bayes Learning)

The Bayes estimator is obtained by not only given data but also prior knowledge. The maximum likelihood estimator was obtained to maximize the likelihood of given data by the model, while instead the Bayes estimation is to estimate the *posterior* distribution from the likelihood of given data and also the prior distribution. We write the idea and method of Bayes estimation mathematically and also very intuitively below, following Bayes theorem. Let $\pi(\boldsymbol{\theta})$ be the prior distribution and $L(\boldsymbol{x}|\boldsymbol{\theta})$ be the likelihood of given data $\boldsymbol{x}$ under parameter $\boldsymbol{\theta}$. The posterior distribution can be computed as follows:

$$p(\boldsymbol{\theta}|\boldsymbol{x}) = \frac{\pi(\boldsymbol{\theta})L(\boldsymbol{x}|\boldsymbol{\theta})}{q(\boldsymbol{x})}, \quad \text{(A.153)}$$

where

$$q(\boldsymbol{x}) = \int_{\boldsymbol{\theta}} \pi(\boldsymbol{\theta})L(\boldsymbol{x}|\boldsymbol{\theta})d\boldsymbol{\theta}. \quad \text{(A.154)}$$

We then obtain the posterior distribution which can be estimated from the prior distribution and the likelihood. In the above we assume continuous random variables for parameters, while if they are discrete random variables, instead of integrals, we can use the sum over the values to be taken by the variables.

Algorithm A.2: Gibbs sampling algorithm.

Function `Gibbs_Sampling_Algorithm(`$\boldsymbol{X}$`)`

  Data: $\boldsymbol{X}$

  Result: $\hat{\Phi}$

  Initialization: Set initial parameter values.;

  $t = 0.$;

  repeat

    foreach $j \in \{1, \ldots, M\}$ **do**

      $x_j^{(t+1)} \sim p(X_j = x_j | X_1 = x_1^{(t+1)}, \ldots, X_{j-1} = x_{j-1}^{(t+1)}, X_{j+1} = x_{j+1}^{(t)}, \ldots, X_M = x_M^{(t)}).$;

    $t \leftarrow t + 1.$;

  until *convergence*;

Notes

Conjugate priors: (Prior and posterior) distributions are usually parametric, in the sense that each of them individually belongs to a certain type of distribution with hyperparameters, which control the distributions. However for computational efficiency, we often use the same type of distribution for both prior and posterior distributions. In other words, we use some type of distribution for priors which is the same as that for posteriors. These priors are called *conjugate priors*. In LDA for topic model in Section 4.2.1, Dirichlet distributions are used as conjugate priors, as shown in (4.7).

Computing marginalized probability: The denominator of (A.153), i.e. (A.154), can be given by a marginalized probability, which is obtained by taking the integral over joint probabilities including latent variables. For example, in LDA of Section 4.2.1, the probability of an instance is given by (4.9), where two integrals are taken over the joint probability given by (4.7).

Computing this marginalized probability is often hard, because of combinatorial explosion of a possible number of values taken by latent variables. Thus we need some strategy to solve/avoid this problem, and a general solution is to approximate the computation of marginalized probability by sampling or a variational method.

A.8.11 Gibbs Sampling

The posterior probability in Bayes estimation is in general obtained by computing the marginalized probability, which is hard to compute, because the number of possible values taken by hidden variables can be huge, and taking the integral (or sum) over all possible values is not possible. We then need to approximate the marginalized probability, and one possible approach to solve this problem is sampling, more generally *Markov chain Monte Carlo (MCMC)* sampling. *Gibbs sampling* is the most typical and well-used method of MCMC.

Gibbs sampling has the following two points:

1. **conditional distribution** The unique feature of Gibbs sampling is that Gibbs sampling considers, instead of joint distribution, as an approximation, conditional distribution of one variable, fixing values of all other variables. That is, the value to be taken by a random variable is estimated from the conditional distribution of the corresponding variable given all current other variables and values.

2. **Markov property** Gibbs sampling is a Markov chain-based approach, and so each time of consecutive time-series sampling depends upon a limited number of past sampling, particularly the first order of Markov chain, meaning that the current sampling depends upon the last time.

We can summarize the above two points into an algorithm. First we define notations below: let $X_i (i = 1, \ldots, M)$ be M random variables and $x_i (i = 1, \ldots, M)$ be their values, particularly $x_i^{(t)} (i = 1, \ldots, M)$ be the value of the t-th iteration of the repetition algorithm of Gibbs sampling.

Then after setting up some initial values for random variables, we can perform an iterative approach. That is, at each iteration, we can estimate the value x_j of the j-th random variable X_j, according to its conditional distribution. Also due to the Markov property, x_j is estimated from only the latest values of all other variables. These two points can be resulted in the following sampling rule:

$$x_j^{(t+1)} \sim$$
$$p(X_j = x_j | X_1 = x_1^{(t+1)}, \ldots, X_{j-1} = x_{j-1}^{(t+1)}, X_{j+1} = x_{j+1}^{(t)}, \ldots, X_M = x_M^{(t)}). \tag{A.155}$$

Algorithm A.2 shows a pseudocode which implements the above ideas of Gibbs sampling.

A.9 Derivations in the Main Text

A.9.1 Derivation of (3.9) in K-means

The objective function of K-means is (3.8):

$$J = \sum_{k=1}^{K} \sum_{i=1}^{N} \boldsymbol{Z}_{ik} ||\boldsymbol{x}_i - \boldsymbol{\mu}_k||_2^2. \tag{3.8}$$

We first think about the following, skipping cluster k:

$$\sum_{i=1}^{N} ||\boldsymbol{x}_i - \boldsymbol{\mu}||_2^2. \tag{A.156}$$

This is equivalent to the variance of $\boldsymbol{x}_i (i = 1, \ldots, N)$, and the following is already known, regarding computing the variance of $\boldsymbol{x} = (x_1, \ldots, x_N)^\top$ (see Section A.9.8):

$$\frac{1}{N}\sum_{i=1}^{N} x_i^2 - \mu^2 \;\Rightarrow\; \frac{1}{2N^2}\sum_{i=1}^{N}\sum_{j=1}^{N}(x_i - x_j)^2. \tag{A.157}$$

We can then see the following:

$$\sum_{i=1}^{N} ||\boldsymbol{x}_i - \boldsymbol{\mu}||_2^2 = \frac{1}{2N}\sum_{i=1}^{N}\sum_{j=1}^{N} ||\boldsymbol{x}_i - \boldsymbol{x}_j||_2^2. \tag{A.158}$$

Thus for each cluster, we can conduct this computation. Then the objective function is as follows:

$$J = \sum_{k=1}^{K}\sum_{j=1}^{N}\sum_{i=1}^{N} \boldsymbol{Z}_{ik}\boldsymbol{Z}_{jk}\frac{1}{2N}||\boldsymbol{x}_i - \boldsymbol{x}_j||_2^2 = \sum_{k=1}^{K}\sum_{j=1}^{N}\sum_{i=1}^{N} \boldsymbol{Z}_{ik}\boldsymbol{Z}_{jk}\boldsymbol{E}_{ij},$$

where $\boldsymbol{E}_{ij} = \frac{1}{2N}||\boldsymbol{x}_i - \boldsymbol{x}_j||_2^2$.

A.9.2 Derivation of (3.51) and (3.52) in Matrix Factorization

The objective function of matrix factorization is given as follows:

$$f(\boldsymbol{U}, \boldsymbol{V}) = \frac{1}{2}||\boldsymbol{X} - \boldsymbol{U}\boldsymbol{V}^\top||_2^2 + \lambda(||\boldsymbol{U}||_2^2 + ||\boldsymbol{V}||_2^2). \tag{A.159}$$

The first term (L^2 norm) can be written in two ways (and note that the L^2 norm can be written as a trace form in two ways), and so the partial derivative of either of the two ways can be used):

1.

$$(\boldsymbol{X} - \boldsymbol{U}\boldsymbol{V}^\top)(\boldsymbol{X} - \boldsymbol{U}\boldsymbol{V}^\top)^\top = (\boldsymbol{X} - \boldsymbol{U}\boldsymbol{V}^\top)(\boldsymbol{X}^\top - \boldsymbol{V}\boldsymbol{U}^\top) \tag{A.160}$$
$$= \boldsymbol{X}\boldsymbol{X}^\top - 2\boldsymbol{X}\boldsymbol{V}\boldsymbol{U}^\top + \boldsymbol{U}\boldsymbol{V}^\top\boldsymbol{V}\boldsymbol{U}^\top. \tag{A.161}$$

The second term can be $2\boldsymbol{U}\boldsymbol{V}^\top\boldsymbol{X}^\top$ as well.

2.

$$(\boldsymbol{X} - \boldsymbol{U}\boldsymbol{V}^\top)^\top(\boldsymbol{X} - \boldsymbol{U}\boldsymbol{V}^\top) = (\boldsymbol{X}^\top - \boldsymbol{V}\boldsymbol{U}^\top)(\boldsymbol{X} - \boldsymbol{U}\boldsymbol{V}^\top) \tag{A.162}$$
$$= \boldsymbol{X}^\top\boldsymbol{X} - 2\boldsymbol{X}^\top\boldsymbol{U}\boldsymbol{V}^\top + \boldsymbol{V}\boldsymbol{U}^\top\boldsymbol{U}\boldsymbol{V}^\top. \tag{A.163}$$

Again the second term can be $2\boldsymbol{V}\boldsymbol{U}^\top\boldsymbol{X}$.

Update of U Using (A.161),

$$\frac{\partial f(\boldsymbol{U},\boldsymbol{V})}{\partial \boldsymbol{U}} = -2\boldsymbol{X}\boldsymbol{V} + 2\boldsymbol{U}\boldsymbol{V}^\top\boldsymbol{V} + 2\lambda\boldsymbol{U}. \tag{A.164}$$

(we used $\frac{\partial \text{trace}(\boldsymbol{A}\boldsymbol{X}^\top)}{\partial \boldsymbol{X}} = \boldsymbol{A}$ for $\boldsymbol{X}$ and $\boldsymbol{A}$)
From $\frac{\partial f(\boldsymbol{U},\boldsymbol{V})}{\partial \boldsymbol{U}} = 0$,

$$\boldsymbol{U}(\boldsymbol{V}^\top\boldsymbol{V} + \lambda\boldsymbol{I}) = \boldsymbol{X}\boldsymbol{V}. \tag{A.165}$$

The update rule can be given as:

$$\boldsymbol{U} \leftarrow \boldsymbol{X}\boldsymbol{V}(\boldsymbol{V}^\top\boldsymbol{V} + \lambda\boldsymbol{I})^{-1}. \tag{A.166}$$

Update of V Using (A.163),

$$\frac{\partial f(\boldsymbol{U},\boldsymbol{V})}{\partial \boldsymbol{V}} = -2\boldsymbol{X}^\top\boldsymbol{U} + 2\boldsymbol{V}\boldsymbol{U}^\top\boldsymbol{U} + 2\lambda\boldsymbol{V}. \tag{A.167}$$

From $\frac{\partial f(\boldsymbol{U},\boldsymbol{V})}{\partial \boldsymbol{V}} = 0$,

$$\boldsymbol{V}(\boldsymbol{U}^\top\boldsymbol{U} + \lambda\boldsymbol{I}) = \boldsymbol{X}^\top\boldsymbol{U}. \tag{A.168}$$

Then the update rule can be given as follows:

$$\boldsymbol{V} \leftarrow \boldsymbol{X}^\top\boldsymbol{U}(\boldsymbol{U}^\top\boldsymbol{U} + \lambda\boldsymbol{I})^{-1}. \tag{A.169}$$

A.9.3 Derivation of (3.118) in Logistic Regression

$o(\boldsymbol{x})$ has parameter $\boldsymbol{w}$, and to clarify this point, hereafter we write $o(\boldsymbol{x}) = \sigma(\boldsymbol{x}, \boldsymbol{w})$. Log-likelihood is given as follows:

$$L_i(\boldsymbol{w}) = \log \sigma(\boldsymbol{x}_i, \boldsymbol{w})^{y_i}(1 - \sigma(\boldsymbol{x}_i, \boldsymbol{w}))^{1-y_i} \tag{A.170}$$

$$= y_i \log \sigma(\boldsymbol{x}_i, \boldsymbol{w}) + (1 - y_i)\log(1 - \sigma(\boldsymbol{x}_i, \boldsymbol{w})). \tag{A.171}$$

The gradient of the above log-likelihood with respect to $\boldsymbol{w}$ is given as follows:

$$\frac{dL_i(\boldsymbol{w})}{d\boldsymbol{w}} = \frac{dL_i(w)}{d\sigma(\boldsymbol{x}_i, \boldsymbol{w})}\frac{d\sigma(\boldsymbol{x}_i, \boldsymbol{w})}{d\boldsymbol{w}} \tag{A.172}$$

$$= \frac{y_i}{\sigma(\boldsymbol{x}_i, \boldsymbol{w})}\frac{d\sigma(\boldsymbol{x}_i, \boldsymbol{w})}{d\boldsymbol{w}} - \frac{1 - y_i}{1 - \sigma(\boldsymbol{x}_i, \boldsymbol{w})}\frac{d\sigma(\boldsymbol{x}_i, \boldsymbol{w})}{d\boldsymbol{w}}. \tag{A.173}$$

$\sigma(\boldsymbol{x}_i, \boldsymbol{w}) = \frac{1}{1+\exp(-(\boldsymbol{x}_i^\top\boldsymbol{w}+\epsilon))}$ and the derivative of $\sigma(\boldsymbol{x}_i, \boldsymbol{w})$ with respect to $\boldsymbol{w}$ is given as follows:

$$\frac{d\sigma(\boldsymbol{x}_i, \boldsymbol{w})}{d\boldsymbol{w}} = \sigma(\boldsymbol{x}_i, \boldsymbol{w})(1 - \sigma(\boldsymbol{x}_i, \boldsymbol{w}))\boldsymbol{x}_i. \tag{A.174}$$

We substitute this for the corresponding part of (A.173), and we can have the following:

$$\begin{aligned}\frac{dL_i(\boldsymbol{w})}{d\boldsymbol{w}} &= \frac{y_i}{\sigma(\boldsymbol{x}_i,\boldsymbol{w})}(\sigma(\boldsymbol{x}_i,\boldsymbol{w})(1-\sigma(\boldsymbol{x}_i,\boldsymbol{w}))\boldsymbol{x}_i) \\ &- \frac{1-y_i}{1-\sigma(\boldsymbol{x}_i,\boldsymbol{w})}(\sigma(\boldsymbol{x}_i,\boldsymbol{w})(1-\sigma(\boldsymbol{x}_i,\boldsymbol{w}))\boldsymbol{x}_i) && \text{(A.175)} \\ &= (y_i(1-\sigma(\boldsymbol{x}_i,\boldsymbol{w}))-(1-y_i)\sigma(\boldsymbol{x}_i,\boldsymbol{w}))\boldsymbol{x}_i && \text{(A.176)} \\ &= (y_i - y_i\sigma(\boldsymbol{x}_i,\boldsymbol{w}) - \sigma(\boldsymbol{x}_i,\boldsymbol{w}) + y_i\sigma(\boldsymbol{x}_i,\boldsymbol{w}))\boldsymbol{x}_i && \text{(A.177)} \\ &= (y_i - \sigma(\boldsymbol{x}_i,\boldsymbol{w}))\boldsymbol{x}_i && \text{(A.178)} \\ &= (y_i - o(\boldsymbol{x}_i))\boldsymbol{x}_i. && \text{(A.179)}\end{aligned}$$

A.9.4 Derivation of (3.148) and (3.152) in Layered Neural Network

The objective function is the squared error loss between $\hat{y}$ and y:

$$e = \frac{1}{2}||o^{(2)}(\boldsymbol{x}) - y||_2^2. \quad \text{(A.180)}$$

The derivative of the error with respect to $o^{(2)}(\boldsymbol{x})$ is given as follows:

$$\frac{\partial e}{\partial o^{(2)}(\boldsymbol{x})} = (o^{(2)}(\boldsymbol{x}) - y). \quad \text{(A.181)}$$

1. $w_k^{(2)}$

   Using (3.141)–(3.143),

$$\begin{aligned}\frac{\partial o^{(2)}(\boldsymbol{x})}{\partial i^{(2)}(\boldsymbol{x})} &= o^{(2)}(\boldsymbol{x})(1-o^{(2)}(\boldsymbol{x})). && \text{(A.182)} \\ \frac{\partial i^{(2)}(\boldsymbol{x})}{\partial w_k^{(2)}} &= o_k^{(1)}(\boldsymbol{x}). && \text{(A.183)}\end{aligned}$$

   Using all above three,

$$\begin{aligned}\frac{\partial e}{\partial w_k^{(2)}} &= \frac{\partial e}{\partial o^{(2)}(\boldsymbol{x})}\frac{\partial o^{(2)}(\boldsymbol{x})}{\partial i^{(2)}(\boldsymbol{x})}\frac{\partial i^{(2)}(\boldsymbol{x})}{\partial w_k^{(2)}} && \text{(A.184)} \\ &= (o^{(2)}(\boldsymbol{x}) - y)o^{(2)}(\boldsymbol{x})(1-o^{(2)}(\boldsymbol{x}))o_k^{(1)}(\boldsymbol{x}). && \text{(A.185)}\end{aligned}$$

2. $\boldsymbol{w}_k^{(1)}$

Using (3.138)–(3.143),

$$\frac{\partial o^{(2)}(\boldsymbol{x})}{\partial i^{(2)}(\boldsymbol{x})} = o^{(2)}(\boldsymbol{x})(1-o^{(2)}(\boldsymbol{x})). \tag{A.186}$$

$$\frac{\partial i^{(2)}(\boldsymbol{x})}{\partial o_k^{(1)}} = w_k^{(2)}. \tag{A.187}$$

$$\frac{\partial o_k^{(1)}(\boldsymbol{x})}{\partial i_k^{(1)}(\boldsymbol{x})} = o^{(1)}(\boldsymbol{x})(1-o^{(1)}(\boldsymbol{x})). \tag{A.188}$$

$$\frac{\partial i_k^{(1)}(\boldsymbol{x})}{\partial \boldsymbol{w}_k^{(1)}} = \boldsymbol{x}. \tag{A.189}$$

Using (A.181) and the above four, we can obtain below:

$$\frac{\partial e}{\partial \boldsymbol{w}_k^{(1)}} = \frac{\partial e}{\partial o^{(2)}(\boldsymbol{x})}\frac{\partial o^{(2)}(\boldsymbol{x})}{\partial i^{(2)}(\boldsymbol{x})}\frac{\partial i^{(2)}(\boldsymbol{x})}{\partial o_k^{(1)}(\boldsymbol{x})}\frac{\partial o_k^{(1)}(\boldsymbol{x})}{\partial i_k^{(1)}(\boldsymbol{x})}\frac{\partial i_k^{(1)}(\boldsymbol{x})}{\partial \boldsymbol{w}_k^{(1)}} \tag{A.190}$$

$$= (o^{(2)}(\boldsymbol{x})-y)o^{(2)}(\boldsymbol{x})(1-o^{(2)}(\boldsymbol{x}))w_k^{(2)}o^{(1)}(\boldsymbol{x})(1-o^{(1)}(\boldsymbol{x}))\boldsymbol{x}. \tag{A.191}$$

A.9.5 Derivation of (3.209) in Support Vector Machine

From (3.206),

$$L(\boldsymbol{w}, b, \boldsymbol{\alpha}) = \frac{1}{2}\boldsymbol{w}^\top \boldsymbol{w} - \sum_i \alpha_i(y_i(\boldsymbol{w}^\top \boldsymbol{x}_i + b) - 1). \tag{3.206}$$

We obtained by setting the partial derivative of the Lagrangian equal to zero, as follows:

$$\boldsymbol{w} = \sum_i \alpha_i y_i \boldsymbol{x}_i, \qquad \sum_i \alpha_i y_i = 0. \tag{A.192}$$

Substituting these two for the corresponding parts of (3.206), we extend that further as follows:

$$\frac{1}{2}\boldsymbol{w}^\top \boldsymbol{w} - \sum_i \alpha_i(y_i(\boldsymbol{w}^\top \boldsymbol{x}_i + b) - 1) \tag{A.193}$$

$$= \frac{1}{2}\sum_i \alpha_i y_i \boldsymbol{x}_i^\top \sum_j \alpha_j y_j \boldsymbol{x}_j - \sum_i \alpha_i (y_i(\sum_j \alpha_j y_j \boldsymbol{x}_j^\top \boldsymbol{x}_i + b) - 1) \tag{A.194}$$

$$= \frac{1}{2}\sum_i\sum_j \alpha_i\alpha_j y_i y_j \boldsymbol{x}_i^\top \boldsymbol{x}_j - \sum_i\sum_j \alpha_i\alpha_j y_i y_j \boldsymbol{x}_i^\top \boldsymbol{x}_j + b\sum_i \alpha_i y_i + \sum_i \alpha_i \tag{A.195}$$

$$= \sum_i \alpha_i - \frac{1}{2}\sum_i\sum_j \alpha_i\alpha_j y_i y_j \boldsymbol{x}_i^\top \boldsymbol{x}_j. \tag{A.196}$$

This is equal to (3.209).

A.9.6 Derivation of (3.253) in Kernel K-means

The objective function in hyperspace is given in (3.249):

$$J = \sum_{k}^{|C_k|} \sum_{i}^{N} \boldsymbol{Z}_{ik} || \phi(\boldsymbol{x}_i) - \frac{1}{|C_k|} \sum_{j}^{N} \boldsymbol{Z}_{jk} \phi(\boldsymbol{x}_j) ||_2^2. \tag{3.249}$$

This can be developed as follows:

$$\begin{aligned} J &= \sum_{k}^{|C_k|} \sum_{i}^{N} \boldsymbol{Z}_{ik} (\phi(\boldsymbol{x}_i)\phi(\boldsymbol{x}_i) - \frac{2}{|C_k|} \sum_{j}^{N} \boldsymbol{Z}_{jk} \phi(\boldsymbol{x}_i)\phi(\boldsymbol{x}_j) \\ &+ \frac{1}{|C_k|^2} \sum_{j}^{N} \sum_{l}^{N} \boldsymbol{Z}_{jk} \boldsymbol{Z}_{lk} \phi(\boldsymbol{x}_j)\phi(\boldsymbol{x}_l)). \end{aligned} \tag{A.197}$$

The first term can be a constant, and so we ignore the first term hereafter. Then by using $K(\boldsymbol{x}_i, \boldsymbol{x}_j) = \phi(\boldsymbol{x}_i)\phi(\boldsymbol{x}_j)$, this can be modified as follows:

$$\sum_{k}^{|C_k|} \sum_{i}^{N} \boldsymbol{Z}_{ik} (-\frac{2}{|C_k|} \sum_{j}^{N} \boldsymbol{Z}_{jk} K(\boldsymbol{x}_i, \boldsymbol{x}_j) + \frac{1}{|C_k|^2} \sum_{j}^{N} \sum_{l}^{N} \boldsymbol{Z}_{jk} \boldsymbol{Z}_{lk} K(\boldsymbol{x}_j, \boldsymbol{x}_l)). \tag{A.198}$$

Note that the third term has nothing to do with i, and so for the third term, $\sum_i^N \boldsymbol{Z}_{ik} = |C_k|$. Thus the above can be developed into the following:

$$\sum_{k}^{|C_k|} (-\frac{2}{|C_k|} \sum_{i}^{N} \boldsymbol{Z}_{ik} \sum_{j}^{N} \boldsymbol{Z}_{jk} K(\boldsymbol{x}_i, \boldsymbol{x}_j) + \frac{1}{|C_k|} \sum_{j}^{N} \boldsymbol{Z}_{jk} \sum_{l}^{N} \boldsymbol{Z}_{lk} K(\boldsymbol{x}_j, \boldsymbol{x}_l)) \tag{A.199}$$

$$= \sum_{k}^{|C_k|} (-\frac{1}{|C_k|} \sum_{i}^{N} \boldsymbol{Z}_{ik} \sum_{j}^{N} \boldsymbol{Z}_{jk} K(\boldsymbol{x}_i, \boldsymbol{x}_j)). \tag{A.200}$$

Thus, the objective function is given as follows:

$$J = -\sum_{k}^{|C_k|} \frac{1}{|C_k|} \sum_{i}^{N} \boldsymbol{Z}_{ik} \sum_{j}^{N} \boldsymbol{Z}_{jk} K(\boldsymbol{x}_i, \boldsymbol{x}_j). \tag{3.253}$$

A.9.7 Derivation of (3.267) in Kernel Ridge Regression

Lagrange function is given as follows:

$$L(\boldsymbol{w}, \boldsymbol{\xi}) = \frac{1}{2} \sum_{i} \xi_i^2 + \frac{\lambda}{2} \boldsymbol{w}^\top \boldsymbol{w} + \sum_{i} \alpha_i (y_i - \boldsymbol{w}^\top \boldsymbol{x}_i - \xi_i). \tag{3.264}$$

Substituting the following two for the corresponding part of the above Lagrange function:

$$\boldsymbol{w} = \frac{1}{\lambda}\sum_i \alpha_i \boldsymbol{x}_i. \tag{A.201}$$

$$\xi_i = \alpha_i. \tag{A.202}$$

Then we can have the following for the right-hand side of the Lagrange function:

$$\frac{1}{2}\sum_i \alpha_i^2 + \frac{\lambda}{2}(\frac{1}{\lambda^2})\sum_i\sum_j \alpha_i\alpha_j \boldsymbol{x}_i^\top \boldsymbol{x}_j + \sum_i \alpha_i(y_i - \frac{1}{\lambda}\sum_j \alpha_j \boldsymbol{x}_j^\top \boldsymbol{x}_i - \alpha_i) \tag{A.203}$$

$$= \frac{1}{2}\sum_i \alpha_i^2 + \frac{1}{2\lambda}\sum_i\sum_j \alpha_i\alpha_j \boldsymbol{x}_i^\top \boldsymbol{x}_j + \sum_i \alpha_i y_i - \frac{1}{\lambda}\sum_i\sum_j \alpha_i\alpha_j \boldsymbol{x}_j^\top \boldsymbol{x}_i - \alpha_i^2 \tag{A.204}$$

$$= -\frac{1}{2}\sum_i \alpha_i^2 - \frac{1}{2\lambda}\sum_i\sum_j \alpha_i\alpha_j \boldsymbol{x}_i^\top \boldsymbol{x}_j + \sum_i \alpha_i y_i. \tag{A.205}$$

(A.205) is equal to (3.267). Thus the Lagrange function is given as follows:

$$L(\boldsymbol{\alpha}) = -\frac{1}{2}\sum_i \alpha_i^2 - \frac{1}{2\lambda}\sum_{i,j} \alpha_i\alpha_j \boldsymbol{x}_i^\top \boldsymbol{x}_j + \sum_i \alpha_i y_i. \tag{3.267}$$

A.9.8 Derivation of (A.11) of Variance

$$s = \frac{1}{N}\sum_{i=1}^N x_i^2 - \mu^2 \tag{A.206}$$

$$= \frac{1}{N}\sum_{i=1}^N x_i^2 - (\frac{\sum_{i=1}^N x_i}{N})^2 \tag{A.207}$$

$$= \frac{1}{N^2}(N\sum_{i=1}^N x_i^2 - (\sum_{i=1}^N x_i)^2). \tag{A.208}$$

The second term of (A.208) is, by using two indices i and j, which is equal to $\sum_{i=1}^N x_i \sum_{j=1}^N x_j$. This is equal to the sum of all elements of $\boldsymbol{x}\boldsymbol{x}^\top$, and below we consider to compute this sum. The $\boldsymbol{x}\boldsymbol{x}^\top$ is symmetric $(N \times N)$-matrix, and so we divide this matrix into three parts: diagonal, upper triangle and lower triangle, where upper triangle and lower triangle are the same, since $\boldsymbol{x}\boldsymbol{x}^\top$ is symmetric.

$$(\sum_{i=1}^{N} x_i)^2 = \sum_{i=1}^{N} x_i \sum_{i=1}^{N} x_i \tag{A.209}$$

$$= \sum_{i=1}^{N} x_i \sum_{j=1}^{N} x_j \tag{A.210}$$

$$= \sum_{i=1}^{N} x_i \sum_{j=i}^{i} x_j + \sum_{i=1}^{N} x_i \sum_{j<i}^{N} x_j + \sum_{i=1}^{N} x_i \sum_{j>i}^{N} x_j \tag{A.211}$$

$$= \sum_{i=1}^{N} x_i^2 + \sum_{i=1}^{N} x_i \sum_{j<i}^{N} x_j + \sum_{i=1}^{N} x_i \sum_{j>i}^{N} x_j \tag{A.212}$$

$$= \sum_{i=1}^{N} x_i^2 + 2\sum_{i=1}^{N} x_i \sum_{j>i}^{N} x_j. \tag{A.213}$$

We substitute the above for the second term of (A.208) and develop that further as follows:

$$s = \frac{1}{N^2}(N\sum_{i=1}^{N} x_i^2 - (\sum_{i=1}^{N} x_i)^2) \tag{A.214}$$

$$= \frac{1}{N^2}(N\sum_{i=1}^{N} x_i^2 - (\sum_{i=1}^{N} x_i^2 + 2\sum_{i=1}^{N} x_i \sum_{j>i}^{N} x_j) \tag{A.215}$$

$$= \frac{1}{N^2}((N-1)\sum_{i=1}^{N} x_i^2 - 2\sum_{i=1}^{N} x_i \sum_{j>i}^{N} x_j). \tag{A.216}$$

Here, we can see the following:

$$(N-1)x_1^2 + (N-2)x_2^2 + \cdots + 2x_{N-2}^2 + x_{N-1}^2 = \sum_{i=1}^{N} x_i^2 \sum_{j>i}^{N} 1. \tag{A.217}$$

Also in the opposite direction, we can have the following:

$$(N-1)x_N^2 + (N-2)x_{N-1}^2 + \cdots + 2x_3^2 + x_2^2 = \sum_{i=1}^{N} \sum_{j>i}^{N} x_j^2. \tag{A.218}$$

Summing the above two equations, we can have the following:

$$(N-1)\sum_{i=1}^{N} x_i^2 = (N-1)x_1^2 + (N-1)x_2^2 + \cdots + (N-1)x_{N-1}^2 + (N-1)x_{N-1}^2$$

$$= \sum_{i=1}^{N} \sum_{j>i}^{N} (x_i^2 + x_j^2). \tag{A.219}$$

We substitute this for the corresponding term of (A.216) and have the following:

$$
\begin{aligned}
s &= \frac{1}{N^2}(\sum_{i=1}^{N} x_i^2 - 2\sum_{i=1}^{N} x_i \sum_{j>i}^{N} x_j) && \text{(A.220)}\\
&= \frac{1}{N^2}(\sum_{i=1}^{N}\sum_{j>i}^{N}(x_i^2 + x_j^2) - 2\sum_{i=1}^{N}\sum_{j>i}^{N} x_i x_j) && \text{(A.221)}\\
&= \frac{1}{N^2}\sum_{i=1}^{N}\sum_{j>i}^{N}(x_i^2 + x_j^2 - 2x_i x_j) && \text{(A.222)}\\
&= \frac{1}{N^2}\sum_{i=1}^{N}\sum_{j>i}^{N}(x_i - x_j)^2 && \text{(A.223)}\\
&= \frac{1}{2N^2}\sum_{i=1}^{N}\sum_{j=1}^{N}(x_i - x_j)^2. && \text{(A.224)}
\end{aligned}
$$

Note that if $i = j$, $x_i - x_j = 0$.

A.9.9 Derivation of (A.40) in Covariance

$$
\mathrm{cov}(\boldsymbol{X}, \boldsymbol{Y}) = \sum_i \frac{(x_i - \mu_X)(y_i - \mu_Y)}{N}, \tag{A.225}
$$

where

$$
\mu_X = \frac{\sum_j x_j}{N}, \quad \mu_Y = \frac{\sum_k y_k}{N}. \tag{A.226}
$$

Using these three, we have the following:

$$
\begin{aligned}
\mathrm{cov}(\boldsymbol{X}, \boldsymbol{Y}) &= \sum_i \frac{(x_i - \mu_X)(y_i - \mu_Y)}{N} && \text{(A.227)}\\
&= \frac{1}{N}\sum_i (x_i - \frac{\sum_j x_j}{N})(y_i - \frac{\sum_k y_k}{N}) && \text{(A.228)}\\
&= \frac{1}{N}(\sum_i x_i y_i - \frac{\sum_i y_i \sum_j x_j}{N} - \frac{\sum_i x_i \sum_k y_k}{N} + \sum_i \frac{\sum_j x_j \sum_k y_k}{N^2}). && \text{(A.229)}
\end{aligned}
$$

For the fourth term, we use $\sum_i 1 = N$:

$$\begin{aligned}
\mathrm{cov}(\boldsymbol{X}, \boldsymbol{Y}) &= \frac{1}{N}(\sum_i x_i y_i - \frac{\sum_i y_i \sum_j x_j}{N} - \frac{\sum_i x_i \sum_k y_k}{N} + N\frac{\sum_j x_j \sum_k y_k}{N^2}) && \text{(A.230)} \\
&= \frac{1}{N}(\sum_i x_i y_i - \frac{\sum_i y_i \sum_j x_j}{N} - \frac{\sum_i x_i \sum_k y_k}{N} + \frac{\sum_j x_j \sum_k y_k}{N}) && \text{(A.231)} \\
&= \frac{1}{N}(\sum_i x_i y_i - \frac{\sum_i x_i \sum_j y_j}{N}) && \text{(A.232)} \\
&= \frac{1}{N}(\frac{\sum_j \sum_i x_i y_i}{N} - \frac{\sum_i \sum_j x_i y_j}{N}). && \text{(A.233)}
\end{aligned}$$

In the last line, we use $\sum_j 1 = N$ again. We develop this further as follows:

$$\begin{aligned}
\mathrm{cov}(\boldsymbol{X}, \boldsymbol{Y}) &= \frac{1}{N}(\frac{\sum_j \sum_i x_i y_i}{N} - \frac{\sum_i \sum_j x_i y_j}{N}) && \text{(A.234)} \\
&= \frac{1}{N^2}(\sum_i \sum_j x_i y_i - \sum_i \sum_j x_i y_j) && \text{(A.235)} \\
&= \frac{1}{2N^2}(2\sum_i \sum_j x_i y_i - 2\sum_i \sum_j x_i y_j) && \text{(A.236)} \\
&= \frac{1}{2N^2}(\sum_i \sum_j x_i y_i + \sum_i \sum_j x_j y_j - \sum_i \sum_j x_i y_j - \sum_i \sum_j x_j y_i) && \text{(A.237)} \\
&= \frac{1}{2N^2}\sum_i \sum_j (x_i y_i - x_i y_j - x_j y_i + x_j y_j) && \text{(A.238)} \\
&= \frac{1}{2N^2}\sum_i \sum_j (x_i(y_i - y_j) - x_j(y_i - y_j)) && \text{(A.239)} \\
&= \frac{1}{2N^2}\sum_i \sum_j (x_i - x_j)(y_i - y_j). && \text{(A.240)}
\end{aligned}$$

Finally this is equal to (A.40).

Bibliography

[1] Abe, N. & Mamitsuka, H. (1997). *Machine Learning,* **29** (2), 275–301.

[2] Agrawal, R. & Srikant, R. (1994). In: *Proceedings of the 20th International Conference on Very Large Data Bases* VLDB '94 pp. 487–499, San Francisco, CA, USA: Morgan Kaufmann Publishers Inc.

[3] Altschul, S. F., Gish, W., Miller, W., Myers, E. W., & Lipman, D. J. (1990). *Journal of Molecular Biology,* **215** (3), 403 – 410.

[4] Altschul, S. F., Madden, T. L., Schffer, A. A., Zhang, J., Zhang, Z., Miller, W., & Lipman, D. J. (1997). *Nucleic Acids Research,* **25** (17), 3389–3402.

[5] Ana, L. N. F. & Jain, A. K. (2003). In: *2003 IEEE Computer Society Conference on Computer Vision and Pattern Recognition, 2003. Proceedings.* volume 2 pp. II–128–II–133 vol.2, Washington, DC, USA: IEEE Computer Society.

[6] Avis, D. & Fukuda, K. (1996). *Discrete Applied Mathematics,* **65** (1), 21 – 46.

[7] Barabasi, A. L. & Oltvai Z. N. (2004). *Nat. Rev. Genet.* **5** (2), 101–113.

[8] Basu, S. (2005). *Semi-supervised Clustering: Probabilistic Models, Algorithms and Experiments.* PhD thesis University of Texas at Austin Austin, TX, USA.

[9] Belkin, M. & Niyogi, P. (2003). *Neural Comput.* **15** (6), 1373–1396.

[10] Bento, A. P., Gaulton, A., Hersey, A., Bellis, L. J., Chambers, J., Davies, M., Krüger, F. A., Light, Y., Mak, L., McGlinchey, S., Nowotka, M., Papadatos, G., Santos, R., & Overington, J. P. (2014). *Nucleic Acids Research,* **42** (D1), D1083–D1090.

[11] Blei, D. M., Ng, A. Y., & Jordan, M. I. (2003). *J. Mach. Learn. Res.* **3**, 993–1022.

[12] Borg, I. & Groenen, P. (1997). *Modern Multidimensional Scaling: Theory and Applications.* Springer series in statistics. Berlin, Heidelberg: Springer.

[13] Borgwardt, K. M. & Kriegel, H.-P. (2005). In: *Proceedings of the Fifth IEEE International Conference on Data Mining* ICDM '05 pp. 74–81, Washington, DC, USA: IEEE Computer Society.

[14] Breiman, L. (2001). *Machine Learning,* **45** (1), 5–32.

[15] Breiman, L., Friedman, J., Olshen, R., & Stone, C. (1984). *Classification and Regression Trees.* Monterey, CA: Wadsworth and Brooks.

[16] Burges, C., Shaked, T., Renshaw, E., Lazier, A., Deeds, M., Hamilton, N., & Hullender, G. (2005). In: *Proceedings of the 22Nd International Conference on Machine Learning* ICML '05 pp. 89–96, New York, NY, USA: ACM.

[17] Cadez, I., Heckerman, D., Meek, C., Smyth, P., & White, S. (2003). *Data Mining and Knowledge Discovery,* **7** (4), 399–424.

[18] Cai, X., Bazerque, J. A., & Giannakis, G. B. (2013). *PLoS Computational Biology,* **9** (5), 1–13.

[19] Chapelle, O., Metlzer, D., Zhang, Y., & Grinspan, P. (2009). In: *Proceedings of the 18th ACM Conference on Information and Knowledge Management* CIKM '09 pp. 621–630, New York, NY, USA: ACM.

[20] Cheng, J. & Greiner, R. (1999). In: *Proceedings of the Fifteenth Conference on Uncertainty in Artificial Intelligence* UAI'99 pp. 101–108, San Francisco, CA, USA: Morgan Kaufmann Publishers Inc.

[21] Choi, H. & Baraniuk, R. G. (2001). *IEEE Transactions on Image Processing,* **10** (9), 1309–1321.

[22] Chomsky, N. (1956). *IRE Transactions on Information Theory,* **2**, 113–124.

[23] Chung, F. R. K. (1997). *Spectral Graph Theory.* Providence, RI USA: American Mathematical Society.

[24] Cover, T. M. & Thomas, J. A. (2006). *Elements of Information Theory (Wiley Series in Telecommunications and Signal Processing).* New York, NY, USA: Wiley-Interscience.

[25] Dhillon, I. S., Guan, Y., & Kulis, B. (2004). In: *Proceedings of the Tenth ACM SIGKDD International Conference on Knowledge Discovery and Data Mining* KDD '04 pp. 551–556, New York, NY, USA: ACM.

[26] Diligenti, M., Frasconi, P., & Gori, M. (2003). *IEEE Transactions on Pattern Analysis and Machine Intelligence,* **25** (4), 519–523.

[27] Drozdetskiy, A., Cole, C., Procter, J., & Barton, G. J. (2015). *Nucleic Acids Res.* **43** (W1), W389–394.

[28] duVerle, D. A. & Mamitsuka, H. (2012). *Brief. Bioinformatics,* **13** (3), 337–349.

[29] Eddy, S. R. (1995). *Proc Int Conf Intell Syst Mol Biol,* **3**, 114–120.

[30] Eisen, M. B., Spellman, P. T., Brown, P. O., & Botstein, D. (1998). *Proceedings of the National Academy of Sciences,* **95** (25), 14863–14868.

[31] Finn, R. D., Coggill, P., Eberhardt, R. Y., Eddy, S. R., Mistry, J., Mitchell, A. L., Potter, S. C., Punta, M., Qureshi, M., Sangrador-Vegas, A., Salazar, G. A., Tate, J., & Bateman, A. (2016). *Nucleic Acids Res.* **44** (D1), D279–285.

[32] Friedman, J., Hastie, T., & Tibshirani, R. (2000). *Ann. Statist.* **28** (2), 337–407.

[33] Grauman, K. & Darrell, T. (2007). *J. Mach. Learn. Res.* **8**, 725–760.

[34] Griffiths, T. L. & Steyvers, M. (2004). *Proceedings of the National Academy of Sciences,* **101** (suppl 1), 5228–5235.

[35] Hagen, L. & Kahng, A. B. (2006). *Trans. Comp.-Aided Des. Integ. Cir. Sys.* **11** (9), 1074–1085.

[36] Han, J., Pei, J., & Yin, Y. (2000). *SIGMOD Rec.* **29** (2), 1–12.

[37] Hand, D. J. & Till, R. J. (2001). *Machine Learning,* **45** (2), 171–186.

[38] Hanley, J. A. & McNeil, B. J. (1982). *Radiology,* **143** (1), 29–36.

[39] Hashimoto, K., Aoki-Kinoshita, K. F., Ueda, N., Kanehisa, M., & Mamitsuka, H. (2006). In: *Proceedings of the 12th ACM SIGKDD International Conference on Knowledge Discovery and Data Mining* KDD '06 pp. 177–186, New York, NY, USA: ACM.

[40] Hashimoto, K., Aoki-Kinoshita, K. F., Ueda, N., Kanehisa, M., & Mamitsuka, H. (2008). *ACM Trans. Knowl. Discov. Data,* **2** (1), 6:1–6:30.

[41] Hauser, A. S., Chavali, S., Masuho, I., Jahn, L. J., Martemyanov, K. A., Gloriam, D. E., & Babu, M. M. (2018). *Cell,* **172** (1), 41 – 54.e19.

[42] Hoerl, A. E. & Kennard, R. W. (1970). *Technometrics,* **12** (1), 55–67.

[43] Hofmann, T. (1999). In: *Proceedings of the 22Nd Annual International ACM SIGIR Conference on Research and Development in Information Retrieval* SIGIR '99 pp. 50–57, New York, NY, USA: ACM.

[44] Hopcroft, J. E., Motwani, R., & Ullman, J. D. (2006). *Introduction to Automata Theory, Languages, and Computation (3rd Edition).* Boston, MA, USA: Addison-Wesley Longman Publishing Co., Inc.

[45] Horton, P., Park, K. J., Obayashi, T., Fujita, N., Harada, H., Adams-Collier, C. J., & Nakai, K. (2007). *Nucleic Acids Res.* **35** (Web Server issue), W585–587.

[46] Hubert, L. & Arabie, P. (1985). *Journal of Classification,* **2** (1), 193–218.

[47] Jaccard, P. (1912). *New Phytologist,* **11** (2), 37–50.

[48] Jebara, T., Kondor, R., & Howard, A. (2004). *J. Mach. Learn. Res.* **5**, 819–844.

[49] Karasuyama, M. & Mamitsuka, H. (2013a). *IEEE Transactions on Neural Networks and Learning Systems,* **24** (12), 1999–2012.

[50] Karasuyama, M. & Mamitsuka, H. (2013b). In: *Advances in Neural Information Processing Systems 26,* (Burges, C. J. C., Bottou, L., Welling, M., Ghahramani, Z., & Weinberger, K. Q., eds) pp. 1547–1555. Curran Associates, Inc.

[51] Karasuyama, M. & Mamitsuka, H. (2017). *Machine Learning,* **106** (2), 307–335.

[52] Katoh, K. & Standley, D. M. (2013). *Molecular Biology and Evolution,* **30** (4), 772–780.

[53] Kim, S., Thiessen, P. A., Bolton, E. E., Chen, J., Fu, G., Gindulyte, A., Han, L., He, J., He, S., Shoemaker, B. A., Wang, J., Yu, B., Zhang, J., & Bryant, S. H. (2016). *Nucleic Acids Res.* **44** (D1), D1202–1213.

[54] Koller, D. & Friedman, N. (2009). *Probabilistic Graphical Models: Principles and Techniques - Adaptive Computation and Machine Learning.* Cambridge, MA, USA: The MIT Press.

[55] Kvalseth, T. O. (1987). *IEEE Transactions on Systems, Man, and Cybernetics,* **17** (3), 517–519.

[56] Lafferty, J. D., McCallum, A., & Pereira, F. C. N. (2001). In: *Proceedings of the Eighteenth International Conference on Machine Learning* ICML '01 pp. 282–289, San Francisco, CA, USA: Morgan Kaufmann Publishers Inc.

[57] LeCun, Y. & an Geoffrey Hinton, Y. B. (2015). *Nature,* **521**, 436–444.

[58] Leemis, L. M. & McQueston, J. T. (2008). *The American Statistician,* **62** (1), 45–53.

[59] Leslie, C., Eskin, E., & Noble, W. S. (2002). *Pac. Symp. Biocomput.* , 564–575.

[60] Lodhi, H., Saunders, C., Shawe-Taylor, J., Cristianini, N., & Watkins, C. (2002). *J. Mach. Learn. Res.* **2**, 419–444.

[61] Luxburg, U. (2007). *Statistics and Computing,* **17** (4), 395–416.

[62] Mamitsuka, H. (2006). *Pattern Recognition,* **39** (12), 2393–2404.

[63] Mamitsuka, H., Okuno, Y., & Yamaguchi, A. (2003). *SIGKDD Explor. Newsl.* **5** (2), 113–121.

[64] Mitchell, T. M. (2017). http://www.cs.cmu.edu/~tom/NewChapters.html.

[65] Neuhaus, M. & Bunke, H. (2006). *Pattern Recognition,* **39** (10), 1852 – 1863.

[66] Ng, A. Y., Jordan, M. I., & Weiss, Y. (2001). In: *Proceedings of the 14th International Conference on Neural Information Processing Systems: Natural and Synthetic* NIPS'01 pp. 849–856, Cambridge, MA, USA: MIT Press.

[67] Nguyen, C. H. & Mamitsuka, H. (2011). *IEEE Transactions on Neural Networks,* **22** (9), 1395–1405.

[68] Pei, J., Han, J., Mortazavi-Asl, B., Pinto, H., Chen, Q., Dayal, U., & Hsu, M. (2001). In: *Proceedings of the 17th International Conference on Data Engineering* pp. 215–224, Washington, DC, USA: IEEE Computer Society.

[69] Ramon, J. & Gärtner, T. (2003). In: *Proceedings of the First International Workshop on Mining Graphs, Trees and Sequences* pp. 65–74 : No Publisher.

[70] Rand, W. M. (1971). *Journal of the American Statistical Association,* **66** (336), 846–850.

[71] Rigoutsos, I. & Floratos, A. (1998). *Bioinformatics,* **14** (1), 55–67.

[72] Rissanen, J. (1978). *Automatica,* **14** (5), 465–471.

[73] Rogers, D. J. & Tanimoto, T. T. (1960). *Science,* **132** (3434), 1115–1118.

[74] Roweis, S. T. & Saul, L K. (2000). *SCIENCE,* **290**, 2323–2326.

[75] Schapire, R. E. (1990). *Machine Learning,* **5** (2), 197–227.

[76] Schapire, R. E. & Freund, Y. (2012). *Boosting: Foundations and Algorithms.* Cambridge, MA, USA: The MIT Press.

[77] Schapire, R. E. & Singer, Y. (1999). *Machine Learning,* **37** (3), 297–336.

[78] Shawe-Taylor, J. & Cristianini, N. (2004). *Kernel Methods for Pattern Analysis.* New York, NY, USA: Cambridge University Press.

[79] Shervashidze, N., Vishwanathan, S. V. N., Petri, T., Mehlhorn, K., & Borgwardt, K. M. (2009). *Journal of Machine Learning Research - Proceedings Track,* **5**, 488–495.

[80] Shi, J. & Malik, J. (2000). *IEEE Trans. Pattern Anal. Mach. Intell.* **22** (8), 888–905.

[81] Shiga, M. & Mamitsuka, H. (2011). *Wiley Interdisciplinary Reviews: Data Mining and Knowledge Discovery,* **1** (6), 496–511.

[82] Shiga, M. & Mamitsuka, H. (2012). *Pattern Recognition,* **45** (3), 1035–1049.

[83] Shiga, M., Takigawa, I., & Mamitsuka, H. (2007). In: *Proceedings of the 13th ACM SIGKDD International Conference on Knowledge Discovery and Data Mining* KDD pp. 647–656, New York, NY, USA: ACM.

[84] Srikant, R. & Agrawal, R. (1996). In: *Proceedings of the 5th International Conference on Extending Database Technology: Advances in Database Technology* EDBT '96 pp. 3–17, London, UK, UK: Springer-Verlag.

[85] Strehl, A. & Ghosh, J. (2003). *J. Mach. Learn. Res.* **3**, 583–617.

[86] Subramanian, A., Tamayo, P., Mootha, V. K., Mukherjee, S., Ebert, B. L., Gillette, M. A., Paulovich, A., Pomeroy, S. L., Golub, T. R., Lander, E. S., & Mesirov, J. P. (2005). *Proceedings of the National Academy of Sciences,* **102** (43), 15545–15550.

[87] Takahashi, K., Takigawa, I., & Mamitsuka, H. (2013). *PLoS ONE,* **8** (12), e82890.

[88] Takigawa, I. & Mamitsuka, H. (2011). *Machine Learning,* **82** (2), 95–121.

[89] Takigawa, I., Tsuda, K., & Mamitsuka, H. (2011). *PLoS ONE,* **6** (2), e16999.

[90] TheGeneOntologyConsortium (2017). *Nucleic Acids Research,* **45** (D1), D331–D338.

[91] Thompson, J. D., Higgins, D. G., & Gibson, T. J. (1994). *Nucleic Acids Research,* **22** (22), 4673–4680.

[92] Vinh, N. X., Epps, J., & Bailey, J. (2010). *J. Mach. Learn. Res.* **11**, 2837–2854.

[93] Vishwanathan, S. V. N , Borgwardt, K. M., & Schraudolph, N. N. (2006). In: *Proceedings of the 19th International Conference on Neural Information Processing Systems* NIPS'06 pp. 1449–1456, Cambridge, MA, USA: MIT Press.

[94] Vishwanathan, S. V. N. & Smola, A. J. (2002). In: *Proceedings of the 15th International Conference on Neural Information Processing Systems* NIPS'02 pp. 585–592, Cambridge, MA, USA: MIT Press.

[95] Vishwanathan, S. V. N. & Smola, A. J. (2004). In: *Kernel Methods in Computational Biology* pp. 113–130. MIT Press Cambridge, MA, USA.

[96] Wagner, S. & Wagner, D. (2007). Technical Report 2006-04 Universität Karlsruhe (TH).

[97] Wishart, D. S., Feunang, Y. D., Guo, A. C., Lo, E. J., Marcu, A., Grant, J. R., Sajed, T., Johnson, D., Li, C., Sayeeda, Z., Assempour, N., Iynkkaran, I., Liu, Y., Maciejewski, A., Gale, N., Wilson, A., Chin, L., Cummings, R., Le, D., Pon, A., Knox, C., & Wilson, M. (2018). *Nucleic Acids Research,* **46** (D1), D1074–D1082.

[98] Xing, E. P., Jordan, M. I., & Karp, R. M. (2001). In: *Proceedings of the Eighteenth International Conference on Machine Learning* ICML '01 pp. 601–608, San Francisco, CA, USA: Morgan Kaufmann Publishers Inc.

[99] Yamaguchi, A., Aoki, K. F., & Mamitsuka, H. (2004). *Information Processing Letters,* **92** (2), 57 - 63.

[100] Yan, X. & Han, J. (2002). In: *Proceedings of the 2002 IEEE International Conference on Data Mining* ICDM '02 pp. 721–724, Washington, DC, USA: IEEE Computer Society.

[101] Yao, Y. Y. (2003). *Information-Theoretic Measures for Knowledge Discovery and Data Mining* pp. 115–136. Berlin, Heidelberg: Springer Berlin Heidelberg.

[102] Yeung, K. Y. & Ruzzo, W. L. (2001). *Bioinformatics,* **17** (9), 763–774.

[103] Yotsukura, S., Karasuyama, M., Takigawa, I., & Mamitsuka, H. (2017). *Briefings in Bioinformatics,* **18** (4), 619–633.

[104] Yu, S., Tranchevent, L.-C., Moor, B., & Moreau, Y. (2013). *Kernel-based Data Fusion for Machine Learning: Methods and Applications in Bioinformatics and Text Mining.* Berlin, Heidelberg: Springer Publishing Company, Incorporated.

[105] Zhang, L., Udaka, K., Mamitsuka, H., & Zhu, S. (2012). *Brief. Bioinformatics,* **13** (3), 350–364.

[106] Zhou, D., Bousquet, O., Lal, T. N., Weston, J., & Schölkopf, B. (2003). In: *Proceedings of the 16th International Conference on Neural Information Processing Systems* NIPS'03 pp. 321–328, Cambridge, MA, USA: MIT Press.

Index

www.ingramcontent.com/pod-product-compliance
Lightning Source LLC
Chambersburg PA
CBHW071538210326
41597CB00019B/3043

* 9 7 8 4 9 9 1 0 4 4 5 0 2 *